PENGUIN REFERENCE BOOKS

THE PENGUIN DICTIONARY OF ASTRONOMY

Jacqueline Mitton cannot remember a time when she was not fascinated by astronomy. After studying physics at Somerville College, Oxford, she went on to Cambridge to do research, and obtained her doctorate. For a number of years she has concentrated on communicating astronomy to a wider audience. She edited the *Journal of the British Astronomical Association* from 1987 to 1993 and has been Public Relations Officer of the Royal Astronomical Society since 1989. Her recent books include *The Great Comet Crash* (with John Spencer), *Pluto and Charon* (with Alan Stern), *The Young Oxford Book of Astronomy* (with Simon Mitton) and a book for younger children about Galileo.

Jacqueline Mitton lives in Cambridge with her husband, Simon, and two Burmese cats (Tristan and Tosca). They have two grown-up daughters, Lavinia and Veronica.

D0313584

00000069186

THE PENGUIN DICTIONARY OF
ASTRONOMY

Jacqueline Mitton

PENGUIN BOOKS

To Veronica

PENGUIN BOOKS

Published by the Penguin Group
Penguin Books Ltd, 27 Wrights Lane, London W8 5TZ, England
Penguin Putnam Inc., 375 Hudson Street, New York, New York 10014, USA
Penguin Books Australia Ltd, Ringwood, Victoria, Australia
Penguin Books Canada Ltd, 10 Alcorn Avenue, Toronto, Ontario, Canada M4V 3B2
Penguin Books (NZ) Ltd, Private Bag 102902, NSMC, Auckland, New Zealand

Penguin Books Ltd, Registered Offices: Harmondsworth, Middlesex, England

First published in Great Britain by Oxford University Press,
in association with Penguin Books Ltd, 1991
Published with revisions in Penguin Books 1993
Third edition, with further revisions,
published in Penguin Books, 1998

3 5 7 9 10 8 6 4 2

Set in 9/10½pt Monotype Bembo
Typeset by Rowland Phototypesetting Ltd, Bury St Edmunds, Suffolk
Printed in England by Clays Ltd, St Ives plc

PREFACE TO THE THIRD EDITION

The world of astronomy has experienced significant change in the five years since the second edition of this dictionary was published. In particular, numerous satellite observatories have been put into Earth orbit, space missions launched to explore the Sun and planets, and a new generation of ground-based telescopes is coming into operation. At the same time, some projects have bitten the dust, due to failure or cancellation. Among notable scientific discoveries requiring new or revised entries are objects in the Kuiper Belt in the outer solar system, none of which were known before 1992, so-called 'EGGS' and 'proplyds' representing stages in the formation of stars and planetary systems, uncovered by the Hubble Space Telescope, the findings of the *Galileo* mission to Jupiter and its moons, and even (in the last few days of the revision work) the first measurement of the distance of a gamma-ray burster. These and other developments, such as a complete change in the nomenclature for comets, are reflected in over 100 new entries and revisions to well over 100 of the existing entries. I have also taken the opportunity to correct a number of errors and omissions, and I am grateful to several readers who have written to me with comments.

JACQUELINE MITTON
Cambridge, May 1997

PREFACE TO THE SECOND EDITION

In the space of just eleven months since the first edition was completed, a number of exciting new results have been announced, notably from the COBE satellite and the *Magellan* probe orbiting Venus. Every effort has been made to incorporate these changes, along with others, such as the consequences of the transformation of the USSR into the Commonwealth of Independent States. At the same time I have taken the opportunity to make a number of minor corrections and add around thirty new entries. I am indebted to Peter Bond, Harold Ridley, Ian Ridpath, Roger Taylor and John Woodruff for new suggestions and information.

JACQUELINE MITTON
Cambridge, June 1992

PREFACE

Astronomy is a branch of science with a particularly rich language of its own. Its vocabulary reflects a heritage accumulated over many centuries but astronomy is also a fast-moving, up-to-the-minute modern science. During the three years in which I compiled this dictionary, it sometimes felt as if astronomers were coining new terms and acronyms, devising new practical techniques, planning new space missions and discovering weird new objects, all at such a pace that it was a real challenge to keep up. And they haven't stopped just because I have finished writing.

My aim has been to collect as many as I could of the terms and names that are particular to astronomy, in both the professional and amateur fields. There is, of course, considerable overlap between astronomy, physics and space science. This book is intentionally a dictionary of *astronomy*, but I have included what I hope is just enough 'pure' physics and space science to make it a useful one-step reference source for a wide range of users.

It is a *dictionary* and not an encyclopedia; the entries could not possibly be encyclopedic in scope in a book of this length. Users who need more detail on a specific subject are recommended to refer to appropriate books listed in the suggestions for further reading at the back.

Wherever possible, I have cross-checked facts with more than one independent reference source and have used primary and authoritative sources. Despite my endeavours to be comprehensive and accurate, I recognize that there are likely to be errors and omissions. For these I apologize. I hope that users will write to me, care of the publisher, with suggestions for improvements.

Tables

Some kinds of information, without which a book like this would certainly not be complete, are best collected in tabular form rather than scattered between individual entries. Data on the planets and their natural satellites, and on bright stars and constellations are collected after the alphabetical entries, together with the Greek alphabet and selected units and conversion factors. Cross-references to these numbered tables appear at the end of certain entries.

Order of alphabetical entries

The alphabetical order takes no account of word breaks or hyphens (i.e. it is the letter-by-letter system). On balance, this system resulted in the least confusion in a subject where, for example, any of the forms 'redshift', red-shift' or 'red shift' might commonly be encountered. The only exception to this scheme is that the entries for H I region and H II region are to be found at the beginning of the H section. Entries beginning with a Greek letter or with a number are alphabetized with that letter or number spelt out. Comets are for most part listed under their formal designations, e.g. Comet Encke. However, the entry for Comet Halley is given under the headword 'Halley's Comet' in recognition of universally common usage. This is one of a number of examples where I have sacrificed absolute consistency in an attempt to predict where users are most likely to look first.

Units of measurement

Appropriate metric units have been used throughout for physical quantities, but imperial equivalents (suitably rounded) are also given in many instances where it seemed helpful to do so. Table 2 at the back of the book lists selected units and conversion factors. Units encountered uniquely in astronomy have their own entries in the alphabetical list, as do a number of other units commonly used by astronomers.

Cross-references

Certain words used in the text that have their own entries are in italics preceded by the symbol '➤'. Not all possible cross-references are highlighted in this way; to do so would make the text almost unreadable in places. Instead, I have indicated cross-references only when I consider that they are most likely to be helpful. Users encountering an unfamiliar technical term in an entry may well find that that term has its own entry, even if it is not specifically indicated.

The symbol '➤' preceding a word or words in italic type means 'see' or 'see also'.

A

AAO   Abbreviation for ►*Anglo-Australian Observatory*.

AAS   Abbreviation for ►*American Astronomical Society*.

AAT   Abbreviation for ►*Anglo-Australian Telescope*.

AAVSO   Abbreviation for ►*American Association of Variable Star Observers*.

Abell Catalogue   A catalogue of 2,712 rich clusters of galaxies drawn up by George Abell from the ►*Palomar Sky Survey* photographs. He demonstrated that there are two types of galaxy cluster: a compact type that is regular in shape and a more spread-out, irregular type.

aberration (1)   An imperfection in the imaging properties of a lens or mirror. The main aberrations are ►*chromatic aberration*, ►*spherical aberration*, ►*coma*, ►*astigmatism*, ►*curvature of the field* and ►*distortion*.

aberration (2)   An apparent displacement in the observed position of a star. It is a result of the finite speed of the light travelling from the star, combined with the motion through space of the observer on Earth relative to the star, etc. Aberration arising from the Earth's orbital motion around the Sun is termed *annual aberration*. The much smaller component that results from the daily rotation of the Earth is called *diurnal aberration*.

ablation   The erosion of a surface through a process such as vaporization or friction. For example, when a meteoroid enters a planetary atmosphere, ablation occurs because of friction between its surface and the gas molecules in the atmosphere.

absolute luminosity   A measure of the actual rate of energy output of a star or other celestial object as opposed to the ►*apparent luminosity*, which depends on the distance to the object.

absolute magnitude (symbol M) of a star.   The ►*magnitude* a star would appear to have if it were at a standard distance of 10 ►*parsecs*. Absolute magnitudes are a method of comparing the actual luminosities of stars on an arbitrary scale.

   For an asteroid or comet, the absolute magnitude is the ►*apparent magnitude* it would have at zero ►*phase angle* and at a distance of 1 AU from both the Sun and Earth.

absolute zero   The point at which all molecular motion ceases and so, theoretically, the lowest possible temperature. It is the zero point of the Kelvin temperature scale used in science. The equivalent on the Celsius scale is −273.16°C.

absorption   The process by which the intensity of radiation decreases as it passes through a material medium. The energy lost by the radiation is transferred to the medium. Many physical processes observed in astronomy involve absorption, including absorption of light from celestial objects in the Earth's atmosphere ➤(*atmospheric extinction*), ➤*interstellar extinction* and absorption within the gaseous layers of a star, resulting in an ➤*absorption line* spectrum.

absorption coefficient   A measure of the ability of a material to absorb radiation that passes through it. A high absorption coefficient indicates that the material absorbs radiation effectively. The absorption coefficient may depend strongly on the wavelength of the radiation, the temperature and other physical conditions.

absorption line   A sharp dip in intensity over a narrow wavelength range in a ➤*continuous spectrum*. In a spectrum produced by a typical ➤*spectrograph*, in which the light passes through a narrow slit before being dispersed, absorption lines have the appearance of dark lines cutting across at right angles to the direction of dispersion.

Absorption lines are a characteristic of the spectra of the majority of stars. In the case of the Sun, they are known as ➤*Fraunhofer lines*. Atoms are able to absorb radiation at a number of precise wavelengths. The wavelengths at which absorption occurs are different for each chemical element, making it possible to identify the elements present in a star, or other celestial body, by analysing which spectral lines are present. The strength of the lines can be used to deduce the abundance of the elements, though not directly since the temperature, density and other physical circumstances greatly influence the strength of absorption lines in a spectrum. ➤*emission line*.

absorption line spectrum   A ➤*continuous spectrum* in which narrow ➤*absorption lines* can be seen.

absorption nebula   A dark interstellar cloud that absorbs the light from bright objects behind it. Absorption nebulae range in size from small ➤*globules* to large clouds visible to the unaided eye, such as the ➤*Coalsack* in the southern Milky Way. Absorption nebulae contain both dust and gas, and the temperatures in them are low enough for simple molecules to form. Much of the knowledge about these nebulae comes from observations of infrared and radio radiation, which, unlike visible light, can pass through them. Strong evidence has been found that the initial stages of star formation take place in such dark nebulae. ➤ *molecular cloud*.

absorption spectrum   A ➤*continuous spectrum* in which absorption features of some kind are seen. They may be ➤*absorption lines* or ➤*diffuse interstellar bands*, for example. ➤*band spectrum*.

abundance   The relative number of atoms of a particular element or isotope in the chemical composition of a single substance or object, or in a structure such as the solar system, or in the universe as a whole.

acceleration   The rate at which velocity changes with time. Any mass with a net force acting on it undergoes acceleration. The *acceleration due to gravity* (symbol g) is the acceleration experienced by an object falling freely at a body's surface, pulled by the force of gravity. On Earth, the average value is around $9.8 \ \text{m/s}^2$.

accretion   The process by which small particles of material coalesce, or accumulate on a larger mass, under the influence of their mutual gravitational attraction or as a result of chance collisions, thus gradually creating larger bodies.

accretion disc   A disc structure that forms around a spinning object, such as a star or ➤*black hole*, when material falls on to it from a close companion in a ➤*binary star* system.

As stars evolve they enter a giant phase, when their size increases dramatically. In a binary system, the gravitational pull of the companion star on the bloated envelope of the giant may be stronger than the force holding the giant star and its envelope together. Under these circumstances, material flows across from one star to the other. The presence of an obscuring accretion disc may cause the star to be variable, as with ➤*Beta Lyrae stars*, and this may be detectable from features in the spectrum.

If the companion is spinning rapidly, as will a collapsed star (a ➤*white dwarf*, ➤*neutron star* or black hole), a swirling accretion disc forms on to which the material falls. The kinetic energy of the falling matter is turned into heat, and X-rays are produced. A mechanism of this kind is thought to be responsible for the production of X-rays in objects such as ➤*Cygnus X-1*. ➤*Roche lobe*, *SS433*.

ACE   Abbreviation for ➤*Advanced Composition Explorer*.

Achernar (Alpha Eridani; α Eri)   The brightest star in the constellation Eridanus. The name, which is of Arabic origin, means 'the end of the river' and the star marks the southern extremity of the constellation at a declination of $-57°$. Achernar is a ➤*B star* of magnitude 0.5. ➤Table 3.

Achilles   Asteroid 588, diameter 116 km. Discovered by Max Wolf in 1906, it was the first of the ➤*Trojan asteroids* to be identified.

achondrite   A type of stony ➤*meteorite* that crystallized from molten rock. The name indicates the absence of ➤*chondrules* in this type of meteorite, in contrast with ➤*chondrites*. Compared with chondrites, they have a greater abundance of calcium-rich minerals and much less of the metal and sulphide minerals.

achromatic lens (achromat)   A composite lens, made up of two elements of different kinds of glass, designed to reduce ➤*chromatic aberration.* ➤*apochromat.*

Acidalia Planitia   A dark plain in the northern hemisphere of Mars, formerly known as Mare Acidalia.

acoustic waves   Pressure waves in a fluid. Acoustic waves can propagate through the gaseous layers of stars (including the Sun). They are produced where pressure is the dominant force restoring the balance when there is a small displacement of the gas.

acronical observation   An observation shortly after sunset of a star that is just rising or just setting. Such observations of bright stars (e.g. Sirius) were used in ancient times for keeping track of the seasons. ➤*heliacal rising.*

Acrux (Alpha Crucis; α Cru)   The brightest star in the constellation Crux. Appearing to the unaided eye as a white star of magnitude 0.9, it is a visual double, the two components being ➤*B stars* of magnitude 1.4 and 1.9. The separation is 4.4 arc seconds. ➤Table 3.

actinometer   An instrument for measuring the intensity of solar radiation. Such instruments are now more usually called pyrheliometers.

active galactic nucleus (AGN)   The central region of an active galaxy in which exceptionally large amounts of energy are being generated from a source other than the normal output of individual stars. This is a common characteristic of various types of object that have been classified differently according to their appearance and the nature of the radiation they emit. ➤*Quasars,* ➤*Seyfert galaxies,* ➤*radio galaxies,* ➤*N galaxies* and ➤*blazars* are all manifestations of the same phenomenon. In every case, the source of power appears to be concentrated within the nucleus; the only mechanism known that might account for the huge amount of energy observed is the presence of a supermassive ➤*black hole,* into which matter is falling with the release of gravitational energy. Variability over relatively short timescales shows that the powerhouse must be concentrated in a small region of space.

Evidence that a highly energetic process is occurring comes both from the sheer luminosity of distant objects like quasars and from the existence of jets of material ejected from the central regions. The giant elliptical galaxy M87, which is also a strong radio and X-ray source, has a spectacular jet consisting of a series of knots of hot gas. With radio galaxies, most of the energy comes from two gigantic clouds of ionized gas, one either side of the galaxy from which they have been ejected.

active optics   A method of maintaining a highly accurate optical surface in a reflecting telescope by means of a computer-controlled feedback system that continually monitors the quality of the image and uses the information to adjust a motorized support system under the mirror.

The development of this technique means that mirrors can be made from thinner, less massive blanks that can be supported in a lighter structure. If the mirror is flexible enough, problems such as ➤*spherical aberration* and ➤*astigmatism* can be greatly reduced by applying constant forces at a large number of positions (perhaps fifty or more) on the back of the mirror. Variations in the forces on the mirror, such as the flexure under gravity as the telescope moves, can be corrected at intervals as short as a few minutes.

The term has also been used to describe the technique now more commonly called ➤*adaptive optics*.

active region A region in the outer layers of the Sun where ➤*solar activity* is taking place. Active regions develop where strong magnetic fields emerge from the subsurface layers. A variety of features may be observed in the ➤*photosphere*, the ➤*chromosphere* and the ➤*corona*. Phenomena such as ➤*sunspots*, ➤*plages* and ➤*flares* are evidence of an active region. Radiation is enhanced over the whole spectrum, from X-rays to radio waves, though in sunspots the localized decrease in temperature reduces the visual brightness. There is a large variation in the size and duration of active regions: they may last from several hours up to a few months. Electrically charged particles and the enhanced ultraviolet and X-radiation from active regions affects the interplanetary medium and the upper atmosphere of the Earth.

active Sun The Sun during periods of ➤*solar activity*.

activity index Any one of a number of indicators of the level of ➤*solar activity* at a given time. Indices used to measure solar activity include the ➤*Wolf sunspot number*, the total area covered by ➤*sunspots* on the visible hemisphere, and a K-plage index derived from the areas and brightness of ➤*plages*. In addition, the total radio and X-radiation from the Sun are also regarded as indices of solar activity.

adaptive optics A technique for improving the quality of the image produced by an astronomical telescope in which the optical system compensates for changes in the quality of ➤*seeing* – constantly varying distortions produced by refraction in the Earth's atmosphere. The distortion is corrected by the rapid bending of a small, very thin mirror placed a short distance before the focus. To be effective, the system needs an image sensor, a microprocessor and actuators to apply forces to the thin mirror, all of which must have response times shorter than a hundredth of a second.

Adhara (Epsilon Canis Majoris; ε CMa) The second-brightest star, after Sirius, in the constellation Canis Major. It is a giant ➤*B star* of magnitude 1.5

and has an eighth magnitude companion. The Arabic name means 'the virgins'. ►Table 3.

Adonis   Asteroid 2101, diameter 2 km, discovered by E. Delporte in 1936. It is a member of the ►*Apollo* asteroid group and came within 2 million kilometres of the Earth in 1937, but was then lost until recovered in 1977 following recomputation of its orbit.

Adrastea   A small satellite of Jupiter (number XV), discovered by David Jewitt in 1979. ►Table 6.

ADS   Abbreviation for ►*Aitken Double Star Catalogue*.

Advanced Composition Explorer (ACE)   A US spacecraft, launched in August 1997, carrying nine scientific instruments for determining the isotopic and elemental composition of the solar ►*corona*, the ►*interplanetary medium*, the local ►*interstellar medium* and galactic material. It was placed in solar orbit, at one of the Earth's ►*Lagrangian points*, 1.5 million km nearer the Sun than the Earth, where it will maintain an almost constant position relative to the Earth.

Advanced X-ray Astrophysics Facility (AXAF)   An American satellite, scheduled for launch via the Space Shuttle in late 1998, which will provide a national X-ray astronomy facility.

aero lens   A camera lens designed and manufactured for aerial survey work.

aerolite   An alternative, largely obsolete name for a stony ►*meteorite*.

aeronomy   The study of the physical and chemical processes in the upper atmosphere of the Earth, or of other planets.

aerosol   A suspension of particles in a gas, such as mist in an atmosphere.

Ae star   An ►*A star* that shows ►*emission lines* of hydrogen superimposed on the ►*absorption lines* in its spectrum. Such emission may arise from interactions between the two members of a close binary system.

aether   Alternative spelling of ►*ether*.

AGB   Abbreviation for ►*asymptotic giant branch*.

Agena   An alternative name for the star ►*Hadar*.

AGK   Abbreviation for the star catalogue *Astronomische Gesellschaft Katalog*. The first project, for cataloguing all stars between declinations −2° and +80° down to ninth magnitude, was initiated by F. W. Argelander in 1867. Seventeen observatories participated and the fifteen volumes of AGK1 listed 150,000 stars. An extension to declination −23° published in 1887 contained a further 50,000 stars. A second version was started in the 1920s, using photography rather than visual observations, and was published in 1951−8. Measurements of ➤*proper motions* were included in the third version, AGK3. AGK positions are accurate positions, calculated relative to fundamental stars. ➤*fundamental catalogue.*

AGN   Abbreviation for ➤*active galactic nucleus.*

Ahnighito meteorite   The largest ➤*meteorite* displayed in any museum, housed at the Hayden Planetarium in New York City. It was found by Robert Peary in Greenland in 1897 and is of the iron type. Its weight is 31 tonnes.

airglow   The light produced and emitted by the Earth's own atmosphere, excluding ➤*thermal radiation,* ➤*aurorae,* lightning and ➤*meteor* trains. The spectrum of the airglow ranges from 100 nanometres to 22.5 micrometres. A major component is an emission line due to oxygen at 558 nanometres, which originates in a layer 30 to 40 kilometres thick at a typical height of 100 kilometres (60 miles). From space, the airglow appears as a ring of greenish light around the Earth.

Air Pump   English name for the constellation ➤*Antlia.*

air shower   A proliferation of high-energy charged particles in the atmosphere, triggered by the collision of a ➤*cosmic ray* particle with an atomic nucleus in the atmosphere. The first collision produces a number of secondary particles, all of which have considerable energy and may subsequently undergo further collisions.

Airy disc   The smallest image a particular telescope can make of a point source of light, such as a star. Diffraction as the light passes through the telescope aperture causes the image of even a point source to have a finite size. The diameter of the Airy disc is smaller for larger apertures. In practice, it is rarely possible to achieve images as small as the Airy disc because of turbulence in the Earth's atmosphere, which distorts and enlarges the perceived image.

Aitken Double Star Catalogue (ADS)   A catalogue of over 17,000 binary stars compiled by Robert G. Aitken and published in 1932. Its formal title is *New General Catalogue of Double Stars.*

AI Velorum star   ➤*Delta Scuti star.*

AJ   A common abbreviation for the ➤*Astronomical Journal.*

Alauda   Asteroid 702, diameter 202 km, discovered by J. Helffrich in 1910.

albedo   The proportion of the light falling on a body or surface that is reflected. Albedo may be expressed as a fraction between zero (perfectly absorbing) and 1 (perfectly reflecting) or as an equivalent percentage. ➤*Bond albedo, geometric albedo, hemispherical albedo.*

Albert   Asteroid 719, diameter 2.6 km. It was discovered by J. Palisa in 1911, when it made a close approach to the Earth, but is now lost.

Albireo (Beta Cygni; β Cyg)   The second-brightest star in the constellation Cygnus. Visual observers regard it as one of the most beautiful double stars. The primary star is a yellow – orange ►*K star*, a giant of magnitude 3.2, and its companion a bluish ►*B star* of magnitude 5.4. The two components are separated by 35 arc seconds.

Alcaid   Alternative form of ►*Alkaid*.

Alcor (80 Ursae Majoris)   The fourth magnitude ►*A star* that forms a naked-eye double with ►*Mizar* in the bear's 'tail'. The two stars are separated by 11.5 arc minutes on the sky and 10 light years in space.

Alcyone (Eta Tauri; η Tau)   The brightest member of the ►*Pleiades* star cluster in the constellation Taurus. The name is that of the daughter of Atlas and Pleione in Greek mythology. Alcyone is a ►*B star* of magnitude 2.9.

Aldebaran (Alpha Tauri; α Tau)   The brightest star in the constellation Taurus. Its Arabic name means 'the follower'. Aldebaran is a giant ►*K star* of magnitude 0.9. Although it appears in the sky to be part of the Hyades star cluster, it is not in fact a cluster member, lying only half as far away. ►Table 3.

Alderamin (Alpha Cephei; α Cep)   The brightest star in the constellation Cepheus. It is an ►*A star* of magnitude 2.7. The name, which is of Arabic origin, means 'the right arm'.

Alfvén waves   Magnetic waves that can propagate through an electrically conducting fluid, such as an ionized gas, in a magnetic field.

Algenib (Gamma Pegasi; γ Peg)   A ►*B star* of magnitude 2.8 that marks one corner of the ►*Square of Pegasus*. The Arabic name means 'the side'. The name is also sometimes applied to the star Alpha Persei, more usually known as ►*Mirfak*.

Algieba (Gamma Leonis; γ Leo)   A ►*visual binary* star consisting of two yellow giants, separated by 4 arc seconds. The magnitudes are 2.3 and 3.5, and they orbit around each other in a period of 620 years. ►Table 3.

Algol (The Demon Star; Beta Persei; β Per)   Perhaps the best-known variable star, its magnitude varying between 2.2 and 3.5 in a period of 2.87 days. The variation occurs because Algol is an ►*eclipsing binary* system in which the two stars regularly cross in front of each other as viewed from Earth. Algol is regarded as the prototype eclipsing binary.

The brighter of the components of Algol is a ►*B star* and the fainter a ►*G star*. The decline in brightness as the G star cuts off the light from its brighter companion takes four hours, and minimum lasts only twenty minutes. The secondary eclipse, when the fainter star causes a dip in brightness of only 0.06

magnitude, is not detectable by eye. Variations in the spectrum of Algol with a period of 1.862 years reveal the presence of a third star in the system.

There is also evidence in the spectrum for ➤*mass transfer* between the two close companions, supported by observations that Algol is a radio star, erratically flaring up to twenty times its normal radio brightness. The radio emission is attributed to gas streaming on to the primary star.

Algonquin Observatory   A Canadian radio astronomy observatory in Ontario. The main instrument is a 46-metre (150-foot) fully steerable dish.

alidade   A movable sighting arm on an ➤*astrolabe* that is used in making observations of the ➤*altitudes* and ➤*azimuths* of celestial objects.

Alinda   Asteroid 887, diameter 4 km. It was discovered in 1918 by Max Wolf when it made a close approach to the Earth. It is a member of the ➤*Amor* group.

Alioth (Epsilon Ursae Majoris; ε U Ma).   The brightest star in the constellation Ursa Major, the Greek letters in this case being allotted in order of position rather than of brightness. Alioth is an ➤*A star* of magnitude 1.8. ➤Table 3.

Alkaid (Eta Ursae Majoris; η U Ma)   A star in Ursa Major, at the end of the bear's 'tail'. It is a ➤*B star* of magnitude 1.9. The Arabic name means 'chief of the mourners', for the Arabs saw the constellation as a bier rather than a bear. ➤Table 3.

Allan Hills   A region in Antarctica from where large numbers of meteorites have been recovered. The meteorites become concentrated in the area by natural movements in the ice sheet, and are relatively easy to identify against the ice.

Allegheny Observatory   A research observatory of the University of Pittsburgh, Pennsylvania. The present building dates from 1912 but its predecessor was started in 1858 by several Pittsburgh businessmen. Inspired by Donati's Comet of that year, they formed the Allegheny Telescope Association and acquired a 33-centimetre (13-inch) refractor. In 1867, the telescope and observatory were given to Western University in Pennsylvania, the forerunner of the University of Pittsburgh. The first salaried director was Samuel Pierpont Langley. He was succeeded by James E. Keeler, co-founder of the *Astrophysical Journal* and subsequently director of the Lick Observatory.

The 1912 building is equipped with three telescopes. The original 33-centimetre refractor is used primarily for educational and testing purposes. The others are the 76-centimetre (30-inch) Thaw refractor and the 79-centimetre (31-inch) Keeler Memorial Reflector. These continue to be used for research.

Allende meteorite   A meteorite of the ➤*carbonaceous chondrite* type, which fell in Mexico in 1969. More than two tonnes of material fell, scattering over an area 48 by 7 kilometres, making it one of the most massive carbonaceous chondrites known.

all-sky camera   A camera with a very wide-angle, fish-eye lens, capable of photographing all or most of the visible hemisphere of the sky on one exposure. Cameras of this type have applications in routine surveys for meteors and artificial satellites.

Almagest   A large astronomical treatise by the Greek astronomer Ptolemy (Claudius Ptolemaeus), who worked in Alexandria between about AD 127 and 151. The name is an Arabic corruption of a Greek title, 'The Greatest', though Ptolemy's original title was 'The Mathematical Collection'. It ranks among the most important works on astronomy ever written. It included a star catalogue and dealt with the motion of the Moon and planets. The rules set out for calculating the future positions of the planets, on the basis of an Earth-centred universe, were used for centuries.

almanac   A book of tables giving the future positions of the Moon, planets and other celestial objects, often compiled with additional information of practical value to users. An almanac normally covers the period of one calendar year.

almucantar (1)   A circle on the ➤*celestial sphere* parallel to the horizon.

almucantar (2)   An instrument for measuring ➤*altitude* and ➤*azimuth*.

Alnath   A variant spelling of the star name ➤*Elnath*.

Alnilam (Epsilon Orionis; ε Ori)   One of the three bright stars forming Orion's belt. The Arabic name means 'string of pearls'. Alnilam is a ➤*supergiant* ➤*B star* of magnitude 1.7. ➤Table 3.

Alnitak (Zeta Orionis; ζ Ori)   One of the three bright stars forming Orion's belt. The Arabic name means 'the girdle'. Alnitak is a ➤*supergiant* ➤*O star* of magnitude 1.8. ➤Table 3.

Alpha Centauri (α Cen)   The brightest star in the constellation Centaurus and the nearest bright star to the Sun, at a distance of 4.34 light years. It is a ➤*visual binary* star with an orbital period of 80 years. The two components are of ➤*spectral types* G and K and have a combined magnitude of −0.27. The eleventh magnitude star Proxima Centauri, though two degrees away on the sky, is thought to be associated with this star system because it has a similar motion in space. Proxima, a dim ➤*M star*, is the nearest star to the Sun at a distance of 4.24 light years. Alpha Centauri is also called by the Arabic name Rigil Kentaurus (sometimes Rigel, or shortened to Rigil Kent), which means 'the foot of the Centaur'. An alternative name is Toliman. ➤Table 3.

alpha particle   The nucleus of a helium atom, consisting of two protons and two neutrons. Alpha particles are emitted by many radioactive isotopes and also play an important role in nuclear fusion processes within stars.

Alphard (Alpha Hydrae; α Hya)   The brightest star in the constellation Hydra. Its Arabic name means 'the solitary one of the serpent'. It is a ➤*K star* of magnitude 2.0. ➤Table 3.

Alphekka (Gemma; Alpha Coronae Borealis; α CrB)   The brightest star in the constellation Corona Borealis. It is an ➤*A star* of magnitude 2.2. The Arabic name, also spelt Alphecca, means 'bright one'. This star is sometimes called by the Latin name Gemma, the 'jewel' in the crown.

Alpheratz (Sirrah; Alpha Andromedae; α And)   The brightest star in the constellation Andromeda, marking one corner of the ➤*Square of Pegasus*. It was formerly considered to belong to that constellation, being designated Delta Pegasi. Alpheratz is an ➤*A star* of magnitude 2.1.

Alphonsine Tables   A set of tables giving the positions of the Sun, Moon and planets, published in 1252 under the patronage of King Alphonso X of Castile. They were computed by a team of astronomers using the principles set out by Ptolemy in the ➤*Almagest* but incorporating more recent observations. They were in use in Europe for nearly 400 years, until superseded by the work of Johannes Kepler, during which time they were the best available.

Alphonsus   A lunar crater, 118 kilometres (73 miles) in diameter. A prominent ridge runs across the centre, almost along a north–south line, through a central peak about one kilometre high. Temporary reddish clouds were observed there in 1958 and 1959, possibly due to the release of gas from the rocks. ➤*transient lunar phenomenon*.

Alpine Valley (Vallis Alpes)   A flat-bottomed valley, 150 kilometres (95 miles) long, crossing the lunar Alps and connecting the Mare Frigoris with the Mare Imbrium.

Alps (Montes Alpes)   A range of mountains on the Moon, lying between the Mare Frigoris and the Mare Imbrium.

ALSEP   Abbreviation for Apollo Lunar Science Experiment Package, experimental set-ups that were deployed on the Moon by astronauts during the manned ➤*Apollo programme* (1969–72). One was left by every mission except the first. Each automated laboratory was powered by a small nuclear generator. The packages included seismometers to measure moonquakes and experiments to measure the solar wind, detect any trace atmosphere and measure the heat flow from the Moon's interior. All the experiments were turned off in 1978.

Altair (Alpha Aquilae; α Aql)   The brightest star in the constellation Aquila. The Arabic name means 'the flying eagle'. It is an ➤*A star* of magnitude 0.8

and one of the closest of the brighter stars at a distance of 17 light years. ►Table 3.

Altar   English name for the constellation ►*Ara*.

altazimuth mounting   A form of telescope mounting in which the two independent rotation axes allow movement of the instrument in ►*altitude* and ►*azimuth*. It is the simplest type of telescope mounting but it is necessary to move the telescope about both axes simultaneously in order to track the motion of celestial objects across the sky. For that reason it is not suitable for small, motor-driven telescopes. However, the ability to control the motion of a large telescope by computer has led to altazimuth mountings being the norm for new, large professional instruments. ►*Dobsonian telescope, equatorial mounting*.

altimetry   The measurement of height, including techniques such as the determination of the height of planetary features by means of radar.

altitude   The direct angular distance between a celestial object and the horizon, measured vertically (i.e. along the great circle that passes through the object and the zenith).

aluminizing   The process in which a thin reflecting layer of aluminium is deposited on the optical glass surface of a telescope mirror.

Amalthea   A small satellite of Jupiter (number V), discovered by E. E. Barnard in 1892. Images obtained by the ►*Voyager 1* mission showed it as a red-coloured, potato-shaped object. The surface is heavily cratered, the largest depression, Pan, being 90 kilometres (56 miles) across. The red colour is thought to be due to sulphur compounds blown off the satellite ►*Io*. ►Table 6.

Amazonis Planitia   A light-coloured plain in the north equatorial region of Mars.

American Association of Variable Star Observers (AAVSO)   An organization based in the USA for amateur observers of ►*variable stars*. It was founded in 1911 by William Tyler Olcott, who began making 'volunteer' observations of the brightness of variable stars and sending his results to Edward C. Pickering at Harvard College Observatory after hearing a talk at the American Association for the Advancement of Science meeting in 1909. It has over a thousand members worldwide and a huge database containing millions of individual observations.

American Astronomical Society (AAS)   The main US organization for professional astronomers, which was founded in 1899. There are over six thousand members, most of whom are professionally engaged in astronomy. Its objective is the advancement of astronomy and the promotion of closely related branches of science. It holds two general meetings each year and publishes three scholarly journals. It also has a grants programme and a Visiting Lectureship programme, and offers its members job-finding services.

American Ephemeris and Nautical Almanac An ►*almanac* produced by the ►*United States Naval Observatory*, Washington, DC, between 1855 and 1980. In 1981 it was replaced by the ►*Astronomical Almanac*.

American Indian English name for the constellation ►*Indus*.

Ames Research Center A NASA research establishment located near San Francisco.

AM Herculis star ►*polar*.

Amor Asteroid 1221, diameter 1 km, discovered by E. Delporte in 1932. It is the prototype of the *Amor group* of Earth-approaching asteroids, with perihelion distances between 1.0 and 1.3 AU, taking them inside the main ►*asteroid belt*.

Amphitrite Asteroid 29, diameter 200 km, discovered by A. Marth in 1854.

amplitude The maximum value of the variable quantity or displacement in a wave or oscillation. In astronomy, the term is used particularly to mean the magnitude range of a ►*variable star*.

Am star An ►*A star* that has unusually strong ►*absorption lines* of certain elements in its spectrum, particularly of metals such as iron and nickel, and weak lines of the elements calcium and scandium. These features are thought to result from vertical diffusion that separates out the various elements in the stable outer layers of a slowly rotating star. Am stars are also called *metallic-line A stars*.

AN Common abbreviation for the journal ►*Astronomische Nachrichten*.

analemma The figure-of-eight shape that results if the Sun's position in the sky is recorded at the same time of day throughout the year. The position of the Sun varies because the Earth's rotation axis is not perpendicular to its orbit round the Sun and because the Earth's orbit is elliptical rather than circular.

Ananke A small satellite of Jupiter (number XII), discovered in 1951 by S. B. Nicholson. ►Table 6.

Andromeda A northern constellation among the 48 listed by Ptolemy (*c.* AD 140). In classical mythology, Andromeda was the daughter of Queen Cassiopeia and was condemned to be sacrificed to a sea monster. The figure traditionally associated with the constellation is that of a chained woman. The three brightest stars, Alpha (Alpheratz or Sirrah), Beta (Mirach) and Gamma (Alamak) represent her head, hip and foot, respectively. Andromeda is large but not very conspicuous, known mainly for the ►*Andromeda Galaxy*. ►Table 4.

Andromeda Galaxy (M31; NGC 224)   A large ➤*spiral galaxy,* visible to the unaided eye as a misty patch in the constellation Andromeda. It lies at a distance of 2.3 million light years and is the largest member of the ➤*Local Group,* with a mass of 300 billion solar masses. It is believed that our own Milky Way ➤*Galaxy* is similar to, though it has only half the mass of, the Andromeda Galaxy. The spiral structure is not easy to see since the galaxy is almost edge-on, the disc being tilted at an angle of only 13° to the line of sight. In a small telescope, only the small central nucleus is visible though the fainter spiral arms actually extend over three degrees of sky – more than six times the apparent diameter of the Moon. Several dwarf galaxies are in orbit around the Andromeda Galaxy, notably M32 and NGC 205.

Historically, it was the first object whose extragalactic nature was recognized, and it is the most distant object visible to the naked eye.

Andromedids   A ➤*meteor shower* associated with ➤*Comet Biela* , not observed since 1940. The first recorded appearance of the shower, which radiated from near the star Gamma Andromedae, was in 1741. Spectacular meteor storms were observed in November 1872 and 1885, following the break-up of the associated comet, when the rates were many thousands per hour. The shower is also known as the *Bielids*.

Angelina   Asteroid 64, diameter 60 km, discovered by E. W. Tempel in 1861. It is one of the most strongly reflecting asteroids known, with an albedo of 34 per cent.

Anglo–Australian Observatory (AAO)   An observatory at the ➤*Siding Spring Observatory* site in New South Wales, Australia, funded jointly by the governments of Australia and the UK. The observatory is administered by the Anglo-Australian Telescope Board (AATB), which was set up in the early 1970s when the 3.9-metre (150-inch) ➤*Anglo-Australian Telescope* was constructed. In 1988, the AATB took over operational responsibility for the 1.2-metre (48-inch) UK Schmidt telescope, which has been at the site since 1973, and it also became a fully shared facility at that time. Widely regarded as one the finest instruments of its type, the Schmidt takes high-quality, wide-angle photographs, 6.4° square, and much of its observing time is dedicated to long-term surveys.

Anglo–Australian Telescope (AAT)   A 3.9-metre (150-inch) reflecting telescope, owned and funded jointly by the governments of Australia and the UK. It is situated at the ➤*Siding Spring Observatory* site in New South Wales, Australia. The telescope was constructed in the early 1970s on an equatorial horseshoe mounting, and started scheduled observing in 1975. It was the first telescope to be fully computer controlled.

Many different instruments and techniques are used in conjunction with this versatile, general-purpose telescope, which has yielded numerous important scientific discoveries and spectacular photographs of the southern skies.

ångström (Symbol Å)   A unit of length used particularly for the wavelength of light, equivalent to 10^{-10} metres or 0.1 nanometres. It was formerly used universally for expressing the wavelength of light but, in the SI system of units, the use of nanometres is now preferred.

angular diameter   The apparent diameter of an object in angular measure, i.e. radians, degrees, arc minutes or arc seconds. Angular diameter is determined by the combination of true diameter and distance.

angular distance   The length of an arc expressed in angular measure (i.e. radians, degrees, arc minutes or arc seconds) as the angle subtended by the arc at the observer. The angular distance between two points on the celestial sphere, for example, is in effect the angle between imaginary lines from the observer in the directions of the two points.

angular momentum   A property analogous to ➤*momentum*, that a body or system of bodies possesses by virtue of its state of rotation or orbital motion. An object with angular momentum will continue to rotate at the same rate unless a torque (i.e. a turning force) acts on it. Within a closed system, on which no outside torques act, the total angular momentum remains constant even if there are changes in the way it is distributed internally. It is a vector quantity, the direction of which is along the axis of rotation or, for orbital angular momentum, along a line perpendicular to the orbital plane.

angular velocity   The rate at which a rotating body sweeps out angle. Angular velocity may be measured in radians, degrees or revolutions per unit time.

anisotropy   The absence of ➤*isotropy*. The properties of an anisotropic object or system are dependent on direction.

annual aberration   ➤*aberration* (2).

annual parallax   (heliocentric parallax) The difference between the position of a star as seen from the Earth and as seen by a hypothetical observer at the Sun. The effect of annual parallax is observed as a shift in the positions of relatively nearby stars against the background of more distant ones during the Earth's yearly journey in orbit round the Sun. If the position of a nearby star is plotted over a year, it appears to sweep out an ellipse on the sky, called the *parallactic ellipse*.

The annual parallax is formally defined as the difference in position that would be measured by hypothetical observations made from the centre of the Earth and the centre of the Sun. (See illustration on page 16.)

annular eclipse   A solar ➤*eclipse* in which a ring of the Sun's photosphere remains visible when the Sun, Moon and Earth are aligned. Since the orbits of the Earth around the Sun and of the Moon around the Earth are elliptical, the ➤*angular diameters* of the Sun and Moon vary slightly as their distances from the

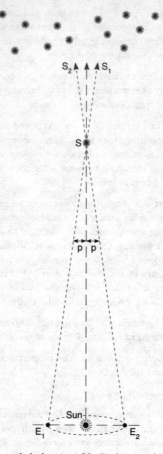

annual parallax. E_1 and E_2 mark the location of the Earth in its orbit on two dates six months apart. Over the six months, the apparent position of star S changes from S_1 to S_2, an angular displacement equal to twice the annual parallax, p. (Not to scale.)

Earth change. A solar eclipse that would otherwise have been total is seen as annular if the Moon's angular diameter at the time is less than the Sun's.

anomalistic month   The time taken by the Moon to orbit the Earth from ➤*perigee* to perigee, equal to 27.554 550 days. ➤*month*.

anomalistic year   The time between successive passages through ➤*perihelion* by the Earth in its orbit around the Sun. Its length, 365.259 64 days, is greater

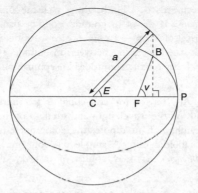

Anomaly. The geometry of true anomaly, v, and eccentric anomaly, E.

than that of the ►*tropical year* by about 27 minutes because of the gradual change in position of the perihelion point.

anomaly   An angle used in describing the motion of a body in an elliptical orbit. The *true anomaly*, v, is the angle between the line joining the body B to the focus of the ellipse, F, and the line joining F to *periapsis* the point on the orbit closest to F (see illustration). The *mean anomaly*, M, is the angle between the line PF and the line joining F to a hypothetical body that has the same orbital period as the real one under consideration but travels at a uniform angular speed. The *eccentric anomaly*, E, is a useful parameter for expressing the variable length of the radius vector, r. The linking equation is $r = a (1 - e \cos E)$ where a is the semimajor axis and e the eccentricity of the elliptical orbit. The relationship between M and E, $M = E - e \sin E$, is known as *Kepler's equation*.

anorthosite   An igneous rock, consisting primarily of the silicate mineral plagioclase. It is an important constituent of the highland regions on the Moon.

ANS   The first Dutch national satellite (Astronomische Nederlands Satelliet), launched on 30 August 1974. It carried experiments in ultraviolet and X-ray astronomy.

ansae (sing. ansa)   The Latin word meaning 'handles', used to describe the appearance created by the protrusion of the rings of ►*Saturn* either side of the planet's disc, as viewed through a telescope.

antapex   The point on the celestial sphere in the diametrically opposite direction to the solar ►*apex*. The whole solar system is moving directly towards the apex and away from the antapex relative to stars in the Sun's vicinity. The antapex lies in the constellation Columba (at about RA 6h, Dec. −30°).

Antares (Alpha Scorpii; α Sco)   The brightest star in the constellation Scorpius. It is a red ➤*supergiant* ➤*M star* of magnitude 1.0. The name is derived from Greek, meaning 'rival of Mars' – a reference to its noticeable colour. It is a ➤*semiregular variable* star, fluctuating between magnitudes 0.9 and 1.1 over a five-year timescale. It has a sixth magnitude blue companion only 3 arc seconds away. ➤Table 3.

antenna (aerial)   Any device for collecting or transmitting radio signals. The design depends on the wavelength and strength of the signal. The simplest form is a straight rod, or dipole; the commonest type used in radio astronomy is a paraboloid dish. A simple telescope may use just a single antenna. ➤*Radio interferometers* may consist of an array of many individual antennas.

Antennae Galaxies   A popular name for the pair of interacting galaxies NGC 4038 and 4039. The name arises from two long, curved streamers of stars, created when they collided. The galaxies are 48 million light years away and the streamers about a hundred thousand light years long.

antenna temperature   A parameter used in radio astronomy as a measure of the power per unit bandwith (p) of the signal received by an antenna after losses in the detector system. It is defined as p/k, where k is ➤*Boltzmann's constant*.

anthropic principle   The recognition that only a limited number of all the theoretically possible universes are favourable to the emergence of life. In theory, a large range of universes with different physical properties and values of the constants of nature could exist. The anthropic principle states that only a restricted class of such models can have intelligent observers. Since we exist, the universe we inhabit must have characteristics within the narrow range that has permitted our evolution. This basic expression of the anthropic principle is not generally regarded as controversial, and is sometimes called the *weak anthropic principle*.

   The so-called *strong anthropic principle*, proposed by Brandon Carter, is more speculative. This asserts that, because there are so many apparently unconnected coincidences in nature, which together have made it possible for life to develop, the universe *must* give rise to observers at some stage in its development.

anticyclone   A region in a planetary atmosphere in which pressure increases towards the centre.

antimatter   Matter composed of elementary particles that have masses and spins identical to those of the particles that make up ordinary matter, but with many other properties, such as electric charge, reversed. Although some *antiparticles* are observed in nature and others are produced in the laboratory, there is no evidence for the existence of large amounts of antimatter, for example, in the form of 'antihydrogen'. If ordinary matter and antimatter were to meet, they would annihilate each other with the release of energy.

Antinoüs   Asteroid 1863, diameter 3 km. It was discovered by A. Wirtanen in 1948, when it made a close approach to the Earth, and was recovered in 1972.

antitail   Part of a comet's ➤*dust tail* which appears to protrude forwards towards the Sun from the comet's head, sometimes like a spike. The effect is one of perspective, due to the relative positions of the Earth and the comet, which determine the angle at which the curving dust tail is viewed.

Antlia (The Air Pump)   A small, faint southern constellation introduced in the mid-eighteenth century by Nicolas L. de Lacaille. It was originally called Antlia Pneumatica. ➤Table 4.

Antoniadi scale   A scale of five points, devised by the French astronomer Eugenios Antoniadi (1870–1944) and widely used by amateur astronomers for describing the quality of ➤*seeing*. The points on the scale are defined as follows: I, perfect; II, slight undulations with periods of virtually perfect seeing lasting for several seconds; III, moderately good seeing, though with noticeable air movements; IV, poor seeing making observations difficult; V, very bad seeing that permits no useful observation.

Apache Point Observatory   An observatory in New Mexico, USA, owned and operated by a consortium of universities: New Mexico State University, the University of Washington, the University of Chicago, Princeton University and Washington State University. The principal instrument is an altazimuth 3.5-metre (138-inch) telescope for both optical and infrared observations. The main mirror is of Pyrex honeycomb produced by ➤*spin casting*, a construction method that results in a mirror five times lighter than a solid one of the same size. A 2.5-metre (98-inch) telescope for the Sloan Digital Sky Survey, together with a 0.6-metre (23½-inch) support telescope for the project, opened in 1997. Its purpose is to collect imaging and spectroscopic data on hundreds of millions of astronomical objects, including faint galaxies. There is also a 1.0-metre (39-inch) telescope at the site, belonging to New Mexico State University. The observatory opened in late 1990.

apastron   The orbital positions of the two members of a ➤*binary star* system when they are furthest apart.

Apennines (Montes Apenninus)   A range of mountains on the Moon, forming part of the eastern boundary of the Mare Imbrium.

aperture (symbol *D*)   The diameter of the main collecting element in a ➤*telescope*. For an optical telescope, the aperture refers to the diameter of the objective lens or mirror; for a radio telescope it is the physical size of the ➤*antenna*. The aperture is one of the most important characteristics of a telescope, since radiation-gathering capacity and resolving power both increase with the size of aperture.

aperture synthesis A technique developed by radio astronomers, more recently applied also in the infrared and optical wavebands, that makes it possible to obtain maps or images with the ➤*resolving power* of a very large aperture by combining the observations made with a number of smaller ➤*antennas* or mirrors.

In the simplest radio astronomy example, two antennas are used as a ➤*radio interferometer*, and the ➤*phase* and ➤*amplitude* of the combined radio signal are measured continuously. As the Earth rotates in the course of a day, one antenna is automatically carried around the other, effectively sweeping out a ring. On successive days the separation between the two antennas is changed, so that a large elliptical area is gradually covered. When the records are combined in a computer, it is possible to produce a radio map of the section of sky under observation with the detail resolved as if the telescope aperture were the size of the total area swept out.

In practice, more than two antennas are normally used to speed up the process and give greater flexibility. It is also possible to combine observations made at different sites, separated by thousands of kilometres, to obtain even better resolution.

Technological developments in the 1990s have made it possible to apply the same physical principle to obtain high-resolution images in the optical and infrared. Pioneering instruments include the Cambridge Optical Aperture Synthesis Telescope (COAST) in the UK and the Navy Prototype Optical Interferometer (NPOI) at the US Naval Observatory's site near Flagstaff, Arizona. ➤*Earth rotation synthesis, interferometer.*

apex (1) The point on the celestial sphere in the constellation Hercules (at about RA 18h, Dec. +30°) towards which the solar system as a whole appears to be moving with respect to other stars in the Sun's vicinity.

apex (2) At any given time, the point on the celestial sphere towards which the Earth appears to be moving due to its orbital motion around the Sun.

aphelion (pl. aphelia) The point furthest from the Sun in the orbit of a body, such as a planet or comet, that is travelling around the Sun.

Aphrodite Terra A highland area on the surface of Venus, roughly equivalent in size to the continent of Africa.

Ap J A common abbreviation for the ➤*Astrophysical Journal.*

APM Machine Abbreviated form of ➤*Automatic Plate Measuring Machine.*

apocentre The point in the orbit of a component of a ➤*binary star* system that is furthest from the ➤*centre of mass* of the system.

apochromat   A composite lens, with three (or more) elements, designed to minimize ➤*chromatic aberration.* ➤*achromatic lens.*

apodization   The use over a telescope aperture of a screen that progressively reduces, from the centre to the edge of the aperture, the amount of light transmitted down the instrument. The purpose is to reduce diffraction effects that tend to blur fine detail in images of the planets. A practical apodizing screen consists of three or four ring-shaped zones.

apogee   The point furthest from the Earth in the orbit of the Moon or of an artificial Earth satellite.

Apollo   Asteroid 1862, diameter 1.4 km, discovered by K. Reinmuth in 1932. It is the prototype of the *Apollo group* of asteroids, whose orbits cross that of the Earth.

Apollo programme   An American space programme, which in 1969 successfully achieved its objective: a manned landing on the Moon. The programme consisted of 17 missions in all. Numbers 1 to 6 were unmanned test flights and *Apollo 13* was aborted following an explosion on board, though the astronauts were returned safely to Earth. Six Moon landings took place between 20 July 1969 and 11 December 1972. The astronauts collected samples of lunar rocks and soils weighing a total of nearly 400 kilograms, and took many photographs both on the surface and from lunar orbit. A variety of scientific experiments were carried out on the surface of the Moon, including ones to detect ➤*cosmic rays* and the ➤*solar wind.*

The Apollo craft consisted of three modules: the Command Module (CM), the Service Module (SM) and the Lunar Module (LM). The Command and

Manned Moon landings

Apollo	Astronauts	Landing date	Landing site
11	Armstrong, Aldrin, Collins	20 July 1969	Mare Tranquillitatis
12	Conrad, Bean, Gordon	19 November 1969	Oceanus Procellarum
14	Shepard, Mitchell, Roosa	5 February 1971	Fra Mauro
15	Scott, Irwin, Worden	30 July 1971	Hadley Rille
16	Young, Duke, Mattingly	21 April 1972	Cayley–Descartes highland region
17	Cernan, Schmitt, Evans	11 December 1972	Taurus–Littrow region

Service Modules (CSM) remained in lunar orbit with one astronaut on board while the other two astronauts made the descent to the Moon's surface in the Lunar Module. The descent stage was left on the Moon when the astronauts returned to lunar orbit by means of the ascent stage, and rejoined the Command and Service Modules. The Service Module was jettisoned shortly before re-entry into the Earth's atmosphere.

Apollo—Soyuz project   A joint US/Soviet space project in July 1975 in which an ►*Apollo programme* Command and Service Module docked with a Soviet Soyuz space station in Earth orbit at an altitude of 225 kilometres. The two teams of astronauts were able to visit each other's craft and they performed a number of experiments jointly.

apparent   An adjective used in conjunction with astrophysical quantities (as in 'apparent magnitude') to mean as perceived by an observer on Earth or at a particular locality. Calculations or corrections may be required to deduce the absolute value.

apparent luminosity   The ►*luminosity* of a star or other astronomical object as it appears to an observer on Earth. The apparent luminosity depends on both the actual energy output of the object and its distance.

apparent magnitude (symbol m)   A measure of the relative brightness of a star (or other celestial object) as perceived by an observer on Earth. Apparent magnitude depends on both the absolute amount of light energy emitted (or reflected) and the distance to the object. The smallest numbers correspond to the greatest brightness.

To encompass very bright objects, the scale is extended to zero and on to negative numbers. For example, the apparent magnitude of the full Moon is -12.6 and that of the planet Venus at its brightest is -4.7.

The term apparent magnitude, without further qualification, is usually taken to mean *apparent visual magnitude*, the relative brightness as seen over the visible part of the spectrum. ►*magnitude, absolute magnitude, photometry.*

apparent place   The position on the celestial sphere an object would be in if it were observed from the centre of the Earth. It is calculated by correcting the observed position for the effects of ►*atmospheric refraction*, diurnal ►*aberration* and ►*diurnal parallax.*

apparent solar time   A measure of time based on the actual daily motion of the real Sun. That motion is not uniform because the Sun's path through the sky is inclined to the ►*celestial equator* and because the Earth's orbit around the Sun is elliptical, not circular. Apparent solar noon is the time when the Sun crosses the ►*meridian* of a particular observer and the apparent solar day is the interval between two successive meridian passages. The difference between apparent solar time and mean solar time, which varies through the year, is called the ➤*equation of time*.

apparition   The period of time during which it is possible to observe a celestial object, such as a planet or comet, that is visible only on a single occasion or that reappears in the sky from time to time.

appulse   A close approach of one celestial object to another, in such a way that the two seem to touch as perceived by the observer, without an ➤*occultation* actually taking place. A typical example is when a planet narrowly misses occulting a star.

April Lyrids   ➤*Lyrids*.

apsidal motion   The rotation of the line of ➤*apsides* of an elliptical orbit, caused by the perturbing gravitational effect of one or more other objects.

apsides (sing. apse or apsis)   The two points in an elliptical orbit lying furthest from and closest to the ➤*centre of mass*. The line joining them, lying along the direction of the major axis of the orbital ellipse, is called the *line of apsides*.

Ap star   An ➤*A star* with an unusual spectrum (p stands for 'peculiar') in which the absorption lines of certain elements are exceptionally strong. The hottest stars of this kind are of ➤*spectral type* B and are designated *Bp stars*. There are several types of Bp and Ap star, each showing different characteristics. The spectral lines enhanced include those of silicon, manganese, mercury, chromium, europium and strontium. Almost all have strong magnetic fields and some show variations in their spectra.

APT   Abbreviation for ➤*automatic photometric telescope*.

Apus (The Bird of Paradise)   A faint constellation near the south celestial pole, introduced probably by sixteenth-century navigators and included by Johann Bayer in his atlas ➤*Uranometria*, published in 1603. ➤Table 4.

Aquarids   Either of two distinct ➤*meteor showers*. The Eta Aquarids are observed between about 24 April and 20 May, peaking around 4 or 5 May. This is a fine southern shower, associated with ➤*Halley's Comet*. The radiant is at RA 22h 20m, Dec. −1°.

  The Delta Aquarids occur between 15 July and 20 August, peaking on 29 July and 7 August. The radiant is double, the two components lying at RA 22h 36m, Dec. −17° and RA 23h 04m, Dec. +2°.

Aquarius (The Water Carrier)   One of the twelve zodiacal constellations listed by Ptolemy (c. AD 140). It is one of the larger constellations but contains no very bright stars. ➤Table 4.

Aquila (The Eagle)   A small but prominent northern constellation, among those listed by Ptolemy (c. AD 140). It is said to represent the eagle of classical mythology sent by Jupiter to carry Ganymede to Olympus. It contains one of the brightest stars, ➤*Altair*. ➤Table 4.

Ara (The Altar) A small, faint, southerly constellation among those listed by Ptolemy (c. AD 140). ➤Table 4.

arachnoid An informal term for a category of volcanic features on ➤*Venus*, revealed by the ➤*Magellan* mission, which resemble spiders connected by a web of fractures.

Arcadia Planitia A plain in the northern hemisphere of Mars.

Arcetri Astrophysical Observatory An observatory in Florence, Italy, founded in 1872 as a memorial to Galileo, who spent his period of house arrest near the site of the observatory. It is now primarily a research institute funded by the Italian government. The instruments at the site are an old 36-cm (14-inch) refractor and a solar tower. A 1.5-metre (60-inch) infrared telescope is located at an altitude of 3,200 metres (10,500 feet) near Zermatt.

archaeoastronomy The study of the practice of astronomy in civilizations and societies of prehistory. In particular, archaeoastronomy is concerned with archaeological evidence for astronomical knowledge rather than written records. Sites that are studied include the stone-age remains in western Europe, ancient meso-America and the classical Mediterranean civilizations.

Archer English name for the constellation ➤*Sagittarius*.

arc minute A unit in which small angles are measured, equal to one-sixtieth of a degree.

arc second A unit in which very small angles are measured, equal to one-sixtieth of an ➤*arc minute*.

Arcturus (Alpha Boötis; α Boo) The brightest star in the constellation Boötes. It is an orange, giant ➤*K star* of magnitude −0.04, the fourth-brightest star in the sky. The name, of Greek origin, means 'bear-watcher'. ➤Table 3.

Arecibo Observatory A radio astronomy observatory in Puerto Rico, where a dish 305 metres (1,000 feet) across has been built into a natural depression in hills south of the city of Arecibo. Completed in 1963, the telescope is operated by the National Ionospheric and Astronomy Center of Cornell University in the USA. The reflecting surface cannot be moved, but radio sources can be tracked by moving the receiver at the focus along a specially designed support structure. A major refurbishment was completed in 1997. The telescope is larger in area than all the other radio telescopes in the world combined. It is used for radar studies of planets, observing ➤*pulsars* and the study of hydrogen in distant galaxies. Because of its large collecting area, it can pick up fainter signals than any other dish.

Arend−Roland, Comet ➤*Comet Arend−Roland.*

areography The study and mapping of the surface features of Mars.

areology The geology of the planet Mars.

Arethusa Asteroid 95, diameter 228 km. It is particularly dark, with an albedo of only 1.9 per cent.

Argo Navis A large constellation of the southern sky, representing the ship of Jason and the Argonauts, listed by Ptolemy (*c.* AD 140). Its excessive size led to difficulties for astronomers and it is no longer recognized officially. Stars that formerly made up Argo are now assigned to three separate constellations: ➤*Carina* (The Keel), ➤*Puppis* (The Stern or Poop) and ➤*Vela* (The Sails).

argument (symbol ω) One of the angles forming the set of ➤*orbital elements* used to define an elliptical orbit. It describes the orientation of an orbit within its plane. The argument of perihelion for the orbits of comets or planets round the Sun is the angle perihelion−Sun−ascending node, measured in the orbital plane in the direction of motion.

Argyre Planitia A circular impact basin, 900 kilometres (550 miles) in diameter, in the southern hemisphere of Mars.

Ariel (1) A series of six satellites, launched between 1962 and 1979, the first four of which were devoted to the study of the ionosphere, and the last two to X-ray astronomy. Ariel 5, launched on 15 October 1974, was very successful and made many important X-ray observations. Several catalogues of sources were produced. The first five satellites were operated jointly by the USA and the UK. Ariel 6 was a purely British X-ray project, but suffered technical problems and produced few results.

Ariel (2) One of the larger satellites of Uranus, discovered by W. Lassell in 1851. Images obtained by the ➤*Voyager 2* mission in 1986 showed the surface to be heavily cratered and crossed by fault scarps and valleys. Its appearance suggests that there has been considerable geological activity in the past. ➤Table 6.

Aries (The Ram) A small zodiacal constellation, listed by Ptolemy (*c.* AD 140). Two thousand years ago it was recognized that the Sun's position in the sky at the vernal ➤*equinox* was in Aries, though the effects of ➤*precession* have since moved this position into Pisces. The constellation is said to represent the ram of the golden fleece sought by Jason in classical mythology. Its brightest star is the second magnitude ➤*Hamal.* ➤Table 4.

Aristarchus A very bright lunar crater, the centre of a conspicuous ray system. It is 45 kilometres (28 miles) across and has multiple terracing on the inner walls. Temporary reddish glows have occasionally been reported, perhaps caused by gas being released from the surface rocks. ➤*transient lunar phenomenon*.

Arizona meteorite crater (Meteor Crater; Barringer Crater; Coon Butte) The best-preserved and most famous of ➤*meteorite* craters on the Earth, 1,200 metres (4,000 feet) in diameter, 183 metres (600 feet) deep and surrounded by a wall 30 to 45 metres (100 to 150 feet) high. It is located between Flagstaff and Winslow in Arizona, USA, and was discovered in 1891. The age is estimated at 50,000 years. Quantities of meteoritic iron have been found in the vicinity; the original meteorite is thought to have been of the iron type and to have weighed more than 10,000 tonnes, but most was destroyed on impact. The many scattered fragments of what is called the Canyon Diablo meteorite total only about 18 tonnes.

Armagh Observatory An observatory in Armagh, Northern Ireland, founded in 1790. The observatory and its associated planetarium, which opened in 1968, are now government funded as an educational establishment and there are no research instruments. The observatory is noted for the compilation there of the ➤*New General Catalogue of Nebulae and Clusters of Stars* (NGC) by J. L. E. Dreyer, Director from 1882 to 1916.

armillary sphere A type of celestial globe in which the sphere of the sky is represented by a skeletal framework of intersecting circles, with the Earth at the centre. The rings in the framework represent important circles on the celestial sphere, such as the ➤*celestial equator* and the ➤*ecliptic*. Some of the rings may be movable so that the sky's appearance at different times and at different latitudes may be reproduced. On some armillary spheres, the positions of the brighter stars are shown by small pointers attached to the fixed rings. The use of armillary spheres, for both observations and demonstrations, dates from at least the third century BC.

Arp Catalogue A catalogue of ➤*peculiar galaxies* (*Atlas of Peculiar Galaxies*) compiled by Halton Arp and published in 1966.

array An arrangement of a number of linked radio ➤*antennas*, which together constitute a ➤*radio telescope*.

Arrow English name for the constellation ➤*Sagitta*.

Arsia Mons One of the large ➤*shield volcanoes* in the ➤*Tharsis Ridge* region of Mars. It is about 350 kilometres (220 miles) in diameter and rises to a height of 27 kilometres (17 miles), 17 kilometres above the level of the surrounding ridge.

ASCA (Advanced Satellite for Cosmology and Astrophysics) A Japanese/

US X-ray astronomical satellite launched in 1993. Before launch it was known as Astro-D.

ascending node   The point where an orbiting object crosses the plane of reference for its orbit, travelling from south to north. The opposite point in the orbit, where the object crosses the plane from north to south, is the *descending node*. For a planet or comet, the reference plane is the ➤*ecliptic*.

Ascraeus Mons   One of the prominent ➤*shield volcanoes* of the ➤*Tharsis Ridge* region of Mars. It is about 250 kilometres (150 miles) in diameter and rises to a height of 27 kilometres (17 miles), 17 kilometres above the level of the surrounding ridge.

ashen light   A dim glow that visual observers occasionally claim to see on the dark part of Venus when the planet is at crescent phase. In the absence of photographic evidence, the reality of the phenomenon is questioned by some, though it has been reported by many different observers. The origin is unknown but, if it is a real physical effect in the atmosphere of Venus, it may be similar to the ➤*airglow* in the Earth's atmosphere.

Asiago Observatory   The main observational facility of the Department of Astronomy at the University of Padua in northern Italy. There are telescopes at two locations. The original instrument, a 1.22-metre (48-inch) reflector inaugurated in 1942, is located in the foothills of the Alps, 90 kilometres (60 miles) to the north of Padua. The 1.82-metre (72-inch) Copernicus Reflector, inaugurated in 1973, is located at a higher site – Cima Ekar, 12 kilometres (7 miles) to the east. There is also a 0.67-metre (26-inch) ➤*Schmidt camera* at this site.

ASP   Abbreviation for ➤*Astronomical Society of the Pacific*.

aspect (1)   The position of a planet or the Moon, relative to the Sun, as viewed from the Earth.

aspect (2)   The angle between the rotation axis of a body in the solar system and the radius vector between that body and the Earth.

association   A loose grouping of young stars, typically with between ten and a hundred members. Stellar associations are found along the spiral arms of the ➤*Galaxy*; about seventy examples are known. All are relatively young in astronomical terms. The member stars are not closely bound by mutual gravitational attraction, and their differing velocities result in the break-up of their association within a few million years. Members of an association must, therefore, have been born together, and relatively recently, from the same star-forming cloud. Associations are always found along with interstellar matter, from which, it may be presumed, they were created.

Two types of association are distinguished. *O* or *OB associations* are made up of massive and luminous ➤*O stars* and ➤*B stars* scattered through a region up to several hundred light years across. *T associations* contain numerous, low-mass ➤*T Tauri stars.*

Association of Universities for Research in Astronomy (AURA)   An incorporated body in the USA, composed of a consortium of twenty universities, which operates the ➤*National Optical Astronomy Observatories* and the ➤*Space Telescope Science Institute.*

A star   A star of ➤*spectral type* A. A stars have surface temperatures in the range 7,500−11,000 K and are white in colour. The most prominent features in their absorption line spectra are the ➤*Balmer lines* due to hydrogen atoms. Lines of heavier elements, such as iron, are also noticeable at the cooler end of the temperature range. Examples of A stars are ➤*Sirius* and ➤*Vega.*

asterism   A prominent pattern of stars, usually with a popular name, that does not constitute a complete ➤*constellation*. Well-known examples of asterisms are the ➤*Plough* (Big Dipper) in Ursa Major and the ➤*Sickle* in Leo.

asteroid (minor planet)   A small rocky object in the solar system. The largest, ➤*Ceres,* is nearly 1,000 kilometres (600 miles) across and they range in size down to dust particles. Many thousands have been individually identified and it is believed that there could be half a million with diameters larger than 1.6 kilometres (1 mile). However, the total mass of all asteroids is less than one-thousandth the mass of the Earth. Most asteroid orbits are concentrated in the ➤*asteroid belt* between Mars and Jupiter at distances ranging from 2.0 to 3.3 AU from the Sun. There are however asteroids in orbits nearer the Sun, such as the ➤*Amor* group, the ➤*Apollo* group and the ➤*Aten* group, and some more distant from the Sun, such as the ➤*Centaurs*. The ➤*Trojan asteroids* share Jupiter's orbit.

Asteroids can be classified according to their spectra of reflected sunlight: 75 per cent are very dark, carbonaceous ➤*C-types*, 15 per cent are greyish, silicaceous (stony) ➤*S-types* and the remaining 10 per cent consist of the ➤*M-types* (metallic) and a number of rare varieties. The classes are linked to the known types of ➤*meteorite*. The evidence suggests that many asteroids and meteorites have similar compositions, so asteroids may be the parent bodies of meteorites. The darkest asteroids reflect only 3−4 per cent of the sunlight falling on them, while the brightest reflect up to 40 per cent. Many vary regularly in brightness as they rotate.

In general, asteroids are irregularly shaped; the smallest asteroids rotate the most rapidly and are the most irregular in shape. The ➤*Galileo* spacecraft, on its way to Jupiter, flew by two asteroids, ➤*Gaspra* (on 29 October 1991) and ➤*Ida* (on 28 August 1993). Detailed images showed their rocky surfaces to be pitted with numerous craters, and that Ida has a small satellite. From the ground, it is possible to obtain information about the three-dimensional structure of asteroids through radar studies using the large radio dish of the ➤*Arecibo Observatory*.

Asteroids are believed to be the remnants of the material from which the solar system formed. This view is supported by the way the predominating asteroid type changes with increasing distance from the Sun within the asteroid belt. High-speed collisions between asteroids are now gradually resulting in their break-up.

asteroid belt   The region of the solar system, between 2.0 and 3.3 AU from the Sun, where the vast majority of ➤*asteroid* orbits lie. Within the belt, there are both concentrations of orbits, forming groups and families, and regions that are avoided, known as the ➤*Kirkwood gaps*. The proportions of different types of asteroid change markedly through the belt. At the inner edge, 60 per cent are silicaceous and 10 per cent carbonaceous; at the outer edge the situation is reversed with 80 per cent carbonaceous and 15 per cent silicaceous. The belt marks the transition zone between the inner and outer solar system.

asthenosphere   A layer in the Earth's ➤*mantle* at a depth of between 100 and 250 kilometres (60 and 150 miles) over which the more rigid plates of the ➤*lithosphere* are in motion.

astigmatism   An imperfection in the imaging properties of a lens or mirror. Light passing through different parts of an astigmatic lens, for example, is focused at different distances beyond the lens. Thus, the image of a point can appear variously as a line or an ellipse. The best image obtainable with such a lens is a small circle called the *circle of least confusion*.

Astraea   Asteroid 5, diameter 120 km, discovered in 1845 by K. L. Hencke.

astration   The cyclic process in which interstellar matter is incorporated into newly formed stars, where it undergoes nuclear processing, is thus enriched with heavier elements, and is then expelled again into the interstellar medium to be used in the next generation of stars. Astration therefore results in a steady increase in the proportion of heavier elements in a galaxy.

Astro-B   A Japanese X-ray astronomy satellite launched on 20 February 1983. It was renamed Tenma ('Pegasus') after launch.

astrobleme   An ancient, greatly eroded impact ➤*crater*.

Astro-C   ➤*Ginga*.

astrochemistry   The study of chemical reactions between atoms, molecules and dust grains in the interstellar medium, including phases of star and planet formation.

Astro-D   ➤*ASCA*.

astrodynamics The science concerned with all aspects of the motion of satellites, rockets and spacecraft.

astrograph An astronomical telescope designed specifically to take wide-angle photographs of the sky for use in the measurement of positions. The term is applied particularly to the refracting telescopes constructed for the ➤*Carte du Ciel* project. These were largely modelled on the 330-mm (13-inch) telescope built at the Paris Observatory in 1886. Astrographic telescopes have now been superseded by the ➤*Schmidt camera*.

Astrographic Catalogue A catalogue of the positions of stars down to twelfth magnitude, drawn up as part of the ➤*Carte du Ciel* project.

astrolabe (1) An ancient instrument for showing the positions of the Sun and bright stars at any time and date. Its invention is credited to Greek astronomers who worked in the second century BC.

The basic astrolabe consists of a circular star map (the 'tablet' or 'tympan') with a graticule (the 'rete') over the top, the two being joined at their common centre so that the rete can rotate over the tablet. Typically, it would be made of brass. Various engraved scales enable the positions of the stars and the Sun to be displayed for any time and date. There may be other scales giving further information. Astrolabes were often fitted with a sight on a movable arm so that they could be used to estimate the ➤*altitudes* of stars, for navigational purposes, for example. However, the use of any particular astrolabe is restricted to places within the range of latitude for which it is constructed.

astrolabe (2) Any instrument for determining accurate positions of celestial objects. ➤*prismatic astrolabe*.

astrology The practice of a tradition that purports to connect human traits and the course of events with the positions of the Sun, Moon and planets in relation to the stars. No theory that would put the claims of astrology on a scientific basis has ever been generally accepted. Reports of anecdotal and circumstantial evidence, and coincidences, continue to attract a following for astrology though most scientists regard it as pure superstition.

In the past, there was less of a clear distinction between astrology and the science of astronomy; many useful astronomical observations were made for astrological purposes.

astrometric binary A ➤*binary star* in which the presence of an unseen companion is revealed by cyclic irregularities in the position of the brighter star.

astrometry The branch of astronomy concerned with the measurement of the positions and apparent motions of celestial objects in the sky and the factors that can affect them.

astronautics The science concerned with all aspects of space travel.

Astronomer Royal   Formerly the title of the director of the Royal Greenwich Observatory in the UK but, since 1972, an honorary title bestowed on a distinguished astronomer who is not necessarily the director of the Royal Observatory.

Astronomers Royal

John Flamsteed 1675–1719	Sir Frank Watson Dyson 1910–33
Edmond Halley 1720–42	Sir Harold Spencer Jones 1933–55
James Bradley 1742–62	Sir Richard Woolley 1956–71
Nathaniel Bliss 1762–4	Sir Martin Ryle 1972–82
Nevil Maskelyne 1765–1811	Sir Francis Graham-Smith 1982–90
John Pond 1811–35	Sir Arnold Wolfendale 1991–5
Sir George Biddell Airy 1835–81	Sir Martin Rees 1995–
Sir William Christie 1881–1910	

Astronomia nova   A book by Johannes Kepler (1571–1630), published in 1609, which included his first two laws of planetary motion. ➤*Kepler's laws.*

Astronomical Almanac   An ➤*almanac* designed primarily for use by professional astronomers, published jointly since 1981 by the United States Naval Observatory, Washington, DC, and the Royal Greenwich Observatory in the UK. It replaced the ➤*American Ephemeris and Nautical Almanac* and the ➤*Astronomical Ephemeris.* The data listed include the phases of the Moon, sunrise and sunset, eclipses, the positions of the Sun, Moon and planets, information about bright stars, locations of observatories and astronomical constants.

astronomical clock   A clock that displays astronomical information, such as the phases of the Moon and sidereal time, in addition to the normal time of day.

Astronomical Ephemeris   A compendium of astronomical data and information, formerly compiled annually at the Royal Greenwich Observatory in the UK, until it was replaced in 1981 by the ➤*Astronomical Almanac.*

Astronomical Journal (*AJ*)   A journal for the publication of astronomical research, established in the USA in 1849 by B. A. Gould. It is published on behalf of the ➤*American Astronomical Society.*

Astronomical Society of the Pacific (ASP)   A society, for both professional and amateur astronomers, which was founded in 1898 and has its headquarters in San Francisco, California. It produces a popular magazine, *Mercury*, and a journal of research papers, *Publications of the Astronomical Society of the Pacific.*

astronomical triangle   A particular spherical triangle on the ➤*celestial sphere* whose three corners are a celestial object, the observer's ➤*zenith* and the north or south ➤*celestial pole.*

astronomical twilight   Formally defined as the interval of time during which the Sun is between 102° and 108° below the ➤*zenith* point. ➤*twilight*.

astronomical unit (AU or a.u.)   A unit of measurement used mainly for distances within the solar system. It is effectively the mean distance between the Earth and the Sun, though it now has a formal definition independent of the Earth's orbit. Its value is 149,597,870 kilometres (92,955,730 miles), which is slightly less than the ➤*semimajor axis* of the Earth's orbit. There are about 63,240 astronomical units in a light year.

Astronomische Nachrichten (*AN*)   A German astronomical journal for the publication of research. First published in 1821, it was one of the first modern-style journals on astronomy.

astronomy   The study of the universe and its contents beyond the bounds of the Earth's atmosphere. Until the twentieth century the term was associated specifically with the study of the movements and positions of celestial objects. ➤*astrophysics*.

Astro-1   An astronomical observatory for operation on board an orbiting ➤*Space Shuttle*. The observatory consists of four telescopes. Three operate in the ultraviolet, specializing in spectroscopy, direct imaging and polarimetry, and one is a broad-band X-ray instrument. The ultraviolet instruments are operated by mission and payload specialists on board the Shuttle. The X-ray telescope is operated from the ground by Goddard Space Flight Center personnel. Astro-1 was flown successfully in December 1990.

astrophotography   A contraction of 'astronomical photography'. The practice of recording celestial objects photographically.

Astrophysical Journal (*Ap J*)   One of the world's foremost journals for the publication of research in astronomy and astrophysics. It was founded in 1895 by George Ellery Hale and is published by the University of Chicago Press on behalf of the ➤*American Astronomical Society*.

astrophysics   The physical theory of astronomical objects and phenomena. The term was introduced in the nineteenth century to draw a distinction between the application of physics to interpret observations and the recording of positions, movements and phenomena that characterized astronomy of an earlier era. Thus it encompasses topics such as the structure and stability of stars, the propagation of electromagnetic radiation in space and the production of spectra, nuclear processes in stars and applications of gravitational theory.

Though the term astrophysics retains its original meaning, 'astronomy' is now generally considered to encompass all aspects of the study of the universe, including astrophysics. The scientific method we use today means that all astronomical work must draw on a knowledge of physics.

astroseismology The study of the propagation of oscillations in stars. In principle, such studies can provide information about the structure of stars in the same way that seismic studies of the Earth reveal details of its structure.

asymptotic giant branch (AGB) A region in the ➤*Hertzsprung–Russell diagram* delineated by points representing stars that initially had masses no greater than ten times the Sun's, and are currently in a late phase of their evolution. For a second time since they ceased to be ➤*main-sequence stars,* vast expansion has resulted in a combination of high luminosity and low temperature that places them among the coolest of the ➤*red giant* stars. All stars of under ten solar masses pass through this phase of ➤*stellar evolution.*

AGB stars have exhausted both hydrogen and helium as nuclear fuels in their cores. The core has become a very hot ➤*white dwarf* composed of carbon and oxygen. A thin shell of helium and carbon overlying the core undergoes recurrent helium 'flash' burning. Between flash episodes, hydrogen burning continues in a shell layer further out. The whole is surrounded by a greatly extended envelope formed from the star's outer layers, with a radius of between a hundred and a thousand times that of the Sun. The internal structure is unstable and these stars form the ➤*long-period variables.* Ultimately, the envelope separates from the core and is ejected to form a ➤*planetary nebula.* A star initially of ten solar masses ejects all but 1.4 solar masses in this way.

ataxite A type of iron ➤*meteorite* with a high nickel content and no obvious structure. The ➤*Widmanstätten figures* associated with other types of iron meteorite are, if present at all, very fine.

Aten Asteroid 2062, diameter 0.8 km, discovered in 1976 by E. Helin. It is the prototype of the *Aten group* of Earth-approaching asteroids that have orbits lying mainly within that of the Earth. They have semimajor axes of less than 1 AU and aphelion distances greater than 0.938 AU.

Atlas (1) The innermost small satellite of Saturn, discovered in 1980 by Richard Terrile during the ➤*Voyager 1* mission. ➤Table 6.

Atlas (2) A third magnitude star in the ➤*Pleiades* cluster.

atmosphere The outermost gaseous layers of a planet, natural satellite or star. Since gas has a natural tendency to expand into space, only bodies that have a sufficiently strong gravitational pull can retain atmospheres. Mercury and the Moon, for example, are not massive enough to hold on to atmospheric gases. Earth, Venus, Mars and Titan are examples of rocky bodies with substantial atmospheres.

The giant planets, Jupiter, Saturn, Uranus and Neptune, have no solid surface; for them the term 'atmosphere' is used to describe the outermost gas envelope.

The outermost layers of a star, where the features observed in the star's spectrum originate, are called the stellar atmosphere.

atmospheric dispersion   The spreading of a star image into a small spectrum as its light travels through the Earth's atmosphere. The atmosphere acts in the same way as a glass prism: the path the light takes depends to a small extent on its wavelength. As a result, the blue light from a star appears to come from slightly nearer the zenith than the red light. ➤*dispersion*.

atmospheric extinction   The reduction in intensity of light from an astronomical object by absorption and scattering in the Earth's atmosphere. Extinction increases the nearer the object is to the horizon due to the greater thickness of atmosphere through which its light must travel. Shorter wavelengths are affected more strongly than longer ones. This has the effect of cutting out more blue light than red light, which results in an apparent reddening of the colour of the object.

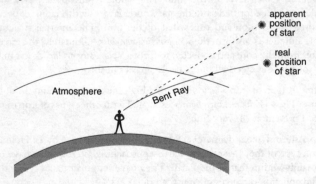

Atmospheric refraction. Due to refraction, starlight follows a curved path through the atmosphere, so a star appears higher than it really is. (Scale of the effect greatly exaggerated here.)

atmospheric refraction   The small change in direction of light rays from astronomical objects as they pass through the Earth's atmosphere, making objects appear to be nearer the ➤*zenith* than they really are. The degree of atmospheric refraction increases the nearer an object is to the horizon.

atmospheric window   A wavelength band in the electromagnetic spectrum that is able to pass through the Earth's atmosphere with relatively little attenuation through absorption, scattering or reflection. There are two main windows: the optical window and the radio window.

In the optical (or visible) region of the spectrum, wavelengths between about 300 and 900 nanometres can pass through the atmosphere. This range

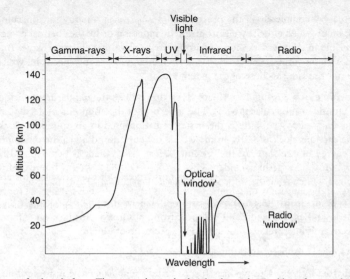

Atmospheric window. The curve shows the height above the Earth's surface at which electromagnetic radiation of different wavelengths is reduced in intensity by a factor of two.

includes near–ultraviolet and infrared radiation, invisible to the human eye. The radio window covers the range between a few millimetres and about 30 metres in wavelength, equivalent to frequencies from 100 GHz to 10 MHz. In addition, there are a number of narrow bands in the infrared (micrometre wavelengths) and submillimetre regions in which the atmosphere is moderately transparent to radiation, particularly in geographical regions where the atmosphere is dry, since the main source of absorption at this wavelength is water molecules.

atom   The smallest stable unit of a chemical element. An atom is typically 0.1 nanometre across. Nearly all the mass is concentrated into a positively charged nucleus, which is a thousand times smaller than the atom as a whole. In a neutral atom the nucleus is surrounded by a cloud of electrons, the number of which matches the positive charge of the nucleus, and is termed the *atomic number*. Each element has a different atomic number, starting with 1 for hydrogen, 2 for helium, and so on.

The electrons can adopt any one of a set of discrete energy levels within an atom. The set of energy levels available is unique to each element. If the electrons make a transition from one energy state to another, electromagnetic energy is either absorbed or emitted. Such transitions result in the production of ➤*absorption line spectra* or ➤*emission line spectra*.

If sufficient energy is absorbed by an atom, one or more electrons may be torn away completely in the process called ►*ionization*. Energy for ionization can be provided either by electromagnetic radiation or by the thermal energy available in a hot gas. An ionized atom carries a positive electric charge. Any ion with one or more electrons has its own energy levels and spectrum, which are not the same as those of the parent atom.

atomic clock   A highly accurate clock that utilizes the regular frequency of an atomic or molecular process. The inversion of the ammonia molecule at a frequency of 23,870 hertz is the basic oscillation used in an ammonia clock. The caesium clock is based on the frequency corresponding to the difference in energy of two states of the caesium ►*atom*; its accuracy is better than one part in 10^{13}. The caesium standard is used to define the SI (Système International) second, which is the basis for ►*International Atomic Time* (TAI).

A-type asteroid   A rare type of asteroid characterized by a moderately high albedo and an extremely red colour. Strong absorption in the near-infrared spectrum is interpreted as indicating the presence of olivine.

AU (a.u.)   Abbreviation for ►*astronomical unit*.

aubrite   A type of stony ►*meteorite*. Aubrites are ►*achondrites* consisting almost entirely of the silicate mineral enstatite.

augmentation   The amount by which the apparent semidiameter of a celestial body is greater, when it is observed from the surface of the Earth, than it would be if observed from the centre of the Earth.

AURA   Abbreviation for ►*Association of Universities for Research in Astronomy*.

Auriga (The Charioteer)   A large and prominent northerly constellation, described from ancient times as representing a charioteer and among those listed by Ptolemy (*c.* AD 140). Its brightest star is ►*Capella,* associated by the Greeks with the mythological she-goat Amalthea, who nurtured the infant Zeus. The nearby triangle of fainter stars, Epsilon (ε), Zeta (ζ) and Eta (η), are collectively called the *Kids*. The star Elnath, formerly designated Gamma (γ) Aurigae and shared with the neighbouring constellation of Taurus, now officially belongs to Taurus as Beta Tauri. ►Table 4.

aurora (pl. aurorae)   A display of rapidly varying, luminous coloured patterns, observed from time to time in the night sky, normally from high-latitude regions of the Earth (both north and south). The coloured lights correspond to the green and red ►*emission lines* from oxygen atoms and nitrogen molecules that have been excited by energetic particles from the Sun. Aurorae occur at heights of around 100 kilometres (60 miles).

A large number of phenomena take place in the ►*ionosphere* during auroral activity, such as pulsations of the ►*geomagnetic field*, electric currents in the ionosphere and the emission of X-rays. Far more energy is emitted in the invisible parts of the spectrum than as visible light.

The occurrence of aurorae is correlated with the ►*solar cycle*, solar rotation, the seasons and magnetic activity.

Aurorae can take several forms. Quiet arcs or bands, several tens of kilometres wide, extend for about 1,000 kilometres in the east−west direction. Bands may be folded into an S-shape or spiral. Rays may be seen, aligned along the magnetic field. Patch or surface aurorae are isolated regions of luminosity with no particular shape. The uncommon veil aurora consists of uniform luminosity covering large areas of the sky.

auroral oval   One of the two oval-shaped belts on the Earth in which aurorae occur. The ovals are disposed asymmetrically around the geomagnetic poles. During the day, they are about 15° of latitude from the poles, increasing to 23° during the night. Their position and width both vary with geomagnetic activity.

auroral zone   The zones on the Earth's surface where the occurrence of observed night-time aurorae is a maximum. The zones are located at latitudes of about 67° north and south, and are about 6° wide.

Australia Telescope   An Australian radio astronomy facility opened in 1988. It consists of a number of antennas at three separate sites in New South Wales and is designed to employ the ►*aperture synthesis* technique of mapping astronomical radio sources. It is the only such array in the southern hemisphere.

The *Compact Array*, located at the Paul Wild Observatory at Culgoora near Narrabri, consists of six antennas, each 22 metres (72 feet) in diameter. Five can be moved along a common east−west track, 3 kilometres (2 miles) long. The sixth is on its own track, a further 3 kilometres to the west.

Greater resolving power is achieved by linking one or more of the antennas in the Compact Array with a seventh new 22-metre dish, 100 kilometres (60 miles) to the south at Mopra, near the Siding Spring optical observatory, and the 64-metre (210-foot) dish at Parkes, which was completed in 1961 and is a further 200 kilometres (120 miles) to the south. Together, these antennas form the *Long Baseline Array*.

A notable feature of the Australia Telescope is the particularly large range of wavelengths at which it can observe. This facility makes it possible to map the emission from interstellar molecules of spectral lines in the radio region of the spectrum.

autoguider   An electronic device that automatically ensures the precise guiding of a telescope during an observation. Even though a telescope may be driven by electric motors to follow the tracks of the stars across the sky, further minor corrections are normally necessary to prevent the field of view drifting somewhat during a long observation. An autoguider contains a photoelectric device that can detect such a drift and activate the drive motors to compensate.

automatic photometric telescope (APT)   A computer-controlled telescope designed to make sequences of observations of the magnitudes of variable stars entirely automatically. From information it has stored, the computer controlling such a telescope can decide when to start observing and which objects to select from the programme. Once an observing programme is set up, the whole process is automated. That includes locating and centring stars in the field of view, taking measurements and storing data, and shutting down if the sky becomes cloudy. With such telescopes, long time-series of variable star observations can be made and plotted as light curves.

Automatic Plate Measuring Machine (APM)   A UK national facility, located at the Institute of Astronomy in Cambridge, which can measure and analyse the images on astronomical photographs. Data on position, brightness and shape are recorded. Typically, half a million images on a 35-centimetre (14-inch) square plate can be processed in 6 hours.

autumnal equinox ➤*equinox.*

averted vision   A technique used by visual observers, making use of the fact that the outer ring of the retina is more sensitive to light than the centre. A deliberate effort is made not to look directly at a faint object, but rather to look to one side of it and be aware of it in the eye's peripheral field of view.

AXAF   Abbreviation for ➤*Advanced X-ray Astrophysics Facility.*

axion   A hypothetical elementary particle, the existence of which has been suggested as a possible explanation for ➤*dark matter* in the universe.

axis   The imaginary line through a body about which it rotates or has rotational symmetry.

azimuth   The bearing of an object measured as an angle round the horizon eastwards starting from north as the zero point. ➤*Altitude* and azimuth are the two coordinates used in the ➤*horizontal coordinate* system.

B

BAA Abbreviation for the ➤*British Astronomical Association*.

Baade's Window An area of sky around the globular cluster NGC 6522 in the constellation Sagittarius, which is particularly rich in stars. The German–American astronomer Walter Baade (1893–1960) drew attention to the area and correctly deduced that it is a region in the disc of our Galaxy relatively clear of interstellar material through which very distant stars in the Galaxy's nuclear bulge can be seen.

background radiation ➤*cosmic background radiation*.

backscattering The scattering of radiation or particles through angles greater than 90° relative to the direction of incidence.

Baikonur The manned space-flight centre established by the former Soviet Union. It is situated north-east of the Aral Sea in Kazakhstan.

Baily A large, highly eroded lunar crater, 298 kilometres (185 miles) in diameter.

Baily's beads A phenomenon observed during the progress towards a total solar eclipse, just before totality and again just after totality. As the Moon gradually obscures the disc of the Sun, the final thin crescent appears to be broken up into a string of bright beads because the mountains and valleys on the Moon make its limb uneven. The English astronomer Francis Baily (1774–1844) drew attention to the phenomenon at the solar eclipse of 1836.

Baker–Nunn camera A form of ➤*Schmidt camera* designed for the photography of artificial satellites.

Baker–Schmidt telescope A form of ➤*Schmidt camera*, incorporating design modifications by J. G. Baker to eliminate aberrations and distortion.

Balmer decrement A marked intensity drop at a wavelength of about 365 nanometres in the ➤*continuous spectrum* of a star (or other astronomical object) in which ➤*absorption* by hydrogen occurs. The individual ➤*Balmer lines* of atomic hydrogen become closer with decreasing wavelength until they merge. The wavelength of 365 nanometres corresponds to the energy required to ionize a hydrogen ➤*atom* when the electron is originally in the second energy level of the atom. At shorter wavelengths (higher energies) the hydrogen effectively causes continuous absorption.

Balmer lines (Balmer series) A series of ➤*spectral lines* in the spectrum of atomic hydrogen. The lines are termed H alpha, H beta, H gamma, and so on, starting with the line of longest wavelength, which is at 656.3 nanometres. The lines become more closely spaced with decreasing wavelength, merging at the series limit of 365 nanometres. ➤*Balmer decrement*.

Bamberga Asteroid 324, diameter 252 km, discovered by J. Palisa in 1892.

band spectrum A spectrum characterized by bands of closely spaced absorption lines in a continuous spectrum. Band spectra result from the presence of molecules rather than single atoms.

bandwidth The range of frequencies or wavelengths to which a detector of electromagnetic radiation is sensitive.

bar A unit in which pressure is measured, particularly planetary atmospheric pressure. One bar is close to the average pressure of the Earth's atmosphere at sea level, and is equal to 10^5 pascals (newtons per square metre). Atmospheric pressures are often quoted in millibars (1,000 mbar = 1 bar).

barium star A giant star with a ➤*spectral type* in the range G2 to K4 that has in its spectrum unusually strong ➤*absorption lines* of the element barium.

Barlow lens A diverging lens used in conjunction with a telescope ➤*eyepiece*. The Barlow lens increases the effective ➤*focal length* of the telescope, which causes the eyepiece to yield a higher ➤*magnification*.

Barnard's Galaxy The galaxy NGC 6822 in Sagittarius.

Barnard's Loop A faint ring of hot gas, forming an ellipse 14° by 10° in the constellation of Orion. It is thought to be the result of ➤*radiation pressure* from the hot stars in the region of Orion's belt and sword acting on interstellar material.

Barnard's Star A ninth magnitude star in the constellation Ophiuchus that has the largest known ➤*proper motion* of any star, a fact discovered by the American astronomer E. E. Barnard in 1916. Its position in the sky changes by 10.3 arc seconds each year as it moves through space relative to the Sun. It is the third-nearest star to the Sun at a distance of 5.88 light years.

Possible 'wobbles' in the motion of Barnard's Star have been interpreted as indicating the presence of unseen planets, but this suspicion has not been confirmed.

barred spiral galaxy A common type of ➤*spiral galaxy* in which the spiral arms apparently emanate from each end of a bright central bar of stars.

barrel distortion ➤*distortion*.

Barringer Crater ➤*Arizona meteorite crater*.

Barwell meteorite   A 46-kilogram stony ➤*meteorite* that fell near the village of Barwell, Leicestershire, UK, in 1965. Though it broke up, it was the largest stony meteorite known to have fallen in the UK.

barycentre   The ➤*centre of mass* of a system of objects moving under the influence of their mutual gravity. The barycentre of the solar system, for example, is constantly moving as the relative positions of the planets (particularly the major ones) change. It lies about a million kilometres from the centre of the Sun.

barycentric coordinates   Coordinates defining the position of a body in the solar system referred to the ➤*barycentre* as origin.

barycentric dynamical time   ➤*dynamical time.*

baryons   A collective name for a group of subatomic particles including protons and neutrons (together termed *nucleons*) and a number of particles of short half-life that produce a nucleon when they decay.

basalt   A volcanic igneous rock, consisting primarily of the silicate minerals pyroxene and plagioclase.

basaltic achondrite   A member of a group of ➤*meteorite* types, including eucrites and howardites, which are similar to terrestrial basalts. ➤*achondrite.*

basin   A large, shallow, circular structure on the surface of a planet, created by the impact of a large ➤*meteorite*. Basins may show concentric rings; some have been filled in by lava in subsequent volcanic activity. ➤*crater.*

Bayer letters   The letters of the Greek alphabet (➤Table 1), used in conjunction with constellation names (as in Alpha Leonis), to identify the brighter stars. Johann Bayer (1572–1625) was responsible for compiling the first complete star atlas, called *Uranometria*, which was published in 1603. In it he introduced the system of naming the brighter stars in each constellation by Greek letters, which he allocated approximately according to brightness or, in some instances, in order of position on the sky. The system was rapidly adopted and is still in use today.

BD   Abbreviation for ➤*Bonner Durchmusterung.*

beam   The area of sky being observed at any one time by a ➤*radio telescope.*

beamwidth   The angular extent of the ➤*beam* of a radio telescope. The half-power beamwidth is the angular extent over which the power received is at least half the maximum measured when the radio telescope points directly at a point source. It gives a measure of the resolving power of the antenna.

Bear Claw Nebula (Bear Paw Nebula)   A name sometimes given to the galaxy NGC 2537.

Bear Driver   English name for the constellation ➤*Boötes*.

Becklin–Neugebauer Object   One of the brightest of all astronomical sources of infrared radiation. It was discovered by Eric Becklin and Gerry Neugebauer in 1967 and is located in the Kleinmann–Low Nebula, within the Orion Nebula. It is thought to be a very young, massive star of ➤*spectral type* B, hidden behind so much dust that it appears very feeble in visible light. There are other infrared sources near by in what is believed to be a region of very active star formation.

Beehive   English name for the open star cluster ➤*Praesepe*.

Beijing Observatory   The astrophysical research institute of the Chinese Academy of Sciences, founded in 1958. Facilities for radio astronomy, optical astronomy, solar observations and time-keeping are located at five observing stations.

Belinda   One of the small satellites of Uranus discovered during the ➤*Voyager 2* encounter with the planet in 1986. ➤Table 6.

Bellatrix (Gamma Orionis; γ Ori)   A giant ➤*B star* of magnitude 1.6. The name, of Latin origin, means 'female warrior'. ➤Table 3.

Belt of Orion   The three stars Delta (δ), Epsilon (ε) and Zeta (ζ) Orionis, forming the belt of the mythological figure of the constellation ➤*Orion*.

Bennett, Comet   ➤*Comet Bennett*.

BeppoSAX   An Italian/Dutch gamma- and X-ray satellite launched on 30 April 1996. Observations it made in 1997 led to the first optical identification of a ➤*gamma-ray burster*.

Berenice's Hair   English name for the constellation ➤*Coma Berenices*.

Besselian year   A concept used in a convention for expressing time in dynamical calculations. The length of the Besselian year was originally defined as the period taken for the Sun's ➤*right ascension* to increase by 24 hours, and is almost equal to the length of a ➤*tropical year*. In 1976, it was redefined as the length of the tropical year in 1900. The Besselian year is defined to commence at the instant when the Sun's mean ➤*longitude* is 280°. In practice, this falls close to the beginning of the calendar year. The simpler ➤*Julian year* system is now usually preferred.

Be star   A ➤*B star* that shows ➤*emission lines* of hydrogen superimposed on the ➤*absorption lines* in its spectrum. ➤*Supergiants*, which may also show such emission, are excluded from this classification.

Beta Canis Majoris star (β CMa star)   A type of giant ➤*B star* that shows short-period variations in brightness and in its spectrum. These stars are variable because they pulsate. Their periods are under seven hours, and the light variation is no more than a tenth of a magnitude. The first such star to be discovered was Beta Cephei, and members of this class of variables are also known as *Beta Cephei stars*.

Beta Cephei star (β Cep star)   ➤*Beta Canis Majoris star.*

beta decay   Decay of a radioactive isotope resulting in the emission from the parent atomic nucleus of either an electron and an antineutrino, or a positron and a neutrino.

Beta Lyrae star (β Lyr star)   A member of a class of variable, ➤*binary stars* of which Beta Lyrae is the prototype. Both members of the system are massive, but one has expanded to fill its ➤*Roche lobe*. This causes material to flow towards the other star, which becomes surrounded by an obscuring ➤*accretion disc.*

beta particle   An electron or positron (i.e. a particle with the same mass as an electron but opposite electric charge) emitted from an atomic nucleus as a result of a nuclear reaction or in the course of radioactive decay.

Beta Pictoris (β Pic)   A fourth magnitude ➤*A star,* which is surrounded by a disc of material. Attention was focused on the star when it was found to be emitting strongly in the infrared. Optical observations confirmed the presence of the disc with a diameter about ten times the size of Pluto's orbit round the Sun. It is believed that planetary systems form from such discs. Warping of the Beta Pictoris disc suggests the presence of at least one planet.

Beta Regio   A highland region on the surface of Venus, dominated by two shield-shaped areas, Theia Mons and Rhea Mons, both of which rise to heights of over 4.5 kilometres (3 miles).

Betelgeuse (Betelgeux; Alpha Orionis; α Ori)   A red supergiant ➤*M star,* one of the largest known. By means of ➤*speckle interferometry* and other interference techniques, the star's diameter has been measured directly and found to be about 1,000 times that of the Sun. The presence of large bright 'star-spots' has also been revealed. Ultraviolet observations made with the Hubble Space Telescope showed Betelgeuse to be surrounded by an extended ➤*chromosphere.* Its mass is about twenty times the Sun's. The brightness varies irregularly between magnitudes 0.4 and 0.9 with a rough period of around five years. ➤Table 3.

Bethe–Weizsäcker cycle   An alternative term for the ➤*carbon cycle,* named after the physicists who, in the early 1930s, first proposed this sequence of nuclear processes as a source of stellar energy.

Bettina   Asteroid 250, diameter 128 km, discovered in 1885 by J. Palisa.

Betulia   Asteroid 1580, diameter 60 km, discovered in 1950 when it made a close approach to the Earth.

Bianca   One of the small satellites of Uranus discovered during the ➤*Voyager 2* encounter with the planet in 1986. ➤Table 6.

Biela's Comet   ➤*Comet Biela*.

Bielids   ➤*Andromedids*.

Big Bang   A ➤*model* for the history of the universe, according to which it began in an infinitely compact state and has been expanding ever since. This apparent beginning occurred between 12 and 15 billion years ago and has come to be known as the Big Bang. The theory is now widely accepted since it explains the two most significant observations in cosmology: the ➤*expanding universe* and the existence of the ➤*cosmic background radiation*.

The known laws of physics can be used to project backwards and calculate what the universe was like at various stages of its development since 10^{-43} seconds after the Big Bang. For the first million years, the matter and energy in the universe formed an opaque ➤*plasma*, sometimes called the *primeval fireball*. By the end of this period, the expansion of the universe caused the temperature to fall below 3,000 K so that protons and electrons could combine to form hydrogen atoms. At this stage, the universe became transparent to radiation. The density of matter then exceeded that of radiation, where previously the situation was the reverse, and so dictated the rate of expansion of the universe. The microwave background is all that remains of the greatly cooled radiation from the early universe. The first stars did not begin to form from the primordial clouds of hydrogen and helium until one or two billion years later.

The term 'big bang' may be applied to any model of an expanding universe that had a hot, dense past. ➤*steady-state theory*.

Big Bear   English name for the constellation ➤*Ursa Major*.

Big Bear Solar Observatory   A solar observatory located at an altitude of 2,000 metres (6,600 feet) on an island in Big Bear Lake in California. The site was chosen to eliminate the distortion caused by turbulence over land heated by the Sun. Formerly owned and operated by the California Institute of Technology, it was transferred in 1997 to the New Jersey Institute of Technology.

There are three main telescopes on the same mount: a 65-centimetre (26-inch) reflector, a 25-cm (10-inch) refractor and a 15-cm (6-inch) refractor. The smallest telescope is used to monitor the whole Sun. The 25-cm is equipped with a magnetograph, and there is a spectrograph fed by light from the 65-cm. An additional instrument is dedicated to the study of ➤*helioseismology*.

Big Crunch   The hypothetical total inward collapse of the universe. If its present expansion were to slow down sufficiently, the universe could enter a phase of contraction that would end with the so-called Big Crunch. ➤*Big Bang*, *oscillating universe*.

Big Dipper   North American name for the ➤*asterism* in the constellation Ursa Major, also called the ➤*Plough*.

Big Dog   English name for the constellation ➤*Canis Major*.

billion   In scientific usage, a thousand million (10^9). Formerly, 'billion' was used in the UK to mean 'a million million', so care with interpretation is needed if there is any possibility of confusion.

BIMA Array   A millimetre-wave telescope at ➤*Hat Creek Observatory* in California, operated by the Berkeley – Illinois – Maryland Association. It consists of ten 6 metre (20 foot) dishes operating in the $1 - 3$ mm wavelength region of the spectrum.

binary star   A pair of stars in orbit around each other, held together by their mutual gravitational attraction. About half of all 'stars' are in fact binary or multiple, though many are so close that the components cannot be seen individually. The presence of more than one star is inferred from the appearance of the combined spectrum.

The two components in a binary system each move in an elliptical orbit around the common ➤*centre of mass*. The further apart they are, the slower they move. Pairs in which the ➤*separation* is great enough for the two stars to be distinguished, or 'split', in a telescope often have orbital periods as long as 50 or 100 years. Such pairs are called *visual binaries*.

If one star is much fainter than the other, its presence may be revealed only by the obvious orbital motion of its brighter companion. Pairs of this type are called *astrometric binaries*.

As the members of a binary system move in orbit, their velocities towards or away from the Earth change in a regular repeating pattern. Through the ➤*Doppler effect*, these velocity variations are reflected as wavelength changes in the features of the combined spectrum. Study of such a spectrum can reveal details about the nature of the stars and their orbits. Binary stars recognized as such only by means of spectroscopy are called *spectroscopic binaries*. Their periods usually lie in the range from a day to a few weeks.

Some binary components are so close that the pull of gravity distorts the individual stars from their normal spherical shape. They may exchange material and be surrounded by a common envelope of gas. An ➤*accretion disc* may develop as material streams towards a compact, spinning star in a binary system. The energy released results in the emission of X-rays. ➤*Novae* are another consequence of mass transfer in binary stars.

If the orbits of a binary pair are oriented in space so that one star has to pass in front of the other as seen from the Earth, the system is described as *eclipsing*. Such a system is observed to be variable since one star periodically blots out light from the other. The best-known eclipsing binary is ➤*Algol*.

binoculars An optical instrument consisting essentially of two small tele-scopes, mounted side by side, one for each eye. The relatively compact size of binoculars is achieved by the use of prisms to reflect the light internally. At the same time, the reflection by the prisms makes the image upright, rather than inverted as it is in an astronomical telescope. The size and magnifying power of binoculars is usually given in the form $A \times B$, where A is the linear magnification and B is the diameter of each objective lens in millimetres (e.g. 10 × 40). Binoculars are popular with amateur astronomers. They are particularly useful for certain types of observation, such as comet hunting.

bipolar nebula A luminous nebula consisting of two lobes of emission, apparently the result of an outflow of material being channelled into two opposing directions. The term may be applied to any type of nebula with bipolar structure, but has been associated particularly with a group of nebulae that are intense sources of infrared radiation. It is believed that such a nebula has at its centre a bright star that is completely concealed by a dense ring of dust and gas. The dust is heated to a temperature of a few hundred degrees by the radiation from the star, as a result of which it emits the infrared radiation. The visible starlight is funnelled along the polar directions and illuminates the more tenuous part of the nebula around the star. ➤*bipolar outflow.*

bipolar outflow Gas streaming outwards in two opposing directions from a newly formed star. The flow emerges from the centre of the star's ➤*accretion disc* and is constrained to move along the rotation axis. The stellar wind, blowing out at typical speeds of 200 kilometres per second, sweeps up interstellar material before it, creating a double-lobed shell structure, extending outwards for a distance of about a light year. Bipolar outflows have been detected by the radio emission from the molecules they contain.

Bird Nebula A name given to a nebula in the region of the galactic centre, seen only in radio emission. It is suspected to be a ➤*supernova remnant.*

Bird of Paradise English name for the constellation ➤*Apus.*

BIS Abbreviation for ➤*British Interplanetary Society.*

black body A body that absorbs all the radiation incident on it. Such an object is a theoretical ideal and the nearest approximation achievable in the laboratory is a small hole in an enclosure held at a uniform temperature. The intensity of the radiation emitted by a black body and the way it varies with wavelength depend only on the temperature of the body and can be predicted by quantum theory.

black body radiation The radiation emitted by a ➤*black body.* The intensity of radiation as a function of wavelength depends only on the temperature of the body. The graph showing the distribution of intensity with wavelength is

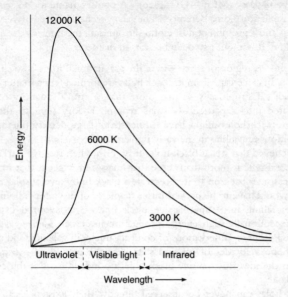

Black body radiation. The distribution with wavelength of the radiation emitted by black bodies at 12,000, 6,000 and 3,000 K. These are typical surface temperatures for stars. The shape of the curves illustrates the colour and luminosity differences between cool and hot stars.

often called the *Planck curve* after the physicist Max Planck, who formulated the relationship. The Planck curves are hill-shaped, with a distinct peak (see illustration). The wavelength of the peak decreases with increasing temperature of the black body in such a way that the product of peak wavelength and absolute temperature is constant. The total amount of energy emitted by a black body is proportional to the product of its surface area and the fourth power of its temperature (T^4).

black drop An illusory effect observed during a transit of Venus or Mercury across the Sun, when the small dark disc of the planet is near the limb of the Sun. When the limbs of the Sun and planet are not quite in contact, a small black spot, or drop, appears to join them.

black dwarf A dead star that is no longer luminous. The final luminous phase in the evolution of a star less massive than about 1.4 solar masses is spent as a ➤*white dwarf*. Since no new energy is generated in a white dwarf, all possible sources having been exhausted, the star's final fate is to gradually cool and fade to a dark stellar 'corpse'. However, the universe is not yet old enough for any black dwarfs to have formed.

Black-eye Galaxy (M64; NGC 4826)   A popular name for a ►*spiral galaxy* in the constellation Coma Berenices. The galaxy is characterized by very smooth spiral arms and a prominent dust cloud surrounding the nucleus that gives rise to the name. It is about 65,000 light years in diameter.

black hole   A region of space where the gravitational force is so strong that not even light can escape from it. Black holes are formed when matter collapses in on itself catastrophically so that more than a critical quantity of mass is concentrated into a particularly small region. Theory suggests that 'mini' primordial black holes might have formed from large density fluctuations in the conditions prevailing in the early universe.

It is believed that stellar black holes may form when massive stars explode, if the central relic is more than three solar masses or exceeds that mass when material cascades back on to it. To create a black hole, several solar masses of material would have to be packed into a diameter of just a few kilometres.

Matter falling into supermassive black holes is a favoured explanation for the exceptionally high energy production in certain ►*active galactic nuclei* and ►*quasars*. Direct observations of compact nuclei in galaxies, and measurements of the velocities of gas and stars near the centres of galaxies, lends weight to the idea that massive black holes do indeed exist at the centres of many galaxies.

Black holes can never be observed directly: their existence can only be inferred from their gravitational effects and the radiation emitted by material falling into them. A number of stellar X-ray sources, such as ►*Cygnus X-1*, are binary star systems in which one component appears, from determinations of its mass and luminosity, to be a black hole. In such systems, observations of the visible star make it possible to compute the orbit and mass of its dark, compact companion. The X-rays result from energy released as matter streams on to the compact star. ►*Kerr metric, Schwarzschild radius.*

blazar   A term applied to ►*BL Lac objects* and ►*quasars* showing violent variations in visible light output.

Blaze Star   A popular name for the ►*recurrent nova* T Coronae Borealis. It is the brightest recurrent nova ever recorded, having reached second magnitude in both 1866 and 1946.

Blazhko variable   ►*RR Lyrae star.*

blink comparator   An instrument for comparing two photographs of a region of sky, usually a pair taken at different times. The purpose of the comparison is to show up any images that are different in position or brightness on the two photographs. This is achieved by means of an optical system that brings images of the two photographs into exact coincidence while they are illuminated alternately. An object whose brightness differs appears to blink on and off as the illumination is changed; an object that is at a different position appears to jump between the two locations.

Blinking Nebula Popular name for NGC 6826, a ➤*planetary nebula* in Cygnus. The name is said to derive from the fact that the central star appears to blink on and off if a visual observer switches rapidly between looking directly at it and looking to one side of it.

BL Lac object (BL Lacertae object; Lacertid) A type of ➤*elliptical galaxy* with a bright, highly variable, compact nucleus. The first such object to be noted was BL Lacertae, which at its discovery in 1929 was thought to be a variable star (hence the form of its name). Their unique characteristics are dramatic short-term variability in light output and a featureless spectrum. The brightness can vary as much as a hundredfold over a period of a month, and day-to-day changes are sometimes observed.

Many BL Lac objects are also radio sources. Intense radio bursts are seen from BL Lacertae itself, but they are apparently not correlated with the light changes.

blue clearing An unusual degree of transparency to blue light in the atmosphere of the planet Mars.

blue moon The origin of this expression, often used just to mean 'a rare event', is obscure. One suggestion is that it refers to the second occurrence of a new Moon in one calendar month. An alternative explanation is that atmospheric effects may occasionally make the Moon appear blue; a possible cause would be dust in the upper atmosphere from volcanoes or forest fires.

Blue Planetary Popular name for NGC 3918, a ➤*planetary nebula* in Centaurus. It derives its name from its visual appearance as a blue, apparently featureless disc.

blueshift The shift in wavelength of ➤*spectral lines* towards shorter wavelengths. Blueshifts arise from the operation of the ➤*Doppler effect* when the source of radiation and its observer are moving towards each other.

Blue Snowball A popular name for the ➤*planetary nebula* NGC 7662 in the constellation Andromeda.

blue straggler A star that appears to belong to a ➤*globular cluster* or an old ➤*open cluster* but has an anomalous blue colour and high luminosity in comparison with other cluster members. When the stars of such a cluster are plotted in a ➤*Hertzsprung–Russell diagram* (see illustration), there is a distinct *turn-off point* on the main sequence. This point marks the lower mass limit of the stars that have evolved into ➤*red giants* so that they lie to the right of the main sequence. In some clusters, a small number of stars appear to lie on the main sequence above this turn-off point – these are the blue stragglers. The reasons for

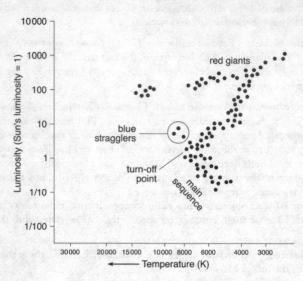

Blue straggler. A schematic Hertzsprung–Russell diagram for a typical globular cluster, showing the location on the diagram of blue stragglers.

their anomalous properties are not fully understood, but a range of explanations might account for the observation, including ►*binary star* membership.

Bode's law ►*Titius–Bode law*.

Bode's Nebula (M81; NGC 3031) A spiral galaxy in the constellation Ursa Major, discovered by J. E. Bode (1747–1826) in 1774. It is now more usually referred to as M81.

Bok globule ►*globule*.

bolide A particularly bright ►*meteor* whose appearance is accompanied by an explosive sound or sonic boom.

bolometer An instrument for measuring the total amount of energy received from a source of electromagnetic radiation, across all wavelengths.

bolometric correction The visual or photovisual magnitude of an object minus its ►*bolometric magnitude*. ►*magnitude*.

bolometric magnitude The magnitude of an object obtained when energy radiated at all wavelengths is included. For an object that emits strongly in the ultraviolet or infrared, for example, the bolometric magnitude may differ greatly from the visual magnitude. ➤*magnitude*.

Bolshoi Teleskop Azimutalnyi ➤*Special Astrophysical Observatory*.

Boltzmann's constant (symbol k) One of the fundamental constants of physics, with the value 1.3806×10^{-23} joules per kelvin. It is the ratio of the gas constant to Avogadro's number. The energy associated with each degree of freedom of an atom or molecule is $\frac{1}{2}kT$ where T is absolute temperature.

Boltzmann's equation An equation that links the relative number of atoms N that will occupy a state of energy E at a particular value of temperature T. It states that N is proportional to $g \exp(-E/kT)$, where g is the statistical weight of the energy level and k is Boltzmann's constant.

Bond albedo The ratio of total reflected light to total incident light for a planetary body. The Bond albedo is generally smaller than the ➤*normal reflectivity*.

Bonner Durchmusterung (BD) A star catalogue (literally the 'Bonn Survey'), prepared by F. W. A. Argelander (1799–1875) at the Bonn Observatory between 1852 and 1868, and extended by E. Schönfeld (1828–91) in 1886. The initial catalogue contained the positions and magnitudes of 324,198 stars down to magnitude 9.5 between the north celestial pole and declination $-2°$. Schönfeld extended the coverage to declination $-22°$ and added a further 133,659 stars. ➤*Córdoba Durchmusterung*.

Boötes (The Herdsman) A constellation of the northern sky, dominated by the bright orange star ➤*Arcturus*. It is one of the ancient constellations listed by Ptolemy (c. AD 140) and is usually said to represent a herdsman driving a bear (the constellation Ursa Major). ➤Table 4.

Boss General Catalogue A catalogue of 33,342 stars, giving positions and proper motions, started by the American astronomer Lewis Boss and completed by his son Benjamin Boss in 1937.

bow shock The boundary around an object's ➤*magnetosphere* where the ➤*solar wind* is deflected and there is a sharp decrease in its velocity. The plasma of the solar wind is heated and compressed at the bow shock.

Boyden Observatory An optical observatory at Bloemfontein, South Africa. Since 1976 it has been operated by the University of The Orange Free State. The original Boyden Station was established in Peru in 1890 by Harvard College Observatory, with funds bequeathed for the purpose in the will of Uriah A. Boyden. It was moved to South Africa in 1926, where it was run by Harvard until 1954. Between 1954 and 1976 it was run by an international consortium of observatories.

Bp star ➤*Ap star.*

Brackett series   A series of ➤*spectral lines* in the infrared region arising from a particular set of transitions in the hydrogen atom.

Brans–Dicke theory   A modification of Einstein's ➤*General Relativity* theory.

breccia   A rock made of broken fragments that have been cemented together by a finer-grained material. Breccias are a common result of impact processes.

bremsstrahlung   Electromagnetic radiation produced when an electron is decelerated by passing close to an atomic nucleus.

brightness   The intensity of radiation from a source. Apparent brightness is the intensity of received radiation, which depends on the source's distance and its true or absolute brightness. The brightness of astronomical objects is measured in ➤*magnitudes*, and the word 'magnitude' is commonly used as if it were synonymous with 'brightness'.

brightness temperature   A measure of the amount of radiation being emitted from a source. It is defined as the temperature a ➤*black body* replacing the source would need to be at in order to produce the level of radiation actually observed.

Brightness temperature is frequently used by radio astronomers as a measure of the intensity of radiation being received. It is a useful concept, since it indicates whether the source is emitting ➤*thermal radiation*, in which case the source's brightness temperature will be measured in hundreds or thousands of degrees, or ➤*synchrotron radiation*, when the brightness temperature will be billions of degrees.

Bright Star Catalogue   A star catalogue published by Yale University Observatory, including stars down to a nominal magnitude limit of 6.5.

British Astronomical Association (BAA)   An organization founded in London in 1890 to promote and encourage popular interest in astronomy and practical work by amateur astronomers in particular. It publishes a bi-monthly journal and an annual handbook, organizes meetings and has a number of sections to coordinate observations undertaken by members. It has an autonomous branch in New South Wales, Australia.

British Interplanetary Society (BIS)   An organization formed in 1933 to promote popular interest in the exploration and use of space. Its activities include publications and conferences.

brown dwarf   A very cool star containing insufficient mass for nuclear reactions to be ignited in its core. A number of plausible candidate brown dwarfs have been identified. One of them, Gliese 229 B, has been shown to have water, methane and ammonia present, all of which would be destroyed in the hot atmosphere of a true star.

Brown–Twiss stellar interferometer An intensity ➤*interferometer* developed by R. Hanbury Brown and Richard Q. Twiss for measuring the angular diameters of bright stars. The first test results were announced in 1956. The only instrument of its type is the one they constructed at Narrabri in Australia. Two ➤*flux collectors*, each 6.5 metres (21 feet) in diameter and formed from several hundred smaller mirrors, were mounted on trolleys on a circular track with a radius of 94 metres (308 feet). The operating wavelength was 433 nanometres and the minimum diameter measurable about 0.0005 arc seconds. Of the order of a hundred stars were accessible, the faintest operating magnitude being 2.5. Diameters were estimated by analysing the correlation between intensity fluctuations at the two light collectors in relation to the separation between them.

B star A star of ➤*spectral type* B. B stars have surface temperatures in the range 11,000–25,000 K and are bluish white in colour. The most prominent features in their spectra are ➤*absorption lines* of neutral helium. The ➤*Balmer lines* of hydrogen are also present, being stronger the cooler the star. Examples of B stars are ➤*Rigel* and ➤*Spica*.

B-type asteroid A subclass of ➤*C-type asteroids*, distinguished by higher albedo.

Bubble Nebula A popular name for a faint, diffuse, luminous nebula (NGC 7635) in the constellation Cassiopeia. Despite its apparently spherical shape, it seems not to have the characteristics of either a planetary nebula or a supernova remnant.

Budrosa Asteroid 338, diameter 80 km, of the rare metallic type. It is the prototype of the *Budrosa family* of unusual asteroids, with six known members. They are grouped at 2.9 AU from the Sun with orbits inclined at 6° to the plane of the solar system.

Bug Nebula A name given to the ➤*bipolar nebula* NGC 6302 in the constellation Scorpius. No central star has been identified, but the central region is hot and active and gas is streaming outwards at velocities up to 400 km/s. The nebula appears red because much of its light is emitted as the red spectral lines of hydrogen and nitrogen.

Bull English name for the constellation ➤*Taurus*.

burst Any sudden emission of unusually strong electromagnetic radiation from a celestial object.

burster ➤*gamma-ray burster*, X-ray burster.

butterfly diagram A representation in graphical form of the way the latitudes at which ➤*sunspots* appear vary throughout the ➤*solar cycle*. It was first plotted

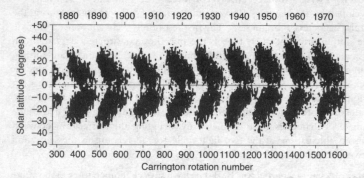

Butterfly diagram. How the distribution of sunspots in latitude changes with time.

in 1922 by E. W. Maunder and is also known as the *Maunder diagram*. On a graph with solar latitude as the vertical axis and time (in years) as the horizontal axis, a vertical line covering one degree in latitude is plotted for each sunspot group centred at that latitude within a ➤*Carrington rotation*. The result (see illustration) is a pattern reminiscent of butterfly wings, which gives the diagram its popular name.

BY Draconis star   A type of ➤*flare star* in which there are small regular variations in brightness during the quiet phase between flares. The group is named after its eighth magnitude prototype.

BY Draconis stars are red dwarfs, of ➤*spectral type* K or M. Of those known, a high proportion are definitely in binary systems. The variability, which amounts to no more than a few hundredths of a magnitude in a maximum period of a few days, is thought to be due to a luminous spot on the surface of the rotating star. ➤*UV Ceti star*.

Byurakan   The location of the Astrophysical Observatory of the Armenian Academy of Sciences. It is situated 40 kilometres (25 miles) to the north of Yerevan at an altitude of 1,500 metres (5,000 feet), where it was founded in 1946 on the initiative of V. Ambartsumian. The main instrument at the observatory is a 2.6-metre (100-inch) reflector.

C

Caelum (The Chisel) A small, insignificant southern constellation, introduced by Nicolas L. de Lacaille in the mid-eighteenth century with the longer name Caela Sculptoris. It has no star brighter than fourth magnitude. ►Table 4.

Calar Alto Location in southern Spain of an observatory constructed by the ►*Max-Planck-Institut für Astronomie* and operated jointly with the Spanish National Commission for Astronomy. The telescopes are 1.2-metre (4-foot), 2.2-metre (7-foot) and 3.5-metre (11½-foot) reflectors, together with a 0.8-metre (2½-foot) Schmidt camera. A 1.5-metre (5-foot) telescope operated by the University of Madrid is also at the site.

caldera A large volcanic crater, usually more than two kilometres in diameter. Calderas are created by collapse or explosive eruption, or a combination of the two, and often have a complex structure.

calendar A practical system for timetabling human affairs, taking into account important natural cycles such as the day, month and year. An ideal practical calendar needs to be organized so that the civil years keep step with the seasons. Since the ►*tropical year* does not contain an integral number of days, devising such a calendar presented great difficulties for earlier civilizations. The calendar now in general use throughout much of the world is the ►*Gregorian calendar*, introduced in 1582. The Gregorian calendar falls out of step with the seasons by only one day in 3,300 years. Other traditional calendars still in use include the Jewish, Muslim and Chinese calendars.

calendar year A year containing 365 or 366 whole days according to whether the year is a leap year or not in the ►*Gregorian calendar*.

California Association for Research in Astronomy (CARA) A non-profit organization formed as a collaboration between the California Institute of Technology and the University of California with the purpose of building and operating the ►*Keck Observatories* in Hawaii.

California Nebula (NGC 1499) A bright gaseous ►*emission nebula* in the constellation Perseus, named for its resemblance to the shape of the US state. It forms the rim of a dark nebula of gas and dust illuminated by the star Xi (ξ) Persei.

Callisto   One of the four large Galilean moons of Jupiter (number IV), discovered in 1610. It is the darkest of the Galilean satellites and also the least dense, which suggests that it contains a high proportion of water, though detailed surface images returned by ➤*Galileo* indicate that the surface at least may contain more rock and dust than previously supposed. ➤*Voyager* and *Galileo* images show that the surface is heavily cratered but has little relief. The most prominent surface feature is a multi-ring structure called Valhalla, which consists of a central bright zone 600 kilometres (375 miles) across, surrounded by numerous concentric rings spaced at intervals of 20 to 100 kilometres (12 to 60 miles). At least seven other multi-ring features have been recognized. ➤Table 6.

Caloris Basin (Caloris Planitia)   A large, multi-ringed, impact basin on Mercury. It is 1,300 kilometres (800 miles) in diameter and the most conspicuous feature on the planet.

Caltech Submillimetre Observatory   ➤*Submillimetre wave astronomy*.

Calypso   A small satellite of Saturn discovered in 1980. It is co-orbital with ➤*Tethys* and ➤*Telesto*. ➤Table 6.

Camelopardalis (Camelopardus; The Giraffe)   A large but inconspicuous constellation occupying a sparse region of the sky near the north celestial pole. It was first recorded in 1624 by the German mathematician Jakob Bartsch, a son-in-law of Johannes Kepler. ➤Table 4.

Camilla   Asteroid 107, diameter 236 km, discovered in 1868 by N. Pogson.

Canada–France–Hawaii Telescope (CFHT)   A 3.6-metre (140-inch) telescope at the ➤*Mauna Kea Observatories* in Hawaii. Commissioned in 1979, it is a major facility for French and Canadian astronomers also used by the University of Hawaii. Its versatile design and location make it suitable for both optical and infrared observations.

canals   Supposed linear features on Mars. The Italian word *canale*, meaning simply 'channel', was used in the nineteenth century by Angelo Secchi to describe linear features he perceived during observations of Mars, and later by Giovanni Schiaparelli. The word was translated into English as 'canal' with the connotation that what had been observed were artificial structures. The notion was elaborated by Percival Lowell, who built an observatory at Flagstaff, Arizona, with the main purpose of observing Mars. His drawings of the planet showed extensive networks of linear 'canals', and he proposed that a civilization of intelligent beings on Mars was responsible for constructing them. Later observers have found little evidence of such markedly linear features, and *Mariner* and *Viking* images show no trace of them; they are now dismissed as optical effects.

Canary Islands Large Telescope   A Spanish national facility, planned for construction at the ➤*Observatorio del Roque de los Muchachos* in the Canary Islands, with completion early in the 21st century. The design is for an optical/infrared reflector, with a segmented mirror of 36 hexagonal components equivalent to a 10-metre (33-foot) single mirror (similar to the telescopes of the ➤*Keck Observatories*).

Cancer (The Crab)   A zodiacal constellation, among those listed by Ptolemy (*c.* AD 140). It is said to represent the crab crushed under the foot of Hercules when he was fighting the Hydra. None of the stars is brighter than fourth magnitude, though the star cluster ➤*Praesepe* at the constellation's centre can be seen with the unaided eye. ➤Table 4.

Canes Venatici (The Hunting Dogs)   A small constellation of the northern sky, lying between Boötes and Ursa Major. It was introduced by Johannes Hevelius in the late seventeenth century and is supposed to represent the dogs Asterion and Chara held on a leash by Boötes. Though small, the constellation contains several interesting objects including the bright star ➤*Cor Caroli*, the fine ➤*globular cluster* M3 and the ➤*Whirlpool Galaxy*. ➤Table 4.

Canis Major (The Greater Dog)   A small constellation, just south of the celestial equator and next to Orion, containing the brightest star in the sky, ➤*Sirius*. It is said to represent one of the dogs following the hunter, Orion, and was listed by Ptolemy (*c.* AD 140). ➤Table 4.

Canis Minor (The Lesser Dog)   A small constellation bordering the celestial equator and close to Orion. With ➤*Canis Major*, it is supposed to represent one of the dogs following Orion, and was listed by Ptolemy (*c.* AD 140). It contains only two bright stars, the brightest being ➤*Procyon*. ➤Table 4.

cannibalism   An informal description sometimes applied to the phenomena of a small galaxy being absorbed by a larger companion galaxy, or a star being merged into another.

Canopus (Alpha Carinae; α Car)   The brightest star in the constellation Carina and the second-brightest star in the sky. Canopus is a supergiant ➤*F star* of magnitude −0.7. The name is that of the pilot of the fleet of King Menelaos of Greek mythology. ➤Table 3.

Cape Canaveral   The location in Florida, USA, of the Kennedy Space Center, from where most of NASA's space missions are launched.

Cape Kennedy   The name by which ➤*Cape Canaveral* was known between 1963 and 1973.

Capella (Alpha Aurigae; α Aur)   The brightest star in the constellation Auriga. It is a ➤*spectroscopic binary,* the primary being a giant ➤*G star* of magnitude 0.1. The name, of Latin origin, means 'little she-goat'. ➤Table 3.

Cape Photographic Durchmusterung (CPD)   A general catalogue of 455,000 southern stars down to tenth magnitude, compiled at the Cape of Good Hope by J. C. Kapteyn (1851–1922) from photographic plates taken by Sir David Gill (1843–1914). The catalogue covers declinations from −19° to the south celestial pole and was produced between 1896 and 1900.

Capricornus (The Sea Goat)   One of the zodiacal constellations recorded by Ptolemy (*c.* AD 140). Its brightest stars are third magnitude. ➤Table 4.

captured atmosphere   A planetary atmosphere that was created by accretion as part of the planetary formation process and subsequently retained.

captured rotation   ➤*synchronous rotation.*

CARA   Abbreviation for ➤*California Association for Research in Astronomy.*

Carafe Galaxy   A peculiar, ringed ➤*Seyfert galaxy* near NGC 1595 and NGC 1598. ➤*Carafe Group.*

Carafe Group   A group of three galaxies, NGC 1595, NGC 1598 and the ➤*Carafe Galaxy*, which share a common motion in space. NGC 1598 has two luminous jets projecting from it, and there is evidence that interaction has occurred between the Carafe Galaxy and NGC 1595.

carbonaceous chondrite   A rare type of stony ➤*meteorite*. Because their average chemical composition is very similar to that of the Sun (apart from the hydrogen and helium) and there is a relatively high abundance of volatile material, carbonaceous chondrites are thought to represent some of the primitive, unprocessed material from which the solar system formed. They are made up of a matrix of carbon-rich minerals in which the ➤*chondrules* are embedded. The water content can be as high as 20 per cent. The largest known example is the ➤*Allende meteorite.*

carbon cycle (carbon–nitrogen (CN) cycle; carbon–nitrogen–oxygen (CNO) cycle; Bethe–Weizsäcker cycle)   A series of nuclear reactions, believed to take place in stars and to be one of the main sources of stellar energy, in which hydrogen is converted to helium.

The six stages in the process are:

$$^{12}C + {}^{1}H \rightarrow {}^{13}N + \text{gamma-ray photon}$$
$$^{13}N \rightarrow {}^{13}C + \text{positron} + \text{neutrino}$$
$$^{1}H + {}^{13}C \rightarrow {}^{14}N + \text{gamma-ray photon}$$
$$^{1}H + {}^{14}N \rightarrow {}^{15}O + \text{gamma-ray photon}$$
$$^{15}O \rightarrow {}^{15}N + \text{positron} + \text{neutrino}$$
$$^{1}H + {}^{15}N \rightarrow {}^{12}C + {}^{4}He$$

Thus, the ^{12}C nucleus reappears at the end. The process cannot take place unless carbon is present, but relatively few ^{12}C nuclei are required. The rate at which the carbon cycle occurs depends very sharply on temperature. A minimum temperature of 14 million K is needed for it to take place at all. Above 16 million K it dominates over the other main hydrogen-burning process, the ►proton–proton chain, and is believed to be the primary source of energy in relatively hot stars with masses greater than two or three times that of the Sun.

Two longer variations on the sequence are also thought to occur, the CNO bi-cycle and the CNO tri-cycle, which result in the creation of ^{14}N and ^{15}N, respectively, rather than the reappearance of carbon. In the shorthand used to represent nuclear reactions, the CNO bi-cycle proceeds as:

$$^{15}N(p,\gamma)^{16}(p,\gamma)^{17}F(,\beta^{\prime}+\nu)^{17}O(p,\alpha)^{14}N$$

The CNO tri cycle includes the additional stages:

$$^{17}O(p,\gamma)^{18}F(,\beta^{+}+\nu)^{18}O(p,\alpha)^{15}N$$

carbon star   A general name for a group of peculiar, red giant stars whose spectra show strong bands of molecular carbon, CN, CH or other carbon compounds, and not the more typical TiO.

In the original Harvard classification system of 1918, the carbon stars were allocated to ►spectral types R and N. It was found that they have temperatures similar to those of the more common ►K stars and ►M stars and that the differences in the spectra arise from differences in the abundance of carbon and oxygen.

The term 'carbon star' was introduced in the 1940s by Morgan and Keenan, who proposed a new sequence of classes from C0 to C7, paralleling the decreasing temperature sequence in normal stars from G4 to M4. Though known carbon stars are rare in our own Galaxy, many thousands have been discovered in the Large and Small Magellanic Clouds.

Some carbon stars contain the unstable element technetium, the longest-lived isotope of which has a half-life of only 210,000 years, a short period on astronomical timescales. A few (less than twenty) of the coolest carbon stars show an extremely strong line of lithium in their spectra.

It is also possible to measure the proportions present of two isotopes of carbon, ^{12}C and ^{13}C. In the carbon stars, particularly the hotter ones, these proportions differ significantly from those encountered in the solar system. A group in which the ratio ^{12}C/^{13}C is unusually low are known as *J stars*.

cardinal points   The four directions north, south, east and west.

Carina (The Keel)   A large constellation in the southern Milky Way, formerly part of ►*Argo Navis*. It contains the second-brightest star in the sky, ►*Canopus*. ►Table 4.

Carina Nebula   ►*Eta Carinae Nebula*.

Carme   A small satellite of Jupiter (number XI), discovered in 1938 by S. B. Nicholson. ➤Table 6.

Carpathian Mountains   A range of mountains on the Moon forming part of the border of the Mare Imbrium.

Carrington rotation number   A number that identifies each rotation of the Sun. The sequence began with rotation number one on 9 November 1853. The system was started by R. C. Carrington, who employed the average synodic rotation rate for sunspots, which he had determined as 27.2753 days. Since the Sun does not rotate as a solid body, the rate actually varies with latitude.

Carte du Ciel   An ambitious project, started in 1887, intended to produce photographic charts of the entire sky, together with an associated star catalogue. The charts were never completed, the standard methods ultimately being overtaken by technical advances in astrophotography.

During the 1880s, the growing importance of photography in astronomy became apparent. In 1885, a 34-cm (13½-inch) photographic refractor was constructed at the Paris Observatory by Paul and Prosper Henry. Impressed by their achievements, the Director, E. B. Mouchez, with encouragement from Sir David Gill and Otto Struve, organized the Astrographic Congress of April 1887. The permanent committee met five times before it was transferred to the auspices of the International Astronomical Union in 1919.

The work was initially divided among eighteen observatories, though notably none in the USA. The Paris instrument was adopted as the standard prototype. Each plate was to be two degrees square with a superimposed grid of fine lines at 5-mm (0.2-inch) intervals. A total of about 22,000 plates was taken, only a quarter of the number originally projected. However, the publication of the associated Astrographic Catalogue was finally completed in 1964.

Cartwheel Galaxy   A popular name for a ➤*peculiar galaxy*, 500 million light years away, otherwise known as A0035. It consists of a circular rim, 170,000 light years in diameter, inside which are a hub and spokes made up of old red stars. It is believed that the galaxy was once a normal large ➤*spiral galaxy* through which a smaller galaxy passed a few hundred million years ago. The intruder can still be seen near by. The shock of the collision caused the formation of large numbers of massive stars in the 'rim'. As a result, the rate at which ➤*supernovae* occur now is about a hundred times greater than in a normal galaxy.

Cassegrain   An adjective used to describe any type of optical, radio or other telescope design which incorporates the use of a central hole in the primary reflecting element. ➤*Cassegrain telescope*.

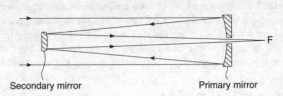

Cassegrain telescope. A schematic diagram of the optical arrangement.

Cassegrain telescope   A reflecting telescope in which the image is brought to a focus just behind a central hole in the primary mirror (see illustration). The design was proposed in about 1672 by Jacques Cassegrain (1652–1712), professor of physics at Chartres in France, some four years after Isaac Newton constructed the first reflecting telescope. It employs a convex secondary mirror, rather than the flat one in Newton's own design. Cassegrain did not build a telescope himself and it was some years before his idea was put into practice. Today, the Cassegrain focus is popular and widely used in both modest amateur instruments and large professional telescopes.

Cassini Division   The conspicuous gap, 2,600 kilometres (1,600 miles) wide, between the two main components (A and B) of the rings of ▶*Saturn*. ▶*planetary rings*, Table 7.

Cassini mission   A joint NASA/ESA mission to explore the Saturnian system, including the planet, its rings, its magnetosphere and several of its moons. Launch was October 1997. Cassini has been designed to use ▶*gravity assist* flybys of Venus (April 1998 and June 1999), Earth (August 1999) and Jupiter (December 2000), resulting in arrival at the Saturnian system in 2004. It is intended to operate in orbit around Saturn for four years.

One of the major objectives of the mission is a study of Saturn's moon ▶*Titan*. Cassini carries the Huygens probe, an instrument package which will parachute down through Titan's atmosphere and land on the surface. The Huygens probe is the ESA contribution to the mission.

Cassini's laws   Three empirical laws describing the rotation of the Moon about its centre of mass, stated by Jacques Cassini (1677–1756) in 1721:

1. The Moon rotates eastwards about an axis fixed within it, with constant angular velocity in a period of rotation equal to the mean sidereal period of revolution of the Moon about the Earth.

2. The inclination of the mean plane of the lunar equator to the plane of the ecliptic is constant.

3. The poles of the lunar equator, the ecliptic and the Moon's orbital plane all lie on one great circle, in that order.

Cassiopeia   A conspicuous W-shaped constellation near the north celestial pole. It is represented by the seated figure of Queen Cassiopeia of classical mythology, and was among the constellations listed by Ptolemy (*c.* AD 140). It is the site of a ►*supernova* observed by Tycho Brahe in 1572 and of the strongest radio source in the sky, a ►*supernova remnant* known as ►*Cassiopeia A*. ►Table 4.

Cassiopeia A (Cas A)   The brightest radio source in the sky (other than the Sun), identified as the remnant of a ►*supernova* that must have occurred around AD 1667. No records exist of any observation of a supernova at this time. It is assumed that it was obscured by large quantities of dust lying in the line of sight to the object, which is 10,000 light years away. The radio emission is concentrated in a ring-like shape 4 arc minutes in diameter, suggesting a shell of material ejected in an explosion, and some faint nebulosity can be detected in optical photographs of that region of the sky. X-ray emission is also detected in the same ring shape.

Castor (Alpha Geminorum; α Gem)   The second-brightest star in the constellation Gemini, after ►*Pollux*. Its magnitude as seen by the naked eye is 1.6, but this is the combined brightness of a multiple system with at least six components. There are two ►*A stars* of magnitudes 2.0 and 2.9 forming a close visual pair, each of which is a ►*spectroscopic binary*, and a more distant ninth magnitude red star, which is an ►*eclipsing binary*. ►Table 3.

cataclysmic variable   A star whose brightness increases dramatically and suddenly in response to an explosive event. The term is applied particularly to ►*novae* and ►*supernovae*. ►*flare star, dwarf nova*.

catadioptric   Describing optical systems that employ a combination of reflecting and refracting elements. The ►*Schmidt camera* is an example of such a system. ►*catoptric, dioptric*.

catalogue equinox   The intersection of the hour circle of zero right ascension of a particular catalogue with the celestial equator. ►*dynamical equinox, equinox*.

catena (pl. catenae)   A chain of craters on a planetary surface.

catoptric   Describing optical systems that employ only reflecting elements (i.e. mirrors). ►*catadioptric, dioptric*.

Cat's Eye Nebula   A popular name for the ►*planetary nebula* NGC 6543. It is about 3,000 light years away and lies in the constellation Draco. Estimated to be about 1,000 years old, its intricate structure suggests that there is a binary star at its centre.

Caucasus Mountains   A range of mountains on the Moon forming part of the border of the Mare Imbrium.

cavus (pl. cavi) A hollow or irregular depression on the surface of Mars.

CCD Abbreviation for charge-coupled device. A CCD is an electronic imaging device widely used in astronomical applications. The CCD consists of semiconducting silicon; when photons of light fall on it, free electrons are released.

To preserve the pattern of light falling on the CCD, the photons are collected in a matrix of small picture elements (pixels), which are defined by means of an array of electrodes, called gates, formed on the surface of the CCD. The electric charge in each pixel is then transferred to the ends of the rows by systematically changing the voltages on each gate so that the charge passes along as if on a conveyor belt. Finally, the electrons in each little packet of charge from individual pixels are counted and converted into a form in which the whole image can be stored in a computer or displayed on a television screen.

CCDs are particularly useful in astronomy because they have all the qualities required in an astronomical detector: a highly efficient response to light so they can be used to pick up very faint objects, sensitivity over a broad spectrum range, low noise levels and a large dynamic range – that is, they can detect bright and faint objects simultaneously. Furthermore, the output is linear – the number of electrons collected is in direct proportion to the number of photons received. This means that the image brightness is a direct measure of the real brightness of the object, a property not shared, for example, by photographic emulsions.

CD Abbreviation for ➤*Córdoba Durchmusterung*.

CDA Abbreviation for ➤*Centre de Données Astronomiques*.

cD galaxy A member of a class of giant elliptical galaxies with an extended halo of stars. The designation was introduced by W. Morgan in the late 1950s. A cD galaxy is often found to be the central galaxy in a rich cluster, and many are also radio emitters. They are five to ten times more luminous than typical elliptical galaxies and their mass may be as high as 10^{13} solar masses. A classic example is NGC 6616, which appears to have multiple nuclei within its envelope and is thought to be 'swallowing' smaller galaxies in its vicinity.

CDM Abbreviation for *cold dark matter*. ➤*dark matter*.

CDS Abbreviation for Centre de Données Stellaires, the former name of the ➤*Centre de Données Astronomiques*.

celestial coordinates Any system of coordinates that can be used to describe the position of an object on the ➤*celestial sphere*. Different types of coordinates are used for different applications in astronomy. Those commonly used are ➤*equatorial coordinates*, ➤*horizontal coordinates*, ➤*ecliptic coordinates* and ➤*galactic coordinates*.

celestial equator   The great circle on the ➤*celestial sphere* marking the boundary between the northern and southern hemispheres, and acting as the zero-mark for ➤*declination*. It is the projection into space of the Earth's equatorial plane.

celestial mechanics   A general term for the branches of astronomy dealing with the movements and consequent positions of astronomical objects, particularly the determination of orbits.

celestial poles   The two points on the celestial sphere about which the sky appears to rotate daily. Their positions are the directions in space towards which the Earth's rotation axis points. The north celestial pole currently lies close to the star Polaris and the south pole is in the constellation Octans, unmarked by any bright star. Because of the effects of ➤*precession*, the positions of the poles are not stationary but sweep out circles with radii of about 23° over a period of 25,800 years.

celestial sphere   The sky considered as the inside of a hollow sphere in order to describe the positions and motions of astronomical objects. Any particular observer is located at the centre of his own celestial sphere. It is easy to imagine the sky as a hemispherical dome. Half the sky is always hidden from an observer on the Earth's surface; which half of the celestial sphere can be seen by an observer depends on his latitude on the Earth and on the date and time. Measurements on the celestial sphere are made in angular measure (degrees) and do not depend on how far away the objects actually are.

Centaur   English name for the constellation ➤*Centaurus*.

Centaurs   A class of asteroids with orbits in the outer part of the solar system. Examples are ➤*Chiron* and ➤*Pholus*. Their orbits are within Neptune's but they approach no closer to the Sun than Jupiter. Their orbits are unstable, and easily perturbed when they pass close to one of the giant planets.

Centaurus (The Centaur)   A large southern constellation, lying in the Milky Way and very rich in stars. It is one of the ancient constellations recorded by Ptolemy (*c.* AD 140). It contains a number of interesting objects, including the nearest star to the solar system, ➤*Proxima Centauri*, and the finest and brightest of all globular star clusters, ➤*Omega Centauri*. ➤Table 4.

Centaurus A (Cen A)   A ➤*radio galaxy*, identified with an elliptical galaxy, NGC 5128. At a distance of 15 million light years, it is the nearest radio galaxy and consequently one of the most studied. The visible galaxy is crossed by a thick dark lane of dust. The radio-emitting lobes jut out at right angles to the dust lane, extending across 7° of sky, equivalent to almost two million light years. It is also a strong source of X-rays.

central engine   An informal expression for the central energy source powering an ►*active galactic nucleus,* ►*radio galaxy* or ►*quasar.* The energy source is generally thought to be a ►*black hole* accreting matter.

central meridian (CM)   The imaginary north−south line bisecting the disc of a planet (or a moon or the Sun) as seen by an observer.

central peak   A mountain formed in the centre of an impact crater as the crust rebounds following the impact explosion. A crater may contain several peaks grouped together.

Centre de Données Astronomiques (CDA)   An institute at the University of Strasbourg in France, established in 1972 as the Centre de Données Stellaires (CDS), which is devoted to the collection, critical evaluation and handling of stellar and other astronomical data.

centre of mass   The balancing point in a system of individual masses or in a solid object through which mass is distributed. ►*barycentre.*

Cepheid variable   A type of pulsating ►*variable star,* named after the group's prototype, Delta Cephei, which varies between magnitudes 3.6 and 4.3 in a period of 5.4 days. Cepheid variables have an unstable structure that causes them to pulse in and out. Their size may change by as much as 10 per cent during a cycle, and the temperature varies too. As pressure builds up inside, the star expands until the pressure is released, rather like it might be through a valve. The star then contracts and the cycle starts again (see illustration (a)).

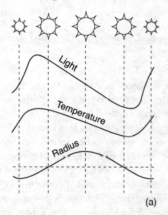

Cepheid variable. (a) The light curve, temperature variation and change in radius represented schematically for the cycle of a typical Cepheid variable.

Cepheids are luminous yellow giant stars that radiate ten thousand times as much energy as the Sun, so they can be seen at very great distances. In 1912, Henrietta Leavitt, working at Harvard College Observatory, noted a number of Cepheids in the Small Magellanic Cloud and plotted their light curves. It became clear to her that there was a relationship between the periods, typically between 3 and 50 days, and the average apparent brightness: the brighter the star, the longer its period. This is called the *period−luminosity relation* (see illustration (b)).

The importance of this discovery lies in the fact that Cepheids can be used

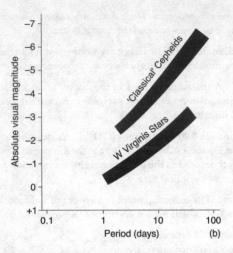

Cepheid variable. (b) The period–luminosity relationship for classical Cepheids and W Virginis stars.

as distance indicators. All the stars in the Small Magellanic Cloud can be considered to be at roughly the same distance (certainly in relation to the distance of the SMC itself), so the apparent magnitudes will differ from the absolute magnitudes by a constant factor. Once the distance to a single Cepheid variable was found by an independent method, the distances to all others could be deduced simply from measuring their periods.

Two distinct varieties of Cepheid variable have been identified: the so-called *classical Cepheids* and 'Population II' Cepheids, also commonly known as *W Virginis stars*. Their period–luminosity relations differ: for a given period, classical Cepheids are about two magnitudes brighter than W Virginis stars. This is a result of differences in mass and chemical composition. The lower mass of W Virginis stars results in a lower luminosity, but the effect is partially offset by the low abundance of elements heavier than helium ('metals') in the stars of the old ➤*Population II*. It is essential to distinguish whether a variable star is a classical Cepheid or a W Virginis star before its distance can be deduced. This is best achieved by determining the metal content from analysis of its spectrum.

Cepheus A constellation close to the north celestial pole, taking its name from the legendary King of Ethiopia, husband of Cassiopeia and father of Andromeda. It is one of the ancient constellations recorded by Ptolemy (*c.* AD 140) but is not conspicuous, having no star as bright as second magnitude. ➤Table 4.

Cerberus   Asteroid 1865, diameter 1.6 km, discovered in 1971 by L. Kohoutek when it made a close approach to the Earth.

Cerenkov radiation   Electromagnetic radiation caused by the shock wave created when electrically charged particles move through a medium at velocities greater than that of light in that medium.

Ceres   The first ►*asteroid* to be discovered, found by Giuseppi Piazzi from Palermo, Sicily, on 1 January 1801. It is by far the largest asteroid, 940 km (585 miles) in diameter, and its orbit lies in the main asteroid belt at a distance of 2.77 AU from the Sun. Its mass of 1.17×10^{21} is about one-third the entire mass of the ►*asteroid belt*. It reaches a maximum magnitude of 6.9, its ►*albedo* being only 9 per cent. Ceres rotates in just over 9 hours during which time its colour and brightness vary only slightly, suggesting that it is almost spherical and uniformly grey. Information from its spectrum indicates that the surface may be similar in composition to the ►*carbonaceous chondrite* meteorites.

Cerro Tololo Inter-American Observatory   An observatory in Chile, forming part of the ►*National Optical Astronomy Observatories* of the USA. The headquarters are at La Serena, 480 kilometres (300 miles) north of Santiago. The mountain site, 70 kilometres (45 miles) inland is at an altitude of 2,200 metres (7,200 feet). The largest instrument, the 4-metre (160-inch) Victor M. Blanco Telescope, is a twin of that at Kitt Peak in Arizona. Among the six other instruments located at the observatory are 1.5-metre (60-inch), 1.0-metre (39-inch) and 92-centimetre (36-inch) reflectors. A 1.2-metre (47-inch) radio telescope of the Universidad do Chile is also at the site.

Cetus (The Whale)   A large constellation in the region of the celestial equator, supposed to represent the sea monster that threatened Andromeda, though normally translated as the 'whale'. It was listed by Ptolemy (*c.* AD 140). It lies in a rather sparse area of the sky and all but one of its stars are fainter than third magnitude. The most notable star is the variable ►*Mira*. ►Table 4.

CFHT   Abbreviation for ►*Canada–France–Hawaii Telescope*.

Chamaeleon (The Chameleon)   A small, faint southern constellation, introduced probably by sixteenth-century navigators and included by Johann Bayer in his atlas ►*Uranometria*, published in 1603. None of its stars is brighter than fourth magnitude. ►Table 4.

Chameleon   English name for the constellation ►*Chamaeleon*.

Champollion   A planned US space mission to Comet 9P/Tempel 1, which would include a lander to retrieve a sample of the comet nucleus and return it to Earth. Launch would be in 2003 for arrival at the comet in December 2005.

Chandler wobble   Small variations in the position of the Earth's geographical poles (the poles of the rotation axis) believed to result from seasonal changes in the distribution of mass on the Earth and movements of material within the Earth. The Earth's rotation axis does, however, maintain the same orientation in space.

Chandrasekhar limit   The upper mass limit for ►*white dwarf* stars of 1.4 solar masses, which was first demonstrated theoretically by the astrophysicist S. Chandrasekhar. When more massive stars exhaust their sources of nuclear energy, they must continue to collapse to a size much smaller than a white dwarf to form a ►*neutron star* or a ►*black hole*.

chaos   A term used in the official names of certain distinctive areas of terrain on planetary surfaces, characterized by an irregular jumble of rugged hills and valleys. On Mars, such chaotic terrain may be associated with the collapse of the surface and the generation of flood water from underground.

chaotic rotation   The rotation of a planetary body without a fixed period. The possibility that rotation of this character could exist arose in the case of one of the small satellites of Saturn, ►*Hyperion*. Because of its eccentric orbit and the many gravitational forces acting on it, there is a constant exchange of energy between Hyperion's orbital motion and its rotation.

CHARA array   ►*Mount Wilson Observatory*.

charge–coupled device   ►*CCD*.

Charioteer   English name for the constellation ►*Auriga*.

Charles's Heart   English name for the star ►*Cor Caroli*.

Charles's Wain   An alternative name for the ►*asterism* in the constellation Ursa Major now more commonly known as the ►*Plough* or the Big Dipper.

Charon   The only known satellite of the remote planet Pluto, discovered by J. Christy in 1978, as a slight elongation in the image of Pluto on a photograph taken at the US Naval Observatory. With a diameter of 1,200 kilometres (740 miles), it is about half the size of Pluto. The orbital period is 6.39 days, which is also the rotation period of both Pluto and Charon. The spectrum of Charon indicates the presence of water ice, but not of methane, which is present on Pluto. ►Table 6.

chasma (pl. chasmae)   A steep-sided canyon on a planetary surface.

chassignite   A type of achondritic ►*meteorite* of which only one example – the Chassigny meteorite – is known. Together with shergottites and nakhlites, chassignites belong to the class of ►*SNC meteorites*, believed to have originated on the surface of Mars.

chemical evolution   The change in the ►*abundances* of the chemical elements as time progresses, as a result of ►*astration*.

Chicxulub crater A large terrestrial impact crater, located around the northern coast of the Yucatan Peninsula in Mexico, where it is now largely buried under sediment. It has been identified as the crater associated with an impact event, 65 million years ago, which appears to be linked with the mass extinction at that time of living species, including the dinosaurs.

child universe A theoretical concept arising from the ►*inflationary universe* cosmology. In mathematical terms, it would be possible to create a density of matter so great in our universe that new inflationary universes, or child universes, would spawn naturally. Although this is consistent with the laws of physics, the densities demanded are far beyond any feasible technology.

Chiron Asteroid 2060, diameter 180 km, discovered in 1977 by Charles Kowal. Its orbit lies between the orbits of the planets Jupiter and Uranus, well outside the main ►*asteroid belt*. It was the first of several asteroids with orbits of this kind, now known collectively as ►*Centaurs*, to be discovered. Infrared observations have shown that it has a moderately dark, rocky or dusty surface and is nearly spherical. In 1989, a ►*coma* was discovered around Chiron, as a result of which it was also designated as a periodic comet (95P/Chiron).

Chisel English name for the constellation ►*Caelum*.

chondrite An abundant type of stony ►*meteorite*, characterized by the presence of ►*chondrules*. About 85 per cent of meteorites are chondrites, as opposed to ►*achondrites*. The name now also implies a chemical composition similar to the Sun's for all but the most volatile elements.

chondrule A small sphere of rapidly cooled silicate minerals (usually olivine and/or pyroxene) found in stony ► *meteorites*. Chondrules range in size from less than 1 mm to more than 10 mm, and are found in abundance in chondritic meteorites. ►*chondrite, achondrite*.

CHON particles Particles composed primarily of light elements and rich in carbon, hydrogen, oxygen and nitrogen. They were found to exist in the vicinity of the nucleus of ►*Halley's Comet*.

chromatic aberration A defect in the imaging properties of a lens that results in the formation of coloured 'fringes' around the image. The problem arises because light of different wavelengths is focused at different distances from the lens. The effect may be countered, but not eliminated completely, by constructing an ►*achromatic lens* of two or more components.

chromosphere The gaseous layer of the Sun (or of another star) above the ►*photosphere*. Through the photospheric layers, the temperature decreases with distance from the centre of the Sun to a temperature minimum. The temperature then gradually rises again to 10,000 K through the overlying chromosphere. The name literally means 'sphere of colour' – the chromosphere is seen as a pinkish glow when the light of the photosphere is hidden at a total solar eclipse.

chronology The study and investigation of the timescales over which events occur, particularly when these are long, as with the geological history of a planet.

chronometer A high-precision clock, particularly one used on board a ship.

Chryse Planitia A circular plain, almost certainly an impact ➤*basin*, in the north equatorial region of Mars. It was the landing site for the ➤*Viking 1* probe.

CH star A giant star of ➤*spectral type* G or K that shows particularly strong bands of the CH molecule in its spectrum.

Circinus (The Compasses) A small and insignificant constellation of the southern sky, introduced by Nicolas L. de Lacaille in the mid-eighteenth century. ➤Table 4.

Circinus X-1 An ➤*X-ray burster* that it is thought to be a binary star system containing a ➤*black hole*.

Circlet An ➤*asterism* in the shape of a ring in the constellation Pisces. It is formed by the stars Gamma (γ), Theta (θ), Iota (ι), Lambda (λ) and Kappa (ϰ) Piscium.

circumpolar star A star that never sets below the horizon as seen by a particular observer. The declination of a star must be greater than 90° minus the latitude of the observer's location for it to be circumpolar. Thus, from the equator no stars are circumpolar whereas at the Earth's poles all stars are circumpolar.

circumstellar matter An envelope of gas or dust surrounding a star.

cirrus ➤*infrared cirrus*.

Cirrus Nebula An alternative name for the Veil Nebula in Cygnus, said to derive from comments made about its appearance by Sir John Herschel. ➤*Cygnus Loop*.

cislunar Pertaining to the region of space between the Earth and the Moon.

civil time The nationally agreed time used throughout a country, or within a ➤*time zone*, for regulating civil affairs. It contrasts with, for example, the ➤*local time* that would be indicated on a sundial.

civil twilight Formally defined as the interval of time when the central point of the Sun's disc is between 90° 50′ and 96° below the ➤*zenith*. ➤*twilight*.

civil year A year of the civil calendar. ➤*calendar year*.

classical Cepheid ➤*Cepheid variable*.

clast   A rock fragment produced by weathering of a larger rock and later included in another rock.

Clavius   A large lunar crater, 225 kilometres (140 miles) in diameter, near the southern limb of the Moon.

C layer   ➤*D layer*.

Clementine   A lunar mission carried out by the US Department of Defense, which returned a large body of scientific data, as well as testing space hardware. Launched on 25 January 1994, it spent two months in orbit around the Moon. The intention had been for Clementine to travel on to a flyby of the asteroid ➤*Geographos*. However, that second part of the mission was cancelled after the spacecraft malfunctioned on 5 May.

Clementine began the task of global digital imaging of the Moon, with ultraviolet-visible and near-infrared cameras. It secured better geological mapping than any previous lunar mission.

clock   Any device that can indicate the passage of equal intervals of time. Things that have been utilized for time-keeping range from the burning down of a candle, through the mechanical oscillations of pendulums and balance wheels, to the frequency associated with vibrations in a quartz crystal or an energy change in caesium atoms. ➤*atomic clock*.

Clock   English name for the constellation ➤*Horologium*.

closed universe   A model of the universe in which the amount of matter present is sufficient to halt its expansion and start it contracting at some finite time far in the future.

In all ➤*Big Bang* models, the only mechanism by which the expansion of the universe currently observed may be slowed down is the mutual gravitational attraction of all the material in it. The universe is closed if the total amount of matter is great enough to reverse the expansion.

Present observations show that there is insufficient luminous matter for closure. We do not know whether non-luminous matter, for example in the form of elementary particles such as neutrinos, might bring the density up to that required for closure. Hypothetical, additional but unseen matter that may exist is termed ➤*missing mass*. ➤*dark matter*.

Cloverleaf   A ➤*quasar* with a redshift of 2.55 which, due to a ➤*gravitational lens* effect, displays a quadruple image resembling a clover leaf.

Cluster   A four-satellite space mission of the European Space Agency designed to study the interaction between the Earth's magnetic field and the ➤*solar wind* and small-scale processes in the Earth's ➤*magnetosphere*. It was destroyed when the first Ariane 5 launcher exploded on 4 June 1996.

cluster of galaxies   A grouping of galaxies in space, linked by their mutual gravitational attraction. The distribution of galaxies in space is not uniform: they tend to clump together on distance scales of millions of light years. Clusters of galaxies take a variety of forms: they can be spherical and symmetrical, or ragged with no particular shape; they may contain a handful of galaxies or thousands; there may or may not be a concentration towards the centre. Regular clusters appear to be populated mainly by elliptical galaxies, while irregular clusters tend to include all galaxy types. Our own Milky Way Galaxy belongs to a small association known as the ►*Local Group*.

Clusters containing a concentration of many large galaxies are described as 'rich'. The nearest rich cluster is the ►*Virgo Cluster*, which has thousands of members. Even larger is the ►*Coma Cluster*, which is at least ten million light years across. The centre of a rich cluster is typically dominated by a giant elliptical galaxy. The most massive galaxies known are at the centres of large rich clusters. It is believed that the largest galaxies tend to coalesce at the centres of clusters in a process called 'galactic cannibalism'. This is consistent with the observation that the second- and third-brightest galaxies in clusters of this sort are fainter than in clusters lacking an exceptionally bright galaxy. The cannibal galaxies are often distended in appearance, and evidence of more than one nucleus is seen in some. They are usually strong sources of radio emission.

A tenuous hot gas pervades the space between the galaxies in rich clusters. Its presence has been discovered through its X-ray emission. In some cases, there is as much matter in the intergalactic gas as in the visible parts of the galaxies. Its presence tends to strip away interstellar gas from the spiral galaxies in the cluster. ►*galaxy, intracluster medium*.

cluster of stars   ►*star cluster*.

cluster variable   An old term for the type of variable star now more commonly called ►*RR Lyrae stars*.

CM (1)   Abbreviation for ►*central meridian*.

CM (2)   Abbreviation for Command Module, part of the craft used by the ►*Apollo programme* of manned Moon landings.

CN cycle   Abbreviation for carbon−nitrogen cycle, an alternative name for the ►*carbon cycle*.

CNO bi-cycle   A sequence of nuclear reactions believed to occur in stellar interiors. ►*carbon cycle*.

CNO cycle   Abbreviation for carbon−nitrogen−oxygen cycle, an alternative name for the ►*carbon cycle*.

CNO tri-cycle   A sequence of nuclear reactions believed to occur in stellar interiors. ►*carbon cycle*.

CN star A star in whose spectra the bands of the CN (cyanogen) molecule are exceptionally strong.

Coalsack An ➤*absorption nebula* of interstellar dust, about 4° across, in the ➤*Milky Way* in the constellation Crux. The nebula reflects most of the light from the stars behind it but a fraction is absorbed, causing the Coalsack to glow feebly with about a tenth of the brightness of the surrounding Milky Way.

COBE ➤*Cosmic Background Explorer.*

COBRAS/SAMBA ➤*Planck Surveyor.*

Cocoon Nebula The diffuse nebula IC 5146 in the constellation Cygnus. It is a region of glowing hydrogen gas surrounding a sparse cluster of twelfth magnitude stars, at an estimated distance of 3,000 light years. It is thought to be similar in nature to the ➤*Orion Nebula,* a complex region of gas and dust in which star formation is taking place.

cocoon star A star surrounded by a shell or 'cocoon' of dust and consequently emitting strong infrared radiation.

coelostat A pair of flat mirrors, mounted and driven in such a way that they compensate for the apparent motion of the stars, constantly reflecting a particular area of sky into a fixed instrument that cannot be conveniently moved. One of the mirrors lies in and rotates about the polar axis so that the image does not rotate; the second mirror directs the image into the instrument.

Coggia's Comet ➤*Comet Coggia.*

colatitude The angle calculated by subtracting latitude from 90°.

cold dark matter ➤*dark matter.*

collapsar A ➤*degenerate star* such as a ➤*white dwarf,* ➤*neutron star* or ➤*black hole.*

colles (sing. collis) Low hills. A term used as part of the name of a feature of planetary topography.

collimation The process of making a beam of light or particles parallel so that it is neither converging nor diverging, or the accurate alignment of the components of a scientific instrument, especially an optical one such as a telescope.

colour ➤*magnitude.*

colour excess The difference between the observed ➤*colour index* of a star and the intrinsic colour index, which may be inferred from the star's ➤*spectral type.* The colour excess is a measure of the degree to which the star's light has been reddened as it travels though interstellar dust. Such reddening occurs because blue light is scattered and absorbed to a greater degree than red light.

colour index The difference between the ➤*magnitude* of a star measured in one wavelength band and its magnitude in another.

When first conceived, the term normally referred to the difference between photographic and visual magnitudes, which were different because the types of photographic plate then used were predominantly sensitive to the blue end of the spectrum, whereas the eye's sensitivity peaks around the green–yellow part. Since the introduction of more precisely defined photometric systems using filters, such as the ➤*UBV photometry* system, colour indices have been formed from different pairs of magnitudes. The quantity *B–V* is one of the most commonly used; *B* and *V* correspond roughly to the old photographic and visual magnitudes.

A colour index such as *B–V* is useful because it can be measured relatively quickly and easily for large numbers of stars and yet is a good indicator of spectral type. It is one of the quantities used in plotting a colour–magnitude diagram. ➤*Hertzsprung–Russell diagram*.

colour–luminosity diagram ➤*Hertzsprung–Russell diagram*.

colour–magnitude diagram ➤*Hertzsprung–Russell diagram*.

colour temperature The temperature of a ➤*black body* that, in the spectral region in which an object is being observed, would radiate the same distribution of intensity with wavelength as in the spectrum of the object.

Columba (The Dove) A small, faint constellation in the southern sky, introduced by A. Royer in 1679. It is sometimes said to represent the dove that followed Noah's Ark. ➤Table 4.

colure A great circle on the ➤*celestial sphere*, passing through the celestial poles and either the ➤*equinoxes* (equinoctial colure) or the ➤*solstices* (solstitial colure).

coma (1) A diffuse envelope of gas surrounding the nucleus of a ➤*comet*. The coma consists of dust together with neutral and ionized gas molecules and radicals. Typically, it reaches its maximum size of up to a million kilometres across when the comet has just passed ➤*perihelion* in its orbit around the Sun.

coma (2) An imperfection in the imaging qualities of an optical system, which results in an image that should be a point instead looking fan-shaped. Coma particularly affects off-axis images, increasing in severity with distance from the optical axis. It may be made worse when the optical parts of a telescope are not correctly aligned.

Coma Berenices (Berenice's Hair) A small and faint constellation adjacent to Boötes, introduced by Tycho Brahe in about 1602. It is supposed to represent the tresses of Queen Berenice of Egypt, who cut off her hair and presented it to the gods in gratitude for the safe return of her husband from battle. The constellation is notable for the large number of galaxies it contains, members of both the Coma and Virgo ►*clusters of galaxies.* ►Table 4.

Coma Cluster   A rich cluster of galaxies in the constellation Coma Berenices, extending over several degrees of sky and containing more than a thousand bright galaxies. One of the nearer rich clusters, it lies at an estimated distance of 300 million light years. It is roughly spherical in shape with most of the galaxies concentrated towards the centre, where the average distance between galaxies is three times smaller than the distance between the Milky Way and the Andromeda Galaxy in the ►*Local Group.*

comes (pl. comites)   The Latin word for 'companion', sometimes used as a term for the fainter component in a ►*binary star* system, especially in older books and catalogues.

comet   An icy body, orbiting in the solar system, which partially vaporizes when it nears the Sun, developing a diffuse envelope of dust and gas and, normally, one or more tails.

Ground-based observations of the behaviour of many comets, together with results from the investigation in 1986 of ►*Halley's Comet* from space probes, support the view first proposed by F. Whipple in about 1949 that the nuclei of comets are essentially 'dirty snowballs' a few kilometres across. They appear to be composed of frozen water, carbon dioxide, methane and ammonia, in which dust and rocky material is embedded. As a comet approaches the Sun, solar heating starts to vaporize the ices, releasing gas that forms a diffuse luminous sphere, called the *coma*, around the nucleus. The coma may be up to a million kilometres across. The nucleus itself is too small to be observed directly. Observations in the ultraviolet region of the spectrum made from spacecraft have shown comets to be surrounded by huge clouds of hydrogen, many millions of kilometres in size. The hydrogen comes from the breakdown of water molecules by solar radiation. In 1996, X-ray emission from ►*Comet Hyakutake* was discovered, and it was subsequently found that other comets are also X-ray sources.

Dust and gas leave the comet nucleus from jets on the side facing the Sun, then stream away under the Sun's influence. Electrically charged ionized atoms are swept away directly by the magnetic field of the ►*solar wind*, forming straight *ion tails* (alternatively called Type I, plasma or gas tails). Variations in the solar wind cause the ion tail to take on structure, or even break off in a ►*disconnection event*. Small neutral dust particles are not carried along by the solar wind but get 'blown' gently away from the Sun by ►*radiation pressure.* ►*Dust tails* (also called Type II tails) are often broad and flat. ►*Comet Hale-Bopp* was found to have a third distinct tail, made of neutral sodium atoms. The tails grow as a comet approaches the Sun and are always directed away from the Sun: they can be as much as a hundred million kilometres long. Large dust particles become strewn along the comet's orbit and form ►*meteor streams.*

Despite their often dramatic appearance, comets contain very little material, perhaps only a billionth the mass of the Earth. The tails are so tenuous that only one five-hundredth of the mass of the nucleus may be lost in a passage round the Sun.

A dozen or so new comets are discovered every year. Some are *short-period comets*, in elliptical orbits that take between about 6 and 200 years to complete. Most are *long-period comets*, in orbits so elongated that the period is many thousands of years. Short-period comets orbit close to the ecliptic plane, but long-period comet orbits are not confined to the main plane of the solar system.

It is now generally accepted that many comets originate in a spherical cloud that surrounds the solar system at a distance of perhaps 50,000 AU. This 'reservoir' of comet nuclei is called the ➤*Oort cloud*. Others appear to come from the ➤*Kuiper Belt,* located beyond the orbit of Neptune. The short-period comets have been captured within the planetary system by the gravitational perturbation of their orbits that can arise from a close encounter with Jupiter.

When a new comet is discovered, or a periodic comet recovered, it is given a designation consisting of the year followed by an upper-case letter indicating the half-month of discovery in that year, e.g. A = 1–15 January, B = 16–31 January . . . Y = 16–31 December. The prefix P/ is added for short-period comets and C/ for long-period comets. Periodic comets that have disappeared or been destroyed are prefixed by D/. New comets are named after their discoverers, no more than three names being permitted when there are several independent discoverers. A few comets have been named after individuals who calculated their orbits (e.g. Halley and Encke) or after observatories or satellites, where discovery was essentially through the efforts of a team. When a short-period comet has been fully established as such, it is also allocated a number (e.g. 1P/Halley).

This system for designating and naming comets was introduced in 1995. Prior to 1995, comets were given designations consisting of the year of discovery, temporarily followed by a lower-case letter indicating the order of discovery in that year. The letters were subsequently replaced with 'permanent' designations in the form of Roman numerals determined by the order of perihelion passage of comets in the particular year.

Authority for naming rests with the ➤*International Astronomical Union.* Its designated centre coordinates discovery reports and observations, and distributes information to subscribers.

Comet Arend—Roland (C/1956 R1)   A bright comet seen in 1957. At one stage it appeared to develop a 'spike' pointing towards the Sun. This was a geometrical effect of viewing angle, caused by sunlight scattering off large dust particles strewn along its path, making them visible as the Earth passed through the plane of the comet's orbit.

cometary globule   A small interstellar cloud the shape of which bears a superficial resemblance to that of a comet. A cometary globule is the remnant of a relatively dense condensation within a cloud of interstellar gas and dust after the more tenuous gas around it has been swept away by the action of strong ultraviolet radiation from nearby stars. The condensation becomes the head of the globule and, in its 'shadow', some of the material of the original cloud is protected from the ultraviolet light, giving rise to the comet-like tail structures.

Cometary globules occur in regions of star formation and it is thought that, over time, they evolve to become Bok ➤*globules*. About forty are known in the ➤*Gum Nebula*. The ➤*Horsehead Nebula* is a cometary globule in the process of formation.

Comet Bennett (C/1969 Y1)   A spectacular comet discovered by J. C. Bennett from South Africa on 28 December 1969. It reached zero magnitude in March 1970 and had a tail 30° long. Observations made with the Orbiting Geophysical Observatory 5 revealed a vast hydrogen cloud surrounding the head and tail measuring 13 million kilometres in the direction parallel to the tail.

Comet Biela (3D/Biela)   A nineteenth-century comet, famous because it split in two before disappearing completely. The comet was originally discovered in 1772 by Montaigne of Limoges. When it was rediscovered by W. von Biela in 1826, the orbit was calculated accurately enough to identify its two previously recorded apparitions. The period was 6.6 years. At its 1846 return it was double. By 1852, the two components were separated by more than two million kilometres but following the same orbit. It was never seen again.

A November meteor shower, the ➤*Andromedids*, is associated with Biela's comet. Occasional brilliant displays have been seen both before and after the break-up of the comet.

Comet Coggia (C/1874 H1)   A bright comet discovered in 1874 by J. E. Coggia from Marseilles. It rapidly moved southwards, developing a tail 40° long. A series of fountain-like envelopes were seen to be thrown off from active regions on its spinning nucleus.

Comet Delavan (C/1913 Y1)   A bright comet discovered by Delavan from La Plata in Argentina in December 1913. It remained visible for many months in 1914.

Comet Donati (C/1858 L1)   A brilliant comet discovered by G. B. Donati of Florence in 1858. Contemporary drawings show it with a broad, curved ➤*dust tail* and two narrow, straight ➤*ion tails*. The head regularly threw off fountain-like envelopes over a period of several weeks.

Comet Encke (2P/Encke)   A periodic comet, first seen from Paris by P.

Méchain (1744–1804) in 1786. It was rediscovered by Caroline Herschel in 1795, by Pons and others in 1805 and by Pons again in 1818. J. F. Encke (1791–1865) computed the orbit of the comet seen in 1818 and linked it with the previous apparitions. He predicted a reappearance in 1822, when it was successfully recovered as a result. Its elliptical orbit has the very short period of 3.3 years.

On subsequent revolutions the comet reached perihelion earlier than predicted by about two hours; the effect has, however, been reducing steadily ever since. These observations can be explained in terms of a 'rocket effect', acceleration caused by the evaporation of gases under the influence of solar radiation, in conjunction with effects of the spin and precession of the comet's nucleus.

The ►*Taurid* meteor shower is associated with Comet 2P/Encke.

Comet Giacobini–Zinner (21P/Giacobini–Zinner) A periodic comet discovered in 1900. Its period of revolution around the Sun is 6.5 years. The Draconids, a ►*meteor shower* seen in October of some years, is associated with the comet, resulting from debris that follows the same orbit as the comet entering the Earth's atmosphere.

In 1985, the American space probe International Sun–Earth Explorer (ISEE) 3, originally launched in 1978 for a different purpose, was directed to pass through the tail of Comet Giacobini–Zinner in a project called ICE (International Cometary Explorer).

comet group A class of comets that have orbits with similar characteristics. Members of a group (or family) of comets are not necessarily close to each other in space. The most notable examples are the group of short-period comets, which have come under the gravitational influence of the planet Jupiter and have orbital periods typically between 6 and 8 years, and the so-called ►*sungrazers*, long-period comets that virtually skim the Sun's outer layers at perihelion passage and share other orbital similarities.

Comet Hale–Bopp (C/1995 O1) One of the brighter comets of the twentieth century, notable for its very large size. Discovered by Alan Hale and Thomas Bopp on 22 July 1995, it reached perihelion on 1 April 1997 and achieved a maximum brightness of around magnitude −1. Its core was estimated to be 40 km (25 miles) across, over twice the diameter of ►*Halley's Comet*.

Comet Halley ►*Halley's Comet*.

Comet Humason (C/1961 R1) A giant comet discovered in 1962. It developed tails while still 5 AU from the Sun, which represents a very unusual degree of activity in a comet still so distant.

Comet Hyakutake (C/1996 B2) A bright comet which reached zero magnitude in March 1996 and developed a tail estimated to stretch at least 7° across the sky. Its brightness was largely due to its proximity to Earth: it passed within 15 million km.

Comet Ikeya–Seki (C/1965 S1)   A particularly brilliant comet, discovered on 18 September 1965 by two Japanese amateur astronomers. It was especially conspicuous in the southern hemisphere after it had passed perihelion. It belonged to the group of comets known as ➤*sungrazers*, characterized by their very small perihelion distances, as a result of which they travel through the Sun's outer layers.

Comet IRAS–Araki–Alcock   ➤*Infrared Astronomical Satellite.*

Comet Klinkenberg   ➤*De Chéseaux's Comet.*

Comet Kohoutek (C/1973 E1)   A comet discovered in 1973, well in advance of its perihelion passage, when it was near the orbit of Jupiter. Suggestions that it would prove spectacular turned out to be incorrect. However, it was subject to an extensive, coordinated professional observing programme, which included observations from the orbiting laboratory ➤*Skylab.* Much new information about comets was obtained, including the first direct proof of the presence of silicates in the ➤*dust tail* of a comet.

Comet Morehouse (C/1908 R1)   A comet discovered from the USA in 1908, the first to be studied extensively by means of photography. Remarkable changes were observed in the tail structure. Throughout 30 September 1908 the tail changed continuously. On 1 October, it broke off and a tail could not be seen visually, though a photograph of 2 October showed three tails. The breaking off and subsequent growth of tails occurred repeatedly. ➤*disconnection event.*

Comet Mrkos (C/1957 P1)   A bright comet of 1957, discovered with the naked eye by a Czech comet-hunter.

Comet Schwassmann–Wachmann 1 (29P/Schwassmann–Wachmann 1)   A periodic comet, discovered from Hamburg in 1927. It is in an almost circular orbit with a period of 16.1 years, which keeps it permanently between Jupiter and Saturn. It can be seen each year at the time of ➤*opposition.* Though it is usually of eighteenth magnitude, it can brighten by as much as 4–8 magnitudes in a few days. The outbursts are accompanied by changes in the nucleus and coma.

Comet Shoemaker–Levy 9 (D/1993 F2)   A comet that crashed into the planet Jupiter in July 1994. When it was discovered photographically on 25 March 1993 by Carolyn and Eugene Shoemaker and David Levy, it was in an elongated 2-year orbit around Jupiter and consisted of a string of about 20 separate fragments. Calculations suggested it had been orbiting Jupiter for several decades, but broke up when subject to tidal forces during a close approach to Jupiter in July 1992. This encounter also set the fragments on their collision course. They struck Jupiter in sequence, between 16 and 22 July 1994. A number of the impacts produced large dark clouds in Jupiter's atmosphere, as well as vivid fireballs observable in the infrared. The dark clouds were discernible for several months before being dispersed by winds and turbulence.

Comet Tebbutt (C/1861 J1)   A brilliant naked-eye comet, discovered by an Australian amateur astronomer in 1861. The Earth passed through the comet's tail on 30 June 1861.

Comet West (C/1975 V1)   A bright comet, visible to the naked eye, which appeared in 1975. The tail covered a large triangular area and the nucleus showed unusual activity, breaking into four pieces shortly after the comet's close perihelion passage.

Command and Service Module   ➤*Apollo programme.*

Command Module   ➤*Apollo programme.*

commensurability   The property shared by two orbits – either actual orbits or ones that are theoretically possible – that the period of revolution in one is equal to or a simple fraction of the period in the other. Many such commensurabilities are observed between the orbits of the planets, their natural satellites and the asteroids. For example, the ➤*Trojan asteroids* share the orbit of Jupiter (1:1), and the periods of Jupiter and Saturn are in the ratio 2:5. Commensurabilities arise from the long-term ➤*resonance* effects that the gravitational interactions between the objects have on their orbits.

   Commensurabilities also occur between rotational and orbital motion. ➤*synchronous rotation.*

common proper motion   A term applied to two or more stars that appear to share similar motion through space but are not otherwise obviously physically linked with each other in a cluster, association or multiple system. ➤*proper motion.*

Compact Array   ➤*Australia Telescope.*

compact object   Any astronomical object that is substantially denser or more compact (typically by a factor of a hundred) than most objects in its class. There is no precise definition of this rather general term.

   White dwarf stars, for example, are compact relative to most stars because they are a million times denser than ➤*main-sequence stars.* An unusually bright galactic nucleus might also be referred to as a compact object if its light comes from a much smaller volume than is generally the case for the observed luminosity. The Swiss–American astronomer Fritz Zwicky made an important catalogue (1971) of compact galaxies that at the time were thought to be intermediate in nature between ordinary galaxies and ➤*quasars.*

comparator An instrument for comparing two similar photographs in order to highlight any small differences between them. A typical application would be to compare two photographs of the same area of sky taken on successive nights in the search for minor planets or comets, which would be observed to move through the fixed pattern of stars. Variable stars can be identified because the size of the photographic image changes as the brightness varies. The most usual sort is the ➤blink comparator.

Compass English name for the constellation ➤Pyxis.

Compasses English name for the constellation ➤Circinus.

Compton effect The interaction (essentially a collision) between a ➤photon and an electrically charged particle, resulting in the photon being scattered with lower energy (i.e. longer wavelength), while the extra energy is given to the particle. ➤comptonization, inverse Compton effect.

Compton Gamma Ray Observatory (Gamma Ray Observatory; GRO) A NASA orbiting observatory carrying four gamma-ray astronomy experiments designed to accomplish mapping, spectroscopy, and the detection and location of bursts over a range of energies. It was launched from the ➤Space Shuttle in April 1991. Originally known simply as the 'Gamma Ray Observatory', it was subsequently named in honour of the American physicist A. H. Compton (1892–1962). Its experiments include EGRET, able to give good resolution of the energy spectrum of high-energy gamma-ray sources, and relatively good source positions. It was used to produce a catalogue of gamma-ray sources, including ➤supernova remnants, stellar OB ➤associations and ➤active galactic nuclei. BATSE (the Burst and Transient Source Experiment) was designed to find ➤gamma-ray bursters and conduct full sky surveys.

comptonization A change in energy brought about by the action of the ➤Compton effect, particularly one in which electrons are accelerated to speeds comparable with that of light by collisions with X-ray and gamma-ray photons.

Cone Nebula A dark cone-shaped dust nebula, forming part of a complex of nebulosity and stars (NGC 2264), near the fifth magnitude star S Monocerotis.

conic section A general term that describes a circle, an ellipse, a parabola or a hyperbola. All four of these mathematically defined curves can be created by slicing through a solid cone in a particular way (see illustration on page 82).

conjunction An alignment of two bodies in the solar system with the Earth, so that they appear to be at the same place (or nearly so) in the sky as seen from the Earth.

A planet is said to be at conjunction when it is at the same ecliptic longitude as (and so is approximately in line with) the Sun. The planets Mercury and Venus can form such a line by being either between the Earth and the Sun,

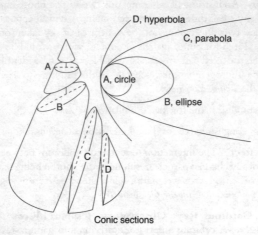

Conic sections

Conic sections.

when they are said to be at *inferior conjunction*, or behind the Sun as seen from Earth, an alignment called *superior conjunction*. The planets further from the Sun than the Earth can only come to superior conjunction.

Conjunctions can also occur between planets or between the Moon and one or more planets but, in that context, the word is often used more loosely to describe an approach within a few degrees. ➤*ecliptic coordinates.*

constellation   One of 88 designated areas in the sky or the pattern of stars within it.

Records show that, from antiquity, civilizations have given names to conspicuous patterns of bright stars. Each culture had its own way of dividing the sky into pictorial elements. Many of those in use today originated in Mesopotamia and were further developed by the Greeks. Ptolemy listed 48 in the second century AD and the rest have been added since about 1600. Some proposed constellations that have never found general acceptance appear on old star maps.

Originally, constellations were regarded simply as star patterns, but they gradually acquired usefulness as a way of specifying stars and their positions. As the science of astronomy developed, the lack of precise standard definitions for the constellations led to confusion in identifying fainter stars in more sparsely populated regions of the sky. This was resolved when, in 1930, there was international agreement among astronomers to define the boundaries of 88 constellations along lines of ➤*right ascension* and ➤*declination.*

Each of the 88 constellations has an entry in this Dictionary under its official Latin name; all the constellations are listed in Table 4.

contact   A significant stage in the course of a solar ➤*eclipse* when the ➤*limbs* of the Sun and Moon appear to be just touching. ➤*First contact* occurs when the east limb of the Moon reaches the opposite side of the Sun. ➤*Second contact* is when the west limbs are touching and ➤*third contact* when the east limbs are touching. Second and third contact mark the beginning and end of totality in a total eclipse. ➤*Fourth contact* is when the Moon finally clears the Sun's disc.

contact binary   A pair of stars in physical contact or surrounded by a common envelope of matter. As stars evolve, they expand. Where two expanding stars are close to each other in a binary system, mass from each is drawn towards the other by the mutual gravitational attraction, until a point of contact is reached. ➤*Mass transfer* between the stars may take place, and the cores of the two stars can share a common envelope as expansion continues.

Such close binary stars often show eclipses as one star crosses in front of the other. The distorted shape of the light curve of star systems such as W Ursae Majoris can be explained if the two stars are in contact. ➤*Roche lobe*.

continental drift   The slow but constant movement of the Earth's continents, relative to the poles and to each other, in response to internal forces.

continuous creation   The notion that matter continuously appears spontaneously in the universe. Continuous creation would need to take place if the expanding universe were always to appear the same, whatever an observer's location in time and space, as proposed in the ➤*steady-state theory*. There is no evidence to suggest that such a process actually occurs.

continuous spectrum   A ➤*spectrum* in which the intensity of radiation changes only gradually with wavelength, in contrast with the sharp peaks of intensity found in an ➤*emission line spectrum*. Any body with a temperature above ➤*absolute zero* emits a continuous spectrum, the character of which depends on its temperature. A continuous spectrum may be crossed by narrow ➤*absorption lines*, as in the Sun's spectrum, for example. ➤*black body radiation*, *thermal radiation*.

convection   The process by which heat is transported in a fluid through the mass movement of the fluid itself. Convection is effective when there is a substantial decrease in temperature with height in a fluid, for example within certain layers in stars. A bubble of gas that is hotter than its surroundings expands and rises. When it has cooled by passing on its extra heat to its surroundings, the bubble sinks again.

convection zone   A layer in a star in which convection currents are the main mechanism by which energy is transported outwards. Energy flows outwards because there is a steady decrease in temperature between a star's core and its surface. There are two possible mechanisms: radiation and convection. Where the temperature gradient becomes steep enough, convection sets in. This occurs where the temperature is low enough for atomic nuclei and electrons to combine into atoms and negative ions. This makes the gas more opaque, so it is harder for radiation to pass through.

In a convection zone, currents of hot gas rise and then fall again after their temperature has dropped. In the Sun, a convection zone extends from just below the ➤*photosphere* for about one-fifth of the solar radius. In ➤*main-sequence stars* cooler and less massive than the Sun, the importance of the convection zone increases with decreasing mass. Stars hotter than ➤*spectral type* F5 have insignificant outer convection zones.

cooling flow   A flow of gas towards the centre of a rich ➤*cluster of galaxies*, which, confined by surrounding gas that is hotter and more diffuse, cools rapidly by emitting energy in the form of X-rays, and condenses into stars.

Coonabarabran   The town in Australia that is nearest to the site of the ➤*Siding Spring Observatory*, 27 kilometres (17 miles) away.

Coon Butte   ➤*Arizona meteorite crater.*

co-orbital satellite   A satellite whose orbit is almost the same as that of another satellite. Saturn's moons ➤*Epimetheus* and ➤*Janus* are examples of co-orbital satellites.

Coordinated Universal Time (UTC)   A modified version of ➤*International Atomic Time* (TAI) on which broadcast time signals are based. ➤*Universal Time.*

coordinate system   A means of specifying the position of a point or object, either in space or on a surface, in terms of its linear or angular distance from some specified plane, line or point. Latitude and longitude, for example, are the coordinates commonly used to specify the location of a place on the surface of the Earth.

In astronomy, several different systems of ➤*celestial coordinates* are used to define positions on the ➤*celestial sphere*, each of which is designed for a particular application. Other coordinates are also used, such as the X, Y, Z three-dimensional spatial coordinates of the planets for defining their positions relative to the Sun and Earth.

Copernican system   A model of the solar system in which the planets orbit around the central Sun, proposed by Nicholas Copernicus and published in his book ➤*De revolutionibus* in 1543. The theory did not find favour at first since it could predict the planetary positions no more accurately than the ➤*Ptolemaic system*, which had been used for hundreds of years, and it displaced the Earth from the centre of the universe, which was regarded as religiously unacceptable by many. In Copernicus's heliocentric model, the basic planetary orbits were circles and he had to invoke ➤*epicycles* to reproduce the observed movements of the planets. Nevertheless, Copernicus's ideas marked a turning point and stimulated work that was later to lead to the development by Johannes Kepler of a more accurate heliocentric model in which the planetary orbits are elliptical rather than circular.

Copernicus (1)   A large and conspicuous crater on the Moon at the centre of a ray system that extends for more than 600 kilometres (370 miles). Terraced walls and multiple central peaks are features of the crater, which is 93 kilometres (58 miles) in diameter.

Copernicus (2)   The name given to the third ➤*Orbiting Astronomical Observatory* (OAO-3).

Cor Caroli (Alpha Canum Venaticorum; α CVn)   The brightest star in the constellation Canes Venatici. The Latin name, meaning 'Charles's heart', is a reference to the execution of King Charles I of England in 1649 and is said to have been given to the star by Charles Scarborough in 1660. It is actually a ➤*visual binary* star, the components having magnitudes of 2.9 and 5.5. The brighter star is an ➤*Ap star*.

Cordelia   A small satellite of Uranus discovered during the ➤ *Voyager 2* encounter with the planet in 1986. Cordelia is one of the two satellites that act as 'shepherds' of the planet's Epsilon ring (the other being Ophelia). ➤Table 6.

Córdoba Durchmusterung (CD)   A catalogue of the positions and magnitudes of 578,802 southern stars down to tenth magnitude prepared under the direction of J. M. Thome (1843–1908) and colleagues at the Córdoba Observatory between 1892 and 1914. The declination range covered, −2° to −22°, was chosen to complement the ➤*Bonner Durchmusterung*.

core   The innermost part of a star or planet. The core of a star may be considered as the region in which thermonuclear reactions are taking place. A marked structural boundary may divide a planetary core from the overlying layers.

Coriolis force   A force that appears to be acting on a moving object observed in a rotating frame of reference. For example, an object thrown northwards from anywhere in the Earth's northern hemisphere will tend to be deflected towards the east by the Coriolis force.

corona (1)   The outermost layers of the Sun, which become visible as a white halo during a total solar ➤*eclipse*. It extends out to many times the Sun's radius, until it merges into the ➤*interplanetary medium*. It consists of a number of components.

The *K corona* (electron corona or continuum corona) is seen as white light scattered from the ►*photosphere* by high-energy electrons, which are at a temperature of a million degrees. The K corona is not uniform, but shows variable structures such as streamers, condensations, plumes and rays. Because the electrons are moving at high speeds, the ►*Fraunhofer lines* are blurred out in the spectrum of the reflected light.

The *F corona* (Fraunhofer corona or dust corona) is light from the photosphere scattered by slower-moving dust particles around the Sun. Fraunhofer lines are seen in its spectrum. The extension of the F corona into interplanetary space is seen as the ►*zodiacal light*.

The *E corona* (emission line corona) consists of the light from discrete ►*emission lines* produced by highly ionized atoms, particularly iron and calcium. It is detected out to two solar radii only. The corona also emits radiation in the extreme ultraviolet and soft X-ray regions of the spectrum.

The extent and shape of the corona changes during the course of the ►*solar cycle*, mainly due to the streamers produced in ►*active regions*.

corona (2)   (pl. coronae) An ovoid-shaped feature on a planetary surface, particularly Venus and Miranda.

Corona Australis (The Southern Crown)   A small and faint but nevertheless quite conspicuous constellation on the southern border of Sagittarius. It is one of the ancient constellations listed by Ptolemy (*c.* A D 140). ►Table 4.

Corona Borealis (The Northern Crown)   A small but quite conspicuous constellation of the northern sky, its main stars forming a semicircular arc. It was listed by Ptolemy (*c.* A D 140). ►Table 4.

coronagraph   An instrument for observing the Sun's ►*corona*, which is normally seen only at a total solar eclipse. Invented by Bernard Lyot in 1930, the coronagraph is a special telescope in which an occulting disc at the prime focus creates an artificial eclipse. This device allows the faint light of the corona, which is normally overwhelmed by light from the disc of the Sun, to be isolated. However, even with a coronagraph located at a site where the sky is very clear, scattering of light by the Earth's atmosphere is a problem. This is partly overcome by the use of special filters or by observing the coronal light with a spectrograph.

coronal hole   An extended region of exceptionally low density and temperature in the solar ►*corona*. Coronal holes are probably linked to areas where the magnetic field diverges outwards from the ►*photosphere* into the corona. They typically last for several rotations of the Sun, and they are sources of strong ►*solar wind*.

coronal mass ejection (CME)   An outward eruption into interplanetary space of material from the solar ►*corona*. CMEs are associated with features in the Sun's magnetic field. During periods of high ►*solar activity*, one or two occur each day, originating over a broad range of solar latitude. In periods when the Sun is quiet, they occur less frequently (around once every 3–10 days), and are confined to lower latitudes. The average ejection speed ranges from 200 km/s at minimum activity to twice that speed at solar maximum. Most occur without an accompanying flare; if a flare does occur, it usually follows the onset of the CME. They are the most energetic of all transient solar phenomena and have a marked effect on the ►*solar wind*. Large CMEs in the plane of the Earth's orbit are responsible for geomagnetic storms.

coronium A hypothetical chemical element to which were attributed unidentified ►*emission lines* in the spectrum of the solar ►*corona*. It was later discovered that the mysterious lines are produced by highly ionized forms of known elements, such as iron.

corrector plate A thin lens with a specially figured surface used in the optical design of certain types of telescope, notably the ►*Schmidt camera* and the ►*Schmidt–Cassegrain* popular with amateur astronomers. The corrector plate serves to eliminate the ►*spherical aberration* that would otherwise affect the image in these designs, which utilize mirrors with a spherical profile.

correlation telescope A ►*radio telescope* in which the voltages from two separate antennas are multiplied and averaged in order to separate the natural cosmic signal from local radio signals. This technique is central to the design of radio interferometers. ►*aperture synthesis, radio interferometer, Earth rotation synthesis*.

correlator An electronic device forming part of the system of a ►*correlation telescope*. It performs the multiplication of signals from each pair of antennas in a ►*radio interferometer*.

Corsa-B A Japanese X-ray astronomy satellite launched in February 1979 and later renamed Hakucho.

Corvus (The Crow) A small constellation on the southern border of Virgo, whose four main stars are of third magnitude and arranged roughly in the shape of a kite. It was listed by Ptolemy (*c.* AD 140) and is sometimes called the 'raven' rather than the 'crow'. ►Table 4.

Cos-B A European gamma-ray astronomy satellite launched on 9 August 1975. It remained in operation until April 1982 and was used to map gamma-ray emission from the sky. About 25 discrete sources were also identified.

Cosmic Background Explorer (COBE) A NASA astronomy satellite launched in 1989 to scan the sky and map diffuse far-infrared and millimetre-wave radiation. It measured a temperature of 2.73 K for the ►*cosmic background radiation* and confirmed that it is entirely ►*thermal radiation* in origin, in accordance with the hot ➤*Big Bang* theory. Infrared detectors mapped the distribution of dust in the Milky Way and uniform diffuse radiation over the whole sky.

cosmic background radiation Diffuse electromagnetic radiation that appears to pervade the whole of the universe. Its discovery in 1964 by Arno Penzias and Robert Wilson (publicly announced in 1965) was of immense importance in ➤*cosmology* because it provided strong evidence in favour of the ➤*Big Bang* theory. It is believed to be the relic of the radiation generated in the event that marked the origin of the universe. The spectrum of the background radiation is characteristic of a ➤*black body* at a temperature of 2.73 degrees above ➤*absolute zero* (2.73 K) and is most intense in the microwave region. The Milky Way Galaxy is travelling through space at 600 km/s relative to the background radiation.

Measurements made by the ➤*Cosmic Background Explorer* satellite (COBE) in 1992 showed for the first time that the radiation is not completely smooth in its distribution across the sky when corrections have been made for expected causes of variation in the basic data. Ripple-like variations amounting to about ten millionths of a degree were discovered. They are believed to be the first signs of structure emerging in the early universe.

cosmic censorship A hypothesis arising from the mathematics of ➤*black holes* which states that ➤*singularities* are always concealed from distant observers by their ➤*event horizon*, and cannot be seen directly from outside their event horizon. If the ➤*Big Bang* began at a singularity, it is a naked singularity that we can, in principle, see because we are on the 'inside'.

cosmic dust ➤*interplanetary dust, interstellar dust.*

cosmic rays Extremely energetic elementary particles travelling through the universe at practically the speed of light. They were discovered by V. F. Hess in 1912 during a balloon flight. Particles beyond the Earth's atmosphere are known collectively as *primary cosmic rays*. When they encounter the atmosphere, their collisions with atomic nuclei produce *air showers* of elementary particles, known as *secondary cosmic rays*.

The chemical composition of atomic nuclei found among cosmic rays mirrors the cosmic abundance as found in stars like the Sun, although there are small differences at the highest energies. Cosmic rays are the only particles we can detect that have traversed the Galaxy. The ones of the highest energy may even have come from ➤*quasars* and ➤*active galactic nuclei*. Lower-energy cosmic rays are generated within the Galaxy in ➤*supernova* explosions, ➤*supernova remnants* and ➤*pulsars*. Solar flares are a source of the lowest-energy cosmic rays, which increase in intensity at times of maximum ➤*solar activity*.

cosmic string A hypothetical line-like defect in the fabric of ➤*spacetime*, in the form of an infinite line or a closed loop, arising as a consequence of ▸*Grand Unified Theories* of elementary particles.

The thickness of cosmic strings would be 10^{-31} metres and their mass about ten million solar masses per light year. These tube-like configurations of energy may have arisen in the early universe and they could have played an important role in the formation of galaxy clusters. There is no observational evidence yet for the existence of cosmic strings.

cosmic year A period of about 220 million years, which is the time taken by the Sun to complete one revolution about the centre of the ▸*Galaxy*.

cosmo- A prefix commonly used to make compound words, meaning 'of the universe' or 'of space' (as in cosmonaut or cosmochemistry).

cosmogony The study of how celestial systems and objects in general are formed but, in particular, the study of the origin of the solar system.

Various theories to account for the formation of the solar system have been proposed since Descartes first attempted in 1644 to apply scientific reasoning to what would now be called ▸*cosmology*. His 'vortex' theory, and more modern versions of it, suppose the existence of turbulence from which planets form but, since there is no known mechanism to drive turbulence, such theories have been rejected. *Tidal theories* invoke the close approach of another star to the Sun, resulting in material being torn out of the Sun, which would then condense into planets, but these too are now regarded as unlikely explanations. Tidal mechanisms in which stars interact with giant molecular clouds are more probable.

The modern view is that the solar system formed from a slowly rotating cloud of gas. As the cloud collapsed, a dense opaque core formed — ultimately to become the Sun – surrounded by a disc of gas and dust. The first such *nebular theories* were suggested by Immanuel Kant (1724–1804) in 1755 and Laplace in 1796. Ideas on how the planets actually formed within the disc have changed greatly in recent years. Current thinking is that they gradually accumulated by ▸*accretion*. The effect of heating by the Sun, which diminishes with distance, accounts for the difference between the inner rocky planets and the outer gas giants.

cosmography A description of how objects are distributed through the universe.

cosmological constant (symbol Λ) A term introduced by Einstein into his gravitational field equations in order to allow a solution corresponding to a static universe. It can be understood as equivalent to an unknown cosmic repulsion that would tend to counter the attractive gravitational force, or an attractive force if its value is negative. Observational evidence is consistent with the value being zero or very small, and modern quantum cosmology attempts to account for its absence.

cosmological distance   The distance to a remote galaxy calculated on the assumption that the measured ➤*redshift* in the galaxy's spectrum is caused by the ➤*Doppler effect* and reflects a true speed of recession consistent with the general expansion of the universe.

cosmological principle   The assertion, central to modern cosmological theory, that all places in the universe are alike, so there are no preferred observers. Apart from local irregularities, it assumes that the universe is both homogeneous (the distribution of matter is the same everywhere) and isotropic (looks the same in all directions). ➤*perfect cosmological principle*.

cosmology   The branch of astronomy concerned with the origin, properties and evolution of the universe. Physical cosmology is about making observations that give information on the universe as a whole, and theoretical cosmology establishes ➤*models* that aim to describe the observed properties of the universe in mathematical terms.

Cosmology in the broadest sense spans physics, astronomy, philosophy and theology because it aims to assemble a world picture explaining why the universe has the properties it has. Ancient cosmologies were simple pictorial models and myths. Greek cosmology sought to build a mathematical model of the motion of the planets. Modern cosmology is firmly rooted in the laws of physics and the constructs of mathematics.

Modern observational cosmology aims to provide data on the universe at large through studies of the distribution of matter at great distances, the velocities of galaxies as a function of their distances from us, and the ➤*cosmic background radiation*. Examples of cosmological investigations are radio ➤*source counts*, which can in principle show whether the universe is evolving; the measurement of the ➤*redshifts* and distances of the furthest galaxies, which gives information on the rate of expansion at great distances; measuring the isotropy of the background radiation to see whether the universe is uniform in all directions.

Theoretical cosmology is normally based on the ➤*General Relativity* theory. Over large distances, gravitation is the dominant force affecting matter and it therefore controls the large-scale structure of the universe. General Relativity is able to describe the relationships between space, time, matter and gravitation; the equations of General Relativity can, in mathematical terms, describe a huge variety of universes. Assumptions such as the ➤*cosmological principle* are used to restrict the range of models. In practice, the ➤*steady-state theory* and ➤*Big Bang* theory have been the major areas of study in the last fifty years. It is now generally accepted that the observational evidence strongly supports a Big Bang cosmology.

cosmos   An alternative word for the universe as a whole or, in a general sense, the realm of space beyond the Earth.

COSMOS   ➤*SuperCOSMOS*.

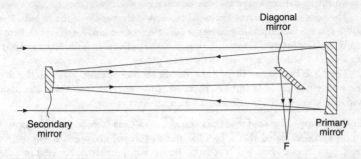

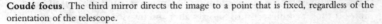

Coudé focus. The third mirror directs the image to a point that is fixed, regardless of the orientation of the telescope.

coudé focus A focal point of a telescope on an ▸*equatorial mounting*, arranged by means of a series of mirrors to lie on the polar axis and so remain stationary regardless of the orientation of the telescope (see illustration). The word 'coudé' is French for 'elbow'. It is necessary to direct the light to a fixed focus when bulky equipment is to be used that cannot be attached directly to the telescope, such as a high-dispersion spectrograph. Large, new-generation telescopes on computer-controlled altazimuth mountings are able to exploit the ▸*Nasmyth focus*, which is also fixed.

coudé spectrograph An astronomical ▸*spectrograph* designed for use at the ▸*coudé focus* of a telescope. Since the coudé focus is fixed, the spectrograph can be very large in order to produce very high dispersion, and may be built into a special room under the telescope.

counterglow ▸*zodiacal light*.

CP catalogue A listing of the first pulsars discovered, compiled in 1968 by radio astronomers at Cambridge. The initials stand for 'Cambridge Pulsar'. This designation has been superseded by PSR (Pulsating Source of Radiation) catalogue numbers.

CPD Abbreviation for ▸*Cape Photographic Durchmusterung*.

Crab English name for the constellation ▸*Cancer*.

Crab Nebula (M1; NGC 1952) The nebulous remnant in the constellation Taurus of a ▸*supernova* that exploded in AD 1054. In coloured photographs it appears as a network of red filaments surrounding an elliptical area of pale white light. This white light is ▸*synchrotron radiation* generated by hot ionized gas in a magnetic field. The filaments are the outer layers of the star that were blown off in the explosion and are travelling outwards at about 1,500 km/s.

The core of the star that exploded remains at the centre of the nebula. It is now a ►*pulsar*. Electrons emitted by the pulsar are responsible for the synchrotron radiation. The interval between flashes from the pulsar is 33 milliseconds; flashes are seen in visible light as well as the radio pulses.

The Crab Nebula is one of the strongest sources of radio waves in the sky; this source was called 'Taurus A' by radio astronomers before it was identified with the known nebula. It is also a source of X-rays.

Crab Pulsar The ►*pulsar* at the centre of the ►*Crab Nebula*. In 1942, astronomers had speculated that a peculiar star in this nebula was a ►*neutron star*, created in the ►*supernova* of AD 1054. Radio astronomers discovered in 1968 that the central star is a pulsar spinning 30 times a second; the period between pulses is thus 33 milliseconds. Shortly after discovery of the radio pulsar, the visible light from the neutron star was also shown to be pulsed. At the time of discovery, this was the fastest pulsar by far and the spin rate indicated unambiguously that the object must be a neutron star: any object with a typical star's mass but larger than a neutron star would fly apart at the rotation rate seen in the Crab Pulsar.

Historically, the discovery is of great importance because theorists subsequently felt much more confident that highly condensed objects really do exist in the universe. This led directly to research on ►*black holes*.

The Crab Pulsar is a source of electrons travelling at almost the speed of light that give rise to strong X-ray and radio emission from the Crab Nebula itself. The pulsar's rotation period is increasing at 36 nanoseconds per day, at which rate its rotation speed will be halved within about 1,200 years. The pulsar has random ►*glitches*, which may be caused by ►*starquakes*. The mass is thought to be about half that of the Sun.

Crane English name for the constellation ►*Grus*.

crater A circular feature on the surface of a planetary body. The word literally means 'cup', and craters are typically shaped like a shallow cup, lower in the centre than the surrounding terrain, with raised walls. The vast majority of craters observed on planets and their satellites are now known to have resulted from the impact of meteorites and the word is often treated as synonymous with 'impact crater'. However, volcanic craters (►*calderas*) also occur, on Mars, for example.

Craters range in size from the smallest discernible to hundreds of kilometres across. The largest impact features are often termed ►*basins*. The detailed forms of craters depend on many factors, including the composition of the planetary surface, the speed, mass and direction of the impacting body and subsequent weathering or geological activity. They may contain a central peak, or a depression.

Crater (The Cup)   A small, faint constellation on the southern border of Virgo. It is one of the ancient constellations listed by Ptolemy (*c.* AD 140) and is said to represent the goblet of Apollo. None of its stars is brighter than fourth magnitude. ➤Table 4.

Crepe Ring   One of the rings of ➤*Saturn* (the C ring), fainter than the prominent A and B rings. It lies within ring B, extending about halfway between the B ring's inner edge and the planet. The name comes from a description of its appearance given by the observer W. Lassell after the announcement of its discovery by W. C. Bond in 1850. ➤Table 7.

crescent   A ➤*phase* of the Moon, Venus or Mercury when less than half the disc is illuminated.

Crescent Nebula   Popular name for NGC 6888, a diffuse shell of gas surrounding the ➤*Wolf–Rayet star* HD 192163. One crescent-shaped segment of the spherical shell is relatively bright.

Cressida   One of the small satellites of Uranus discovered during the ➤*Voyager 2* encounter with the planet in 1986. ➤Table 6.

Crimean Astrophysical Observatory   An observatory of the Russian Academy of Sciences, located near Simeiz in the Crimea. It is an important solar observatory, and there is a 2.6-metre (102-inch) optical telescope and a milli-metre-wave telescope 22 metres (72 feet) in diameter.

critical density   In cosmology, the minimum density of matter that would ensure that the universe could not expand for ever. The observed expansion can be reversed by gravity only if the density is sufficiently high. The critical density is defined as the density that would ensure that both the deceleration and the velocity of expansion would become zero at the same time. Its value is between 10^{-29} and 2×10^{-29} g/cm^3, about ten times larger than the density inferred from visible matter, such as stars and galaxies.

For philosophical and aesthetic reasons many cosmologists like to imagine that the universe is closed. If it is, the amount of so-called ➤*missing mass* is considerable. ➤*Dark matter* in the form of particles other than ➤*baryons* – such as ➤*neutrinos* – would be required in order to match the actual density to the critical density. ➤*closed universe, expanding universe, Big Bang*.

Cross   English name for the constellation ➤*Crux*.

crossing time   The ratio between the diameter of a rich ➤*cluster of galaxies* and the average random velocity for galaxies within the cluster, which is thus a measure of the time for a galaxy to drift through the cluster. Typical crossing times are about one-tenth the age of the universe. This result implies that members of clusters are gravitationally bound together: otherwise they would have dispersed long ago.

Crossley Telescope   A 90-centimetre (36-inch) reflecting telescope at the ►*Lick Observatory*, presented in 1895 by an Englishman, E. Crossley. The mirror was figured more accurately than had been possible previously and the instrument demonstrated the potential for larger reflectors. Its success stimulated work on the construction of instruments of greater size.

Crow   English name for the constellation ►*Corvus*.

crust   The outermost solid layer of a planet or satellite, usually consisting of rock, ice, or a mixture of the two.

Crux (The (Southern) Cross)   The best-known of all southern constellations, formerly included in Centaurus (which surrounds it on three sides) before A. Royer introduced it as a separate constellation in about 1679. It is the smallest constellation by area, yet one of the most distinctive and recognizable. It lies in the ►*Milky Way* and contains a fine star cluster known as the ►*Jewel Box*. ►Table 4.

CSM   Abbreviation for Command and Service Module, part of the craft used by the ►*Apollo programme* of manned Moon landings.

CSO   Abbreviation for Caltech Submillimeter Observatory. ►*Submillimetre wave astronomy*.

C-type asteroid   A category of dark grey asteroids with albedos of about 5 per cent. The 'C' is for 'carbonaceous', and they are believed to consist of the same type of material as the ►*carbonaceous chondrite* class of meteorites. C-type asteroids are common in the outer part of the main belt.

Culgoora   The location in Australia of the Paul Wild Observatory, forming part of the ►*Australia Telescope*. Culgoora, in New South Wales, near the township of Narrabri, was the site of a ►*radioheliograph* that operated between 1968 and 1983.

culmination   The moment when a star or other celestial object reaches its maximum altitude above the horizon as the Earth's rotation takes it across the sky. Culmination occurs when the object crosses the observer's ►*meridian*. Thus, at culmination an object is either due south or due north of the observer. Circumpolar stars cross the meridian both below and above the pole, events called lower and upper culmination, respectively. 'Meridian passage' and 'transit' are alternative terms for culmination.

Cup   English name for the constellation ►*Crater*.

curvature of the field   The imaging by a telescope of objects in the field of view in such a way that the plane in which the images lie is curved rather than flat. Although this may cause minor inconvenience for a visual observer, it is possible to compensate for curvature of the field in a photographic telescope by bending the plate or film to the appropriate shape in a special holder.

curve of growth The relationship between the intensity of a ➤*spectral line* in an ➤*absorption line spectrum* and the effective number of absorbing atoms. The precise shape of the curve depends on the physical conditions (pressure and temperature) prevailing. Plotting the observed line intensities divided by wavelength against ➤*oscillator strength* multiplied by wavelength for a series of lines produced by one ion effectively delineates the curve of growth for that ion, assuming that the curve of growth is the same for all lines. Such a plot can be used to estimate the relative abundance of different chemical elements in a star, though the method takes no account of the variation in physical parameters through the line-forming layers of the star.

cusp The pointed extremity of a crescent-shape.

Cybele Asteroid 65, diameter 308 km, discovered by E. W. Tempel in 1861.

cyclone A converging circulation around a low-pressure region in a planetary atmosphere.

Cyclops An ambitious plan, proposed in the 1970s but never implemented, to deploy large numbers of steerable radio dishes in an attempt to communicate with any civilization that may exist elsewhere in the Galaxy.

cyclotron radiation Electromagnetic radiation emitted by electrons travelling in circular or helical paths in a magnetic field.

Cygnus (The Swan) A conspicuous constellation of the northern ➤*Milky Way*, in the shape of an elongated cross, supposedly like a swan in flight. It was among the 48 listed by Ptolemy (c. AD 140) and is sometimes called the Northern Cross. It contains 11 stars brighter than fourth magnitude, including the first magnitude ➤*Deneb* and a well-known double star, ➤*Albireo*. ➤Table 4.

Cygnus A An active elliptical galaxy that is one of the strongest radio sources in the sky.

Cygnus A (3C 405) is the strongest radio source in the constellation Cygnus and was detected by the first radio telescopes. The source consists of two similar clouds of radio emission, symmetrically located either side of a fifteenth magnitude galaxy at a redshift of 0.057. Cygnus A is the prototype of all powerful radio galaxies. It is one of the largest physical structures in the universe, its clouds spanning a total extent of about 300,000 light years. The energy associated with the radio clouds is about 10^{53} joules, about ten million times more than with normal galaxies such as the Andromeda Galaxy (M31).

The galaxy has strong emission lines in its optical spectrum, indicating that it contains an ➤*active galactic nucleus*. It is widely accepted that the energy generated by Cygnus A can be explained only by a central ➤*black hole* that is releasing very large amounts of energy as matter falls towards it.

Cygnus Loop A circular shell of nebulosity with a diameter of 3° in the constellation Cygnus. The ►*Veil Nebula* forms one section of the circular structure. Radio emission confirms that the shell is in fact a ►*supernova remnant*. It is estimated to be 30,000 years old and to lie at a distance of 2,500 light years. The loop is still expanding at a rate of 6 arc seconds every hundred years, though this must be slower than the original expansion rate because of the braking effect of interstellar material.

Cygnus X-1 An intense source of X-rays in the constellation Cygnus, discovered in 1966. It has since been the subject of much study and speculation, being regarded as a candidate ►*black hole*.

The source has been identified as a ►*binary star* system, the primary of which, designated HDE 226868, is a hot supergiant of ►*spectral type* O or B. The best evidence from the binary orbit of HDE 226868 around its unseen companion suggests that the invisible star has a mass significantly higher than the upper limit for neutron stars. The conclusion is that the invisible star can only be a black hole. The X-ray energy is generated by matter from the primary streaming on to the compact companion.

Cygnus X-3 A source of X-rays in the constellation Cygnus. It is a close binary system in which the orbital period is 4.8 hours. The primary star has a mass about the same as the Sun's. Spectra reveal it has the characteristics of a ►*Wolf–Rayet star*. The secondary is a pulsed source of gamma-rays with a period of 12.6 milliseconds. This star is likely to be a ►*pulsar* whose rate of rotation has been increased through interaction within the binary system.

Cosmic rays have been detected at ground level on Earth from the direction of Cyg X-3. This suggests that Cyg X-3 is emitting the most energetic photons known from any astronomical source, and demonstrates that binary X-rays sources are important generators of high-energy cosmic rays. ►*X-ray astronomy*.

cynthian Pertaining to the Moon. Cynthia is an alternative name for the Roman goddess of the Moon, who is more usually called Diana. The name is derived from Mount Cynthus, her birthplace according to the myths.

cytherean (cytheran) Pertaining to Venus; used as an alternative to 'Venusian'. Cythera is the Ionian island on which the goddess Venus first set foot, according to mythology.

D

Dactyl ►*Ida.*

Daedalus Asteroid 1864, diameter 3.2 km, discovered by T. Gehrels in 1971. Its orbit crosses that of the Earth.

Dall–Kirkham telescope A form of ►*Cassegrain telescope* in which the profile of the primary mirror is ellipsoidal rather than the more usual paraboloid. A spherical mirror is used for the secondary. The resulting field of view is considerably smaller than that of a conventional Cassegrain telescope of the same size.

Damocles Asteroid 5335, discovered in 1991. It is in an unusual, highly elliptical orbit which ranges between 1.6 and 22 astronomical units from the Sun.

Danjon astrolabe ►*prismatic astrolabe.*

Danjon scale A scale devised by the French astronomer André Danjon (1890–1967) to describe the relative darkness of the Moon during a total lunar ►*eclipse.* The scale runs from 0 (zero), for a very dark eclipse, to 4, representing an eclipse in which the Moon is a very bright copper colour or orange.

Daphne Asteroid 41, diameter 204 km, discovered by H. Goldschmidt in 1856.

dark adaption (dark adaptation) Allowing the eye a period of adjustment in the dark during which it becomes more sensitive to faint sources of light as the pupil gets larger. The time taken varies with individuals, but is likely to be at least ten minutes, and improvements may be noticed for up to an hour. Dark adaption is essential for visual astronomical observations. It is instantly destroyed by any exposure to bright lights. For that reason, dim red lights are normally used by astronomers where some lighting is essential, such as for reading charts.

dark matter Matter postulated to exist in the universe but not yet detected. The evidence for dark matter stems primarily from observations of the velocities of galaxies within galaxy clusters. The dynamical behaviour of cluster galaxies strongly suggests that the masses of clusters are about ten times greater than the mass contributed by the luminous parts of galaxies.

For individual galaxies it is possible to estimate the distribution of mass within them from the way the rotational velocity varies between the centre and the edge. Such measurements for giant spiral galaxies show that there is more matter in the galaxy than can be accounted for by glowing stars and gas.

The presence of dark matter is also an important element in theories that seek to account for how galaxies formed in the early universe. There are two main categories of such theories, those calling for *cold dark matter*, and those requiring *hot dark matter*.

The cold dark matter would take the form of exotic elementary particles that interact only weakly with radiation and with the ►*baryons* (neutrons and protons) of ordinary atoms. Such material could start to form into clumps early in the history of the universe following the ►*Big Bang*, when any fluctuations in the density of neutrons and protons would be smoothed out by their interaction with the high density of radiation present. These structures could survive to some extent on a relatively small scale, creating the framework for the formation of galaxies. Clusters and superclusters of galaxies would then be built up through the action of gravity.

The alternative hot dark matter theory postulates dark matter particles with large random velocities at the era in the universe when matter starts to dominate over radiation. Neutrinos would be a possible candidate if they had a small but finite mass. In this scenario, the largest-scale structures form first and then fragment into clusters and galaxies, which is in direct contrast with the prediction of the cold dark matter theory.

Computer simulations suggest that the cold dark matter does not produce as much large-scale structure as is actually observed, while hot dark matter appears to produce too many voids and stringy structures. It is not yet possible to determine whether either of these theories is correct.

Other evidence for dark matter comes from comparing the mass contributed by galaxies with that needed theoretically for a ►*closed universe*. The observed matter is only around 2 per cent of that required cosmologically. There are numerous candidates for the dark matter, massive galactic haloes, ►*brown dwarfs*, very low mass stars, neutrinos and ►*WIMPs* being just a few. ►*critical density, missing mass*.

dark nebula   ►*absorption nebula*.

data centre   An establishment for the collection, evaluation, storage and dissemination of scientific information. ►*Centre de Données Astronomiques*.

Davida   Asteroid 511, diameter 324 km, discovered by R. S. Dugan in 1903. It is one of the largest asteroids.

David Dunlap Observatory   The observatory of the University of Toronto, Canada, located 25 kilometres (15 miles) north of the university campus. It was presented to the university in 1935 by Mrs Dunlap as a memorial to her husband. The main instrument is a 1.88-metre (74-inch) reflector, the largest in Canada. Domes on top of the administration building house 0.5-metre and 0.6-metre reflectors. The observatory is used for training students and fostering public interest in astronomy, as well as research.

Dawes' limit   An empirical formula devised by William Rutter Dawes (1799–1868) for the minimum angular separation of a close pair of star images detectable as double by a telescope of a particular aperture – its resolving power. The rule assumes that observing conditions are good and that the two stars are not of very different brightness. It gives a separation in arc seconds of 11.6/D, where D is the diameter of the telescope aperture in centimetres. If D is in inches, the formula is 4.6/D.

day   In astronomy, a unit of time defined as 86,400 seconds, where the second is in turn defined in terms of the frequency utilized in a caesium ►*atomic clock*. This definition of the day is closely linked with the Earth's rotation period, though that rotation is not completely uniform. ►*solar day, sidereal day*.

daylight saving time   An adjustment to the normal ►*civil time*, introduced in some countries as a matter of convenience, for part or all of the year. One of the chief reasons for the adjustment is to arrange for the habitual working day to occur as far as possible during the hours of daylight, hence its name.

Dec.   Abbreviation for ►*declination*.

decametric radiation   Low-frequency radio waves, with wavelengths of tens or hundreds of metres. Waves of this type are received, for example, from the planet ►*Jupiter*, in bursts that are associated with the interaction between the planet and its satellite ►*Io*.

deceleration parameter (symbol q_0)   A number that indicates the rate at which the expansion of the universe is being slowed down by the mutual gravitational attraction of the matter within it. For a value greater than 0.5 the expansion of the universe will ultimately be halted, followed by contraction and collapse; a value less than or equal to 0.5 indicates that the universe will continue to expand for ever. ►*critical density*.

De Chéseaux's Comet   An exceptionally bright ►*comet* discovered independently by Klinkenberg from Haarlem on 9 December and De Chéseaux from Lausanne on 13 December 1743. It reached magnitude −7 and developed a fan of multiple tails: eleven separate tails were noted.

decimetric radiation   Radio waves in the wavelength range 10 to 300 centimetres. Radio waves of this type are produced, for example, in the radiation belts surrounding ►*Jupiter*, where electrically charged particles are trapped in the planet's magnetic field.

declination(1) (Dec.)   One of the coordinates used to define position on the ►*celestial sphere* in the equatorial coordinate system. Declination is the equivalent of latitude on the Earth. It is the angular distance, measured in degrees, north or south of the ►*celestial equator*. Northerly declinations are positive and southerly ones negative. ►*right ascension*.

declination (2) One of the parameters used to describe the direction of the ►*geomagnetic field*.

declination axis One of the two axes about which an equatorially mounted telescope can rotate (the other being the polar axis). Movement around the declination axis allows the telescope to point to different ►*declinations* while the ►*right ascension* remains constant. ►*equatorial mounting*.

declination circle A graduated disc or circle fitted to an ►*equatorial mounting* in order to indicate the ►*declination* coordinate of the direction in which the telescope is pointing.

decoupling era The era in the evolution of the universe, about 300,000 years after the ►*Big Bang*, when electrons and protons began to combine to form hydrogen atoms. At this point, the radiation in the universe ceased to scatter material particles and began to propagate freely. In other words, the radiation and the matter became 'decoupled'. This phase is also called the *recombination epoch*.

deep sky object An object of astronomical study that does not belong to the solar system. The expression is not precisely defined and is used primarily by amateur astronomers as an umbrella term for both galactic and extragalactic objects, such as ►*nebulae* and ►*galaxies*.

Deep Space Network A ground-based network of radio dishes used by NASA to communicate with space missions to solar system destinations beyond the Moon. The three 70-metre (230-foot) antennas are located at ►*Goldstone* in California, near Madrid in Spain and near Canberra in Australia.

defect of illumination The proportion of the disc of a planetary body, measured as an angle, that is not illuminated as viewed by an observer on Earth.

deferent A basic circular orbit, which, in combination with the ►*epicycle*, was necessary to the theory of planetary motions developed by Ptolemy in the second century A D. The planets were supposed to travel uniformly around a small, circular epicycle, the centre of which in turn moved around a larger circle, called the deferent.

dE galaxy ►*dwarf galaxy*.

degenerate star A term that covers both ►*white dwarfs* and ►*neutron stars*, which are made up of degenerate matter. These stars are in an advanced state of evolution and have suffered extreme gravitational collapse. Normal atoms cannot exist under the conditions of very high pressure.

In white dwarfs, the electrons and atomic nuclei collapse from the normal 'open' atomic structure into a dense, compressed mass. A quantum-mechanical effect called *degeneracy pressure* counters further gravitational collapse. However, if the total mass of the star exceeds 1.4 times the mass of the Sun, the degeneracy pressure is insufficient to balance the gravitational force and a neutron star results. Electrons and nuclei combine into a form of matter consisting of tightly packed neutrons. ►*black hole*.

Deimos One of the two small satellites of Mars, discovered in 1877 by Asaph Hall. Images obtained by the ►*Viking 2* Mars mission show a heavily cratered surface. The low albedo and properties of the surface suggest that both satellites have a composition similar to that of the ►*carbonaceous chondrite* meteorites, and it has been suggested that they are captured ►*asteroids*. ►Table 6.

Delavan's Comet ►*Comet Delavan*.

Delphinus (The Dolphin) A small, faint but distinctive constellation lying in the ►*Milky Way* just north of the celestial equator. It is one of the constellations listed by Ptolemy (*c.* AD 140). ►Table 4.

Delta Aquarids ►*Aquarids*.

Delta Cephei (δ Cep) A yellow giant star in the constellation Cepheus that varies in brightness between magnitudes 3.6 and 4.3 over a period of 5.37 days, and is the prototype of the ►*Cepheid variable* stars.

Delta Delphini star (δ Del star) A member of a group of luminous stars of ►*spectral type* A and F, characterized by weak absorption lines of calcium in their spectra. The star Delta Delphini is the brightest of the group and regarded as the prototype. Some are low-amplitude, short-period variables similar to the ►*Delta Scuti stars*, and they may also be related to the ►*Am stars*.

Delta Scuti star (δ Sct star) A short-period pulsating ►*variable star*. Delta Scuti, found to be variable in 1935, is the prototype of the group. Typically, the periods are less than eight hours and the variation amounts to only a few hundredths of a magnitude, which is imperceptible to the naked eye. Stars of this type are also known as *dwarf Cepheids* or *A1 Velorum stars*.

Dembowska Asteroid 349, diameter 164 km, discovered by A. Charlois in 1892. It belongs to the rare class of ►*R-type asteroids* and is a member of the ►*Budrosa* family.

Demon Star ►*Algol*.

Deneb (Alpha Cygni; α Cyg) The brightest star in the constellation Cygnus. It is a supergiant ►*A star* of magnitude 1.3. The Arabic name means 'tail'. ►Table 3.

Denebola (Beta Leonis; β Leo) An ►*A star* of magnitude 2.1, the third-brightest in the constellation Leo. The name, of Arabic origin, means 'the lion's tail'.

density parameter (symbol Ω_0) In cosmological theory, the ratio of the total density of the universe to the ►*critical density*.

density wave   The passage of a compressed region through a material medium while the elements of the material itself move by only a relatively small amount around their average positions. Sound waves are a common example of density waves. A density-wave theory has been proposed to explain the structure of ▶*spiral galaxies*. According to the theory, spiral arms are not permanent structures but regions where stars and interstellar matter concentrate as the density wave passes through.

De revolutionibus   The shortened form of the title, *De revolutionibus orbium coelestium*, the book by Nicholas Copernicus (1473–1543) in which he set out his heliocentric theory of the solar system. It was published in the year of Copernicus's death.

Copernicus's work marked a turning-point in astronomical thought, which led ultimately to the rejection of Earth-centred cosmology, though it was a century before the idea of a Sun-centred solar system gained general acceptance. The theory assumes that the planetary orbits are circular, when they are in fact elliptical, so it provided little if any improvement on existing methods for calculating the positions of the planets. Copernicus could offer no direct proof of the heliocentric theory and the book was subsequently condemned by the Church authorities, who long regarded the ideas it contains as heretical.

descending node   ▶*ascending node*.

Desdemona   One of the small satellites of Uranus discovered during the ▶*Voyager 2* encounter with the planet in 1986. ▶Table 6.

de Sitter universe   A model of the expanding universe proposed in 1917 in which there is no matter or radiation. This unrealistic hypothesis is historically important because it established the idea that the universe might be expanding rather than static. ▶*Einstein–de Sitter universe*.

Deslandres   A large lunar crater, diameter 234 kilometres (145 miles), on the southern border of the Mare Nubium. Its walls are overlapped by several other craters, including Regiomontanus, Walter and Lexell. The crater Hell lies within it.

Despina   A satellite of Neptune (1989 N3) discovered during the flyby of ▶*Voyager 2* in August 1989. ▶Table 6.

detached system   A ▶*binary star* system in which neither star fills its ▶*Roche lobe*. ▶*semidetached system, contact binary*.

detector   The element of an instrument system that is sensitive to the incoming radiation or particles to be observed.

dew cap   An extension fixed to a telescope tube to inhibit the condensation of water droplets on the optical elements. A simple tube extension works by lowering the cooling effect of direct air currents. A more sophisticated dew cap may incorporate a small electric heating element.

D galaxy   A type of large ►*elliptical galaxy* with a bright nucleus surrounded by an extensive envelope. D galaxies are often radio galaxies. The term comes from a classification system devised by W. W. Morgan. ►*cD galaxy*.

diagonal (1) (star diagonal)   An attachment for a small telescope, containing a small plane mirror or prism, used to turn through a right angle the beam of light in the telescope and the direction of the draw-tube into which the eyepiece is inserted. Use of a diagonal may be helpful when the visual user of a small amateur telescope would have difficulty accessing the eyepiece in its normal tube. However, there is some loss of light when a further optical element is introduced, and there may also be loss of image quality. The image in a diagonal is reversed right-to-left.

diagonal (2)   In a ►*Newtonian telescope*, the flat secondary mirror, which is positioned diagonally (i.e. at an angle of 45°) to the optical axis of the telescope.

Dialogue   Abbreviated form of the title, *Dialogue Concerning the Two Chief World Systems, the Ptolemaic and the Copernican*, a book by Galileo Galilei (1564–1642) published in 1632. Aware of potential opposition from the Church authorities, Galileo had delayed writing such a book at all. Encouraged by a change in the papacy in 1624, Galileo eventually started his work but cast it in the form of a conversation between three men in order to appear as if he were not supporting one argument against another. However, he presented the arguments in favour of a Copernican, Sun-centred solar system so persuasively that the few disclaimers were ineffectual, and Galileo, then aged 68 and infirm, was tried before the Inquisition in Rome. He was forced to abjure his 'heresies' and spent the rest of his life under house arrest. ►*Copernican system*, *Ptolemaic system*.

diamond ring effect   A phenomenon observed at the very beginning and end of totality in a total solar ►*eclipse* when the last or first glimpse of the brilliant ►*photosphere* of the Sun shines through a valley on the limb of the Moon. The visual effect is very much that of a flashing solitaire diamond ring. ►*Baily's beads*.

dichotomy   The time when the Moon, Mercury or Venus is exactly at half phase.

Dicke switch   A technique used in radio astronomy, particularly in microwave work, in which the strength of a cosmic radio signal is compared with the strength of a known terrestrial standard by rapidly switching between the two.

differential geometry   A branch of mathematics that deals with the properties of curved spaces, and is applied in cosmology to the analysis of the geometry of the universe.

differential rotation   The rotation of a gaseous body, such as the Sun or the planet Jupiter, at a rate that varies with latitude, or the rotation of a non–solid, disc-shaped structure, such as a galaxy, at a rate that varies with distance from the centre.

A solid planet like the Earth must rotate so that the ➤*angular velocity* is the same everywhere. However, the equatorial regions of a gaseous planet or star rotate more quickly than regions at higher latitudes, so two features at differing latitudes will move relative to each other.

In a galaxy, the component parts (stars and clouds of interstellar material) are in individual orbits around the centre of the galaxy. The angular velocity varies with radial distance from the centre, so the galaxy does not rotate like a solid disc.

differentiation   The process in which a body such as a terrestrial planet, which is initially homogeneous in composition, becomes stratified into different regions, usually of different densities. In the case of a planet, these may be the core, mantle and crust, for example.

diffraction   The spreading of a beam of light as it passes by the edge of an obstacle, into what geometrically would be the shadow. When diffraction occurs, interference between different parts of the light beam results in a pattern of light and dark areas called a *diffraction pattern*.

diffraction grating   An optical device used to disperse light into a spectrum. It consists of a large number of narrow, closely spaced lines ruled either on glass to form a transmission grating or on polished metal to form a reflection grating. Typically there are several thousand rules per centimetre. Interference between the beams of light created by ➤*diffraction* at each slit result in the ➤*dispersion* of the light with wavelength. Diffraction gratings can produce very high dispersion spectra of good quality, and are used for this purpose in astronomical ➤*spectrographs*.

diffuse cloud   A cold, dark, relatively small cloud of interstellar matter, several light years in diameter. Such clouds are of relatively low density, and contain gas mainly in the form of atoms and atomic ions, with a sparse population of ➤*interstellar molecules*.

diffuse interstellar bands   Features of unknown origin between 440 and 685 nanometres in the ➤*absorption spectrum* of the ➤*interstellar medium*.

diffuse interstellar medium   The general ➤*interstellar medium*, which is not in discrete ➤*nebulae*.

diffuse nebula   A gaseous ➤*nebula*. The use of the adjective 'diffuse' dates from a time when all objects of 'fuzzy' appearance were classed as 'nebulae', and it was necessary to distinguish between them. In current terminology, star clusters and galaxies are no longer called 'nebulae', the word being reserved for diffuse clouds of gas and dust.

diogenite A type of stony ➤*meteorite* composed of the silicate minerals pyroxene and plagioclase.

Dione A medium-sized satellite of Saturn, discovered by G. D. Cassini in 1684. Images from the ➤*Voyager 1* mission show several different types of terrain on Dione: heavily cratered areas, cratered plains with a lower density of craters, and smooth plains with few craters or other features. The largest craters are over 200 kilometres (125 miles) across, and craters over 100 kilometres are common in the heavily cratered areas. Another noticeable feature is an irregular network of light wispy streaks on a dark background which, it has been suggested, may be frosty deposits. ➤Table 6.

dioptric Describing optical systems that employ only refracting elements (e.g. lenses). ➤*catadioptric*, *catoptric*.

Diotima Asteroid 423, diameter 208 km, discovered in 1896 by A. Charlois.

dipole antenna ➤*antenna*.

direct (prograde) Describing motion of an object on the ➤*celestial sphere* in the west-to-east direction, or, for orbital motion or axial rotation in the solar system, motion that is counterclockwise as observed from north of the ➤*ecliptic*. ➤*retrograde*.

dirty-snowball model A popular description of the structure of ➤*comets*, which are believed to consist of water, methane and ammonia snows in which are embedded dust particles of minerals and metals. The model was first proposed by F. Whipple in about 1949.

disc (disk) Any relatively thin, flat, circular structure, particularly the main part of a spiral galaxy containing the arms.

The circular shapes presented by the Sun, Moon and planets are also described as discs.

disc galaxy A ➤*spiral galaxy* that has been stripped of most of its interstellar gas through interaction with the intergalactic medium in a ➤*cluster of galaxies*.

disconnection event A break in the ➤*ion tail* of a ➤*comet* caused when the comet crosses one of four boundaries between sectors of the ➤*solar wind* where the direction of the magnetic field changes.

discordant redshift The ➤*redshift* of a distant ➤*galaxy* or ➤*quasar* when it is markedly different from the redshift of another neighbouring object with which it appears to be physically linked. The redshifts observed in the spectra of galaxies and quasars are generally thought to reflect the overall expansion of the universe that has been continuing since the ➤*Big Bang*. In that interpretation, the magnitude of the redshift is proportional to distance, and distances can be inferred if redshifts are measured. If two galaxies are at the same distance, they would be expected to show the same redshift. However, there are instances where the images of galaxies and quasars that are close together on the sky suggest the existence of a physical link or bridge between them, yet the measured redshifts of the objects are at variance. This situation could arise if the apparent connection is a coincidental alignment or if the redshift did not arise purely from the general expansion of the universe.

disc star   A star located within the disc of a spiral galaxy as opposed to the ➤*galactic halo*.

dispersion   The decomposition or spreading of a beam of electromagnetic radiation in relation to wavelength. A simple example is provided by the dispersion of white light into a coloured spectrum on passing through a glass prism. The light disperses because the wave velocity of light in a medium (which is reflected in the value of the refractive index of the medium) varies with wavelength.

Dispersion is also used to mean a qualitative measure of the spread of a spectrum on a detector, expressed, for example, in ångströms per millimetre.

dispersion measure (DM)   A quantity that indicates the delay between arrival times from a ➤*pulsar* of radio pulses at different frequencies. Arrival times are spread by the presence of electrons in the interstellar medium. If the density of electrons is known from independent measurements, the dispersion measure of a pulsar may be used to calculate its distance.

distance modulus   The difference between the apparent and absolute ➤*magnitudes* of a star (or other astronomical object), which is a direct measure of the object's distance. From the definitions of absolute and apparent magnitude, the distance modulus takes the form $5 \log d - 5$, where d is the distance in ➤*parsecs*.

distortion   An imperfection in the imaging qualities of a lens such that the magnification varies over the field of the lens. The result may be *barrel distortion* or *pincushion distortion* according to whether the magnification decreases or increases towards the edge of the lens.

diurnal   An adjective meaning 'daily'.

diurnal aberration   ➤*aberration* (2).

diurnal motion   The apparent motion of a celestial object around the sky during the course of a ➤*sidereal day*. Diurnal motion results from the rotation of the Earth on its axis.

diurnal parallax (geocentric parallax)   The apparent difference between the position of a celestial object measured from the Earth's surface and the position that would be recorded by a hypothetical observer at the centre of the Earth. The effect of diurnal parallax causes nearby celestial objects to move against the background of more distant stars as the Earth rotates daily on its axis. In practice, the effect is so small as to be detectable only for objects within the solar system. It varies from a maximum when the object is on the horizon to a minimum when it is on the meridian. The maximum value is called the *horizontal parallax*. Since the Earth is not spherical, the value of horizontal parallax varies with latitude. The standard value quoted is normally that at the equator, the *equatorial horizontal parallax*. ➤*parallax*.

D layer A region of the Earth's ➤*ionosphere* at heights between about 50 and 90 kilometres (30 and 55 miles). The lower D layer, between heights of 50 and 70 kilometres, is also known as the *C layer*.

D lines A close pair of strong ➤*spectral lines* in the yellow region of the spectrum of sodium. The name comes from the identifying letters given by Joseph von Fraunhofer to prominent absorption lines in the solar spectrum. The wavelengths are 589.0 and 589.6 nanometres. ➤*Fraunhofer lines*.

Dobsonian telescope A low-cost, large-aperture reflecting telescope on a simple, undriven ➤*altazimuth mounting*. The design is suitable for visual observing by amateurs and is particularly portable.

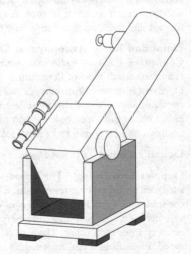

The name and concept derive from the pioneering work in the 1960s and 1970s by John Dobson of the San Francisco Sidewalk Astronomers. The telescope tube is typically square and constructed from plywood. The mount consists of a box, open at the top and on one side, which is mounted on a baseplate in such a way that it can rotate about a vertical axis. Semicircular cut-out yokes on the top of the box take large circular trunnions attached to opposite sides of the telescope tube (see illustration). The non-stick material Teflon is used to obtain smooth motion around the two axes.

Dobsonian telescope. A typical design.

Dobson also demonstrated that a large mirror could be constructed cheaply and successfully from plate glass, which is thinner than a conventional mirror blank. The thin mirror needs to be supported loosely on a bed of carpet or foam rubber to avoid distortion.

Dog Star   A popular name for the star ►*Sirius*.

Dolphin   English name for the constellation ►*Delphinus*.

domain wall   A sheet-like defect in the structure of ►*spacetime* that arises in some ►*Grand Unified Theories* of elementary particles. ►*magnetic monopole*.

dome (1)   A low, rounded hill on a planetary surface, resulting from volcanic activity, often capped by a small crater. The Marius Hills on the Moon are examples of such domes.

dome (2)   A building with a hemispherical roof designed to accommodate a telescope and be used as an observatory. The roof rotates and has a wide slit opening so that the telescope can be conveniently pointed to different parts of the sky, while affording some protection from wind.

Dominion Astrophysical Observatory   The National Research Council of Canada's centre for optical astronomy, located near Victoria, British Columbia. It is part of the Herzberg Institute of Astrophysics. The observatory is noted for the stellar and galactic studies done there, particularly work on binary and multiple stars. It was founded by J. S. Plaskett, and the 1.85-metre (72-inch) telescope began operating in 1918. A 1.2-metre (48-inch) telescope was added in 1962 and the Canadian Astronomy Data Centre established there in 1988.

Dominion Radio Astrophysical Observatory   The National Research Council of Canada's radio astronomy observatory, located 20 kilometres (12 miles) south-west of Penticton in British Columbia. It forms part of the Herzberg Institute of Astrophysics and was founded in 1959. The main instrument is an ►*aperture synthesis* radio telescope consisting of seven 9-metre (30-foot) dishes on an east–west baseline 600 metres (2,000 feet) long. There is also a 26-metre (85-foot) dish and a small solar telescope.

Donati's Comet   ►*Comet Donati*.

Doppler broadening   The spread in wavelength range of a feature in a ►*spectrum* produced by internal motion of the emitting atoms and molecules in the source of light. The ►*Doppler effect* causes a change in the wavelength of a ►*spectral line* when the source and observer are in relative motion, either towards or away from each other. The size of the change increases with the relative speed. Light from a star, for example, consists of individual photons emitted by atoms that are moving rapidly in the hot gas of the star's outer layers. Some atoms will always be moving towards the observer, others away. The result is to broaden the range of wavelength over which the line is spread. The hotter the star, the faster the gas atoms move and the greater the broadening effect.

Doppler effect   The change in the observed frequency of sound or ►*electromagnetic radiation* when the source of waves and the observer are moving towards or away from each other. The operation of the Doppler effect is commonly experienced when an emergency vehicle sounding its siren speeds past: the pitch of the sound drops suddenly the instant the vehicle passes the listener. As the source of sound approaches, the waves are compressed in front of it, raising the frequency and the pitch. As the source recedes, the waves are stretched, lowering the frequency and pitch. A similar effect is observed with the light from astronomical objects, features in the spectra of which may be shifted towards longer or shorter wavelengths according to whether the source of light is receding from or approaching the Earth. ➤*redshift*.

Doppler shift   The magnitude of the wavelength change in a spectrum caused by the ➤*Doppler effect*. A Doppler shift in the spectrum of an astronomical object is commonly described as a ➤*redshift* when it is towards longer wavelengths (object receding) and as a ➤*blueshift* when it is towards shorter wavelengths (object approaching). Its magnitude, z, is quantified in terms of the ratio of the wavelength change, $\Delta\lambda$, to the original wavelength λ, since the theory shows that this quantity is a constant determined by the velocity of separation, v, between the object and its observer. When v is small in relation to the velocity of light, c, $\Delta\lambda/\lambda = v/c$. If v is a significant fraction of c, a formula derived from ➤*Special Relativity* theory must be used:

$$\Delta\lambda/\lambda = [(1+v/c)/\sqrt{(1-v^2/c^2)}] - 1.$$

Dorado (The Goldfish or Swordfish)   A southern constellation, introduced probably by sixteenth-century navigators and included by Johann Bayer in his atlas ➤*Uranometria*, published in 1603. Its stars are inconspicuous but the Large ➤*Magellanic Cloud* lies on its southern border with Mensa. ➤Table 4.

Doris   Asteroid 48, diameter 246 km, discovered in 1857 by H. Goldschmidt.

dorsum (pl. dorsa)   An irregular, wrinkled ridge feature found on planetary surfaces.

Double Cluster in Perseus   ➤*h and χ (chi) Persei*.

double star   ➤*binary star*.

doublet   A lens composed of two elements, each of a different type of glass, which are either air-spaced in a mount or cemented together. With such a compound lens, it is possible to reduce the effects of ➤*chromatic aberration*. For that reason, the expression 'achromatic doublet' is often used. Compound lenses may consist of more than two elements; one with three components is called a *triplet*, for example.

Dove   English name for the constellation ➤*Columba*.

DQ Herculis star   ➤*polar*.

Draco (The Dragon)   An extensive but rather faint constellation stretching halfway round the north celestial pole and enclosing Ursa Minor on three sides. It is one of the ancient constellations listed by Ptolemy (*c.* AD 140). ➤Table 4.

draconic month (draconitic month)   The time interval of 27.212 221 days between successive passages of the Moon through the ascending (or descending) ➤*node* of its orbit, i.e. where its orbit crosses the ➤*ecliptic*. This interval is important in predicting the occurrence of ➤*eclipses*.

Draconids   A ➤*meteor shower*, associated with ➤*Comet Giacobini–Zinner*, seen occasionally around 9 or 10 October. The radiant is near the 'head' of Draco at RA 17h 23m, Dec. +57°. The number of meteors seen per year is very variable. Spectacular displays were witnessed in 1933, when the rate briefly reached 350 per minute, and again in 1946. Moderate showers occurred in 1952 and 1985. The shower is also known as the Giacobinids.

Dragon   English name for the constellation ➤*Draco*.

Drake equation   A mathematical expression, formulated by Frank Drake in 1961, which gives the number of potentially detectable civilizations in space (beyond the solar system) as the product of several distinct factors. Drake originally wrote his expression in the form:

$$N = R f_p n_e f_l f_i f_c L$$

where N is the number of detectable civilizations, R is the rate of star formation, f_p is the fraction of stars forming planetary systems, n_e is the number of planets hospitable to life, f_l is the fraction of those planets where life actually emerges, f_i is the fraction of planets where life evolves into intelligent civilizations, f_c is the fraction of planets with intelligent beings capable of interstellar communication and L is the length of time that such a civilization remains detectable.

draw–tube   The tube at the observing position on a telescope into which ➤*eyepieces* are inserted.

drive   The mechanism that moves a telescope so it can track a star or other astronomical object across the sky. The apparent motion of the sky reflects the rotation of the Earth on its axis relative to the distant stars. The period of one such rotation is a ➤*sidereal day*. Since the sidereal day is shorter than a solar day by about 4 minutes, the telescope needs to be driven so as to make a complete revolution in 23 hours 56 minutes. This is known as the *sidereal rate*. In practice, the drive mechanism is normally an electric motor, appropriately geared. Formerly, mechanical clockwork drives were constructed, powered by falling weights for example. Such drives are still operational on some historical instruments.

D-type asteroid   A reddish type of asteroid, rare in the main belt, but found increasingly at greater distances from the Sun.

Dubhe (Alpha Ursae Majoris; α UMa)   One of the two stars of the ➤*Plough* in Ursa Major (with Merak) called the ➤*Pointers*. It is a giant ➤*K star* of magnitude 1.8 with a fifth magnitude companion that orbits it in 44 years. Dubhe, literally 'the bear', is a shortened version of an Arabic name meaning 'the back of the greater bear'. ➤Table 3.

Dumbbell Nebula (M27; NGC 6853)   A large ➤*planetary nebula*, a quarter of a degree in diameter, in the constellation Vulpecula. It was discovered by Charles Messier in 1764. The name is descriptive of its hourglass shape.

Dunsink Observatory   Now part of the School of Cosmic Physics of the Dublin Institute for Advanced Studies, Eire. It was founded in 1783 as the university astronomical and meteorological observatory and is located at Castleknock, County Dublin. The only remaining instrument at the historical site is a 30-centimetre (12-inch) refractor dating from 1868, which is maintained for demonstrations to the public.

dust grains   Small particles of matter, typically of order 10–100 nanometres in diameter, which co-exist with atoms and molecules of gas in interstellar space. The dust grains are thought to consist mainly of silicates and/or carbon in the form of graphite. They form in the extended atmospheres of ➤*red giant* stars. Dark clouds of dust are evident when they obscure the light from stars and luminous gas clouds, as happens in the plane of the Milky Way. Though tenuous, such clouds are very effective at absorbing visible light, but radiation at millimetre and longer wavelengths can pass through the dust clouds unimpeded. The presence of dust is also revealed by the emission of infrared radiation, generated when the grains are warmed by the absorption of visible and ultraviolet radiation. The temperature of dust is typically in the range 30–500 K.

Dust grains are thought to play an important role in the formation of ➤*interstellar molecules*, acting as host surfaces on which gas atoms can combine. The molecules so formed can then leave the surfaces of the grains.

Dust clouds are also an important constituent of star-forming regions. The dust appears to shield interstellar molecules from the destructive effect of high-energy radiation and provides ➤*protostars* with an efficient means of radiating away surplus energy.

dust lane   A dark streak of obscuring dust observed against the bright background of the Milky Way, or in other galaxies. ➤*dust grains*.

dust tail (type II tail)   One of the two types of tail ➤*comets* develop as they approach the Sun. The dust tail is composed of particles about one micrometre in size that shine by reflected sunlight. Dust tails can be as much as ten million kilometres long. They curve away from the Sun under the influence of solar ➤*radiation pressure*.

dwarf Cepheid star   ➤*Delta Scuti star*.

dwarf galaxy A small elliptical or spheroidal ►*galaxy* containing between a few hundred thousand and a few million stars – far fewer than a typical galaxy. Dwarf galaxies have low luminosity and may be so loose as to resemble large ►*open clusters* of stars. Six dwarf spheroidal galaxies are known that are satellites of the Milky Way Galaxy. Their diameters range between 2 and 7 per cent of the size of the Milky Way. Dwarf ellipticals are referred to as *dE galaxies*.

dwarf nova A type of cataclysmic variable star that exhibits sudden increases in brightness at intervals ranging from several days to a year. The dwarf novae are binary systems consisting of a ►*main-sequence star* and a ►*white dwarf*. Mass transfer takes place from the main-sequence star to the white dwarf via an ►*accretion disc* that builds up around it. Outbursts take place when hot regions form on the accretion disc. Dwarf novae are also known as *U Geminorum stars*.

Several subtypes have been identified. *SS Cygni stars* increase in brightness by between 2 and 6 magnitudes in an outburst lasting several days. *SU Ursae Majoris stars* have occasional supermaxima, two magnitudes brighter and five times longer than normal. *Z Camelopardalis* stars sometimes undergo periods when the eruptions are suspended for weeks or years.

dwarf star A star that is of normal size for its mass. The term is often used to describe any star on the main sequence in the ►*Hertzsprung–Russell diagram*, that is, any star producing energy by converting hydrogen to helium in its core. The name arises from the natural comparison with giant and supergiant stars, which have greatly increased in size as a consequence of the evolutionary process. However, the largest main-sequence stars may be approaching the size of some red giants, so the distinction can become blurred. ►*white dwarf, stellar evolution*.

Dwingeloo 1 A nearby barred spiral galaxy discovered at the ►*Dwingeloo Observatory* in 1994 through its radio emission. It had been overlooked in previous optical images because it lies in the plane of the Milky Way, where it is obscured by dust and gas. Its estimated distance is 10 million light years.

Dwingeloo Observatory A radio astronomy observatory in the Netherlands, established in 1956, and the administrative headquarters of the Netherlands Foundation for Research in Astronomy (NFRA). It also hosts the Joint Institute for VLBI in Europe. The 25-metre (82-foot) Dwingeloo radio telescope is owned and operated by the NFRA. ►*VLBI, Westerbork Observatory*.

dynamical equinox Formally defined as the ascending node of the Earth's mean orbit on the Earth's equator, which is equivalent to the intersection of the ►*ecliptic* with the ►*celestial equator* where the Sun's declination changes from south to north. ►*equinox, catalogue equinox*.

dynamical mean Sun An imaginary object introduced to assist in the defining of ►*mean solar time*. It is in effect a point that moves uniformly around the sky along the ►*ecliptic,* coinciding with the actual position of the Sun at its perihelion passage time.

dynamical parallax   A measure of the distance to a visual ►*binary star* based on estimates of the masses of the stars deduced from their spectra and the observed properties of their orbits around each other.

dynamical time   The concept of time that is used as the variable in gravitational equations of motion. The dynamical time formerly used for the computation of ►*ephemerides* was ephemeris time (ET), but this has been replaced by terrestrial dynamical time (TDT) and barycentric dynamical time (TDB).

TDT is essentially a continuation of ET, intended for use as the timescale in calculations of geocentric ephemerides. No assumption is made about the theory of gravity used. It is based on the SI ►*second* and measured by ►*atomic clocks.* It commenced in 1977, when the relationship between TDT and ►*International Atomic Time* (TAI) was defined to be:

1977 Jan 1, 0 hours TAI = 1977 Jan 1.000 372 5 TDT.

TDB is intended for use in the equations of motion of planetary bodies referred to the ►*barycentre* of the solar system. It is not uniquely defined but depends on the theory of gravitation that is adopted. However, it is stipulated that it differs from TDT by only periodic discrepancies.

dynamics   The study and theory of how and why objects move.

E

Eagle   English name for the constellation ➤*Aquila*.

Eagle Nebula (M16; NGC 6611)   An ➤*emission nebula* surrounding a brilliant cluster of young stars in the constellation Serpens. The nebula glows with the characteristic red colour of hydrogen gas ionized by the radiation from the stars, which are estimated to be only 2 million years old. Detailed observations show it to be a region of ongoing star formation. The nebula lies at a distance of 7,000 light years and is half a degree in diameter.

early-type galaxy   Originally (1926), a term used by Edwin Hubble for tightly wound spiral ➤*galaxies* (types Sa and SBa), which he considered to be the first stage in a progression through 'intermediate' (Sb, SBb) to 'late' (Sc, SBc). It is now thought very unlikely that spiral galaxies evolve in this manner and, in this sense, the nomenclature has generally fallen into disuse, though it is encountered occasionally.

In modern astrophysics the term is increasingly applied indiscriminately to any type of galaxy that is relatively young as judged from its observed properties. Galaxies at very high ➤*redshifts*, or galaxies containing large amounts of dust and exhibiting rapid star formation may be loosely described by the adjective 'early', used essentially as a synonym for 'young'. ➤*Hubble classification*.

early-type star   An obsolete term applied to the hottest and most massive types of stars, usually those of spectral types O, B and A. The expression reflects the belief once held, but now known to be completely wrong, that the sequence of spectral classes – from hottest to coolest – represents an evolutionary progression. Despite its misleading nature, the term is still frequently used.

Earth   The third planet from the Sun. From the astronomical perspective, Earth belongs to the group of terrestrial planets, which also includes Mercury, Venus and Mars. It is with this group, and also the ➤*Moon*, that its origin, structure and evolution are often compared.

Earth has an atmosphere intermediate in density between those of Venus and Mars. It is unique in possessing vast oceans of liquid water. The complex interaction between ocean, atmosphere and planetary surface determines the energy balance and the temperature regime. Cloud cover is typically 50 per cent, and heat trapped within the atmosphere (the ➤*greenhouse effect*) raises the average temperature by more than 30 degrees.

The present composition of the atmosphere is 77 per cent molecular nitrogen,

21 per cent molecular oxygen, 1 per cent water vapour and 0.9 per cent argon. Carbon dioxide is the most important trace constituent. The high concentration of oxygen, which dates from 2,000 million years ago, is a direct result of the existence of plants. The presence of oxygen allowed the formation of the high-level *ozone layer*, which shields the surface from solar ultraviolet radiation damaging to life.

Earth is the only major planet known for certain to be geologically active. Its large-scale features have all been determined by the creation, destruction, relative movement and interaction of a dozen or so crustal plates – comprising the *lithosphere* – which slide over the less rigid *asthenosphere* below. Collisions between plates produce folded mountains and zones of seismic activity are concentrated along the plate boundaries.

Seismic waves generated during earthquakes reveal the internal structure of the Earth by the way they propagate. At the centre, there is a molten metallic core of iron and nickel, possibly with a solid core at the very centre. The central temperature is around 4,000°C. A silicate mantle overlies the core. The outermost crust is about 10 kilometres (6 miles) thick under the oceans and 30 kilometres (20 miles) thick where there are continents.

In planetary terms, the surface of the Earth is very young. The basaltic rocks forming the ocean floors are among the youngest. The Precambrian shields – which occupy about 10 per cent of the surface – are the oldest, and the nearest approximation to the cratered terrain that forms a large part of other planetary surfaces. Weathering has removed all but a few traces of whatever impact craters there were.

The molten metallic core gives rise to the Earth's magnetic field and ►*magnetosphere*. A layer of electrically charged particles between heights of about 50 and 600 kilometres (30 and 400 miles) form the ►*ionosphere*. The funnelling of charged particles by the magnetic field to regions between latitudes of 60° and 75° creates the phenomenon of the ►*aurorae*. Satellite measurements have shown that the Earth is also an intense source of radio waves at kilometre wavelengths, though these are generated high up and are not detectable at ground level. ►Tables 5 and 6.

Earth-grazer A comet or asteroid whose orbit brings it relatively close to the Earth.

Earth rotation synthesis A technique in radio astronomy that uses the rotation of the Earth to enable small ►*radio interferometers* to achieve the resolution of a dish antenna many kilometres in diameter. This form of ►*aperture synthesis* was developed at Cambridge University.

earthshine A faint illumination of what would otherwise be the 'dark' part of the Moon when its phase is a thin crescent. The effect is caused by sunlight reflected by the Earth towards the Moon. A popular expression describes the appearance as 'the new Moon in the old Moon's arms'.

Easel   English name for the constellation ➤*Pictor*.

eccentric   Displaced relative to some specified centre.

eccentric anomaly   ➤*anomaly*.

eccentricity (Symbol *e*)   One of the parameters used to describe the shape of curves belonging to the family known as ➤*conic sections*: circles, ellipses, parabolas and hyperbolas. The orbit of a body moving under the influence of a central gravitational force, such as a planet travelling round the Sun, is necessarily one of the conic section curves. Thus eccentricity is one of the important elements used to describe an orbit.

Circles and ellipses are closed curves. A circle is defined as having $e = 0$. The eccentricity of an ellipse is a measure of how much it deviates from being a circle. If c is the distance from the centre of an ellipse to one of its focal points, and a the semimajor axis of the ellipse, the eccentricity is given by the ratio c/a. The eccentricity of an ellipse must be less than unity.

Parabolas and hyperbolas are open curves. The observed orbits of non-periodic comets are typically parabolic. A parabola has $e = 1$ and a hyperbola $e > 1$. ➤*orbital elements*.

echelle spectrograph   A ➤*spectrograph* in which dispersion of the incident light into a spectrum is achieved by the use of an echelle grating. The particular characteristic of an echelle grating is that the profile of the parallel grooves is step-like or zigzag, and their spacing is relatively wide. The light to be dispersed is directed on to the grating at right angles to the faces of the grooves. The direction of the light is thus at a large angle with the normal to the grating as a whole. This produces a set of many overlapping spectra with a high degree of resolution. A second, low-dispersion grating, or a prism, arranged perpendicular to the echelle is used to separate out the overlapping spectra.

eclipse   A phenomenon in which the light from a celestial body is temporarily cut off by the presence of another. This may be: (1) the passage of a planetary satellite, such as the Moon, into shadow so that the direct illumination from the Sun normally causing it to shine is cut off; (2) the obscuration of all or part of the Sun by the passage of the Moon directly across it (solar eclipse); (3) the passage of a star belonging to a binary system behind its companion so that the total light received from the system is reduced.

There is some confusion of usage between the terms 'eclipse' and ➤*occultation*. If 'eclipse' is reserved for the cutting off of sunlight by shadow, solar 'eclipses' and the phenomena observed in eclipsing binary stars are, strictly speaking, occultations. However, the use of 'eclipse' in these contexts is firmly established and normal.

In the description of the motion of the moons of other planets, such as Jupiter, it is usual to distinguish between eclipses and true occultations.

The Moon's orbit around the Earth is inclined at only 5° to the plane of the Earth's orbit around the Sun. From time to time, the three bodies become aligned and an eclipse of the Sun or Moon occurs (see illustration).

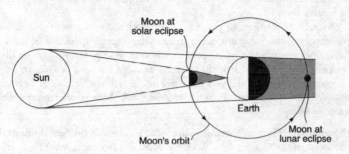

Eclipse. The relative positions of the Sun, Moon and Earth during lunar and solar eclipses (not to scale).

A *solar eclipse* can occur only at or very close to new Moon. Though the Moon is much nearer than the Sun, their apparent diameters are nearly equal at about half a degree. This coincidence makes total solar eclipses possible, with a maximum duration of 7½ minutes. However, there are small variations in the apparent sizes of the Sun and Moon because the orbits of the Moon and Earth are elliptical rather than circular. The ratio of the diameters of the Moon and Sun is described as the 'magnitude' of a solar eclipse. If a solar eclipse that would otherwise have been total occurs when the Moon's diameter appears less than the Sun's, an annulus (ring) of the Sun's disc remains visible when the centres of the two bodies are aligned. Such a solar eclipse is described as 'annular'.

The Moon's shadow on the Earth is only a few hundred kilometres wide. It traces out a curved path as the motion of the three bodies makes the eclipse visible at successive locations. Over a wider region either side of the path of totality, a *partial eclipse* is seen. Partial eclipses may occur when no part of the Earth witnesses a total eclipse.

During the brief moments of a total solar eclipse, darkness falls, and the outer parts of the Sun, the ➤*chromosphere* and the ➤*corona*, whose light is normally swamped by the bright photosphere, become visible.

Lunar eclipses occur when the Moon passes into the shadow of the Earth. They can take place only close to full Moon and can be seen from any location where the Moon has risen. The Moon does not normally disappear completely: its disc is illuminated by light scattered by the Earth's atmosphere. It often takes on a deep reddish hue.

The full shadow (*umbra*) cast by the Earth is surrounded by a region of partial shadow, called the *penumbra*. In the early and late stages of the progress of a lunar eclipse, the Moon enters the penumbra. It is possible for lunar eclipses to occur which are only penumbral. The length of the Moon's path through the umbra, divided by the Moon's apparent diameter, defines the 'magnitude' of a lunar eclipse.

The relative motions of the Sun, Earth and Moon are such that at least two eclipses of the Sun must occur in any year (though most will be partial). The maximum number of eclipses in any one year is seven, two or three of which must be lunar. It is theoretically possible for solar eclipses to occur at successive new Moons, and for there to be a lunar eclipse in between. However, lunar eclipses at two successive full Moons are not possible. ➤*eclipsing binary*.

eclipse year   The time between successive passages of the Sun through the same ➤*node* of the Moon's orbit, which is equal to 346.620 03 days. This period is less than a sidereal year because the changing orientation of the Moon's orbit causes the node to change position in the sky. ➤*saros*.

eclipsing binary   (eclipsing variable) A binary or multiple star system whose total brightness varies on a regular cycle because the orbital motion of the member stars causes them to pass one in front of another as viewed from Earth (see illustration). If the two stars in an eclipsing binary are of unequal luminosity, the light curve displays a *primary minimum* (at A) when the dimmer star passes in front of the brighter one, and a *secondary minimum* (at C) when the opposite occurs. The best-known example of an eclipsing binary is ➤*Algol*.

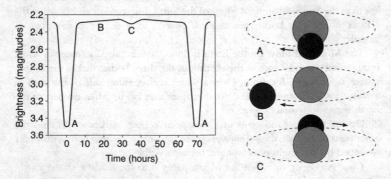

Eclipsing binary. The light curve of the prototype eclipsing binary, Algol. (A) Primary minimum. (C) Secondary minimum. (B) Light of the two components combined.

eclipsing variable ➤*eclipsing binary*.

ecliptic   The mean plane of the Earth's orbit around the Sun. The name arises because eclipses of the Sun or Moon can occur only when the Moon passes through this plane.

From the point of view of an observer on Earth, the relative orbital motion of the Earth and Sun makes it look as if the Sun were travelling around the Earth once a year. The Sun's path on the celestial sphere traces out the ecliptic plane, and is often marked as the ecliptic on celestial charts.

Since the orbits of the other planets are inclined to the ecliptic plane by only very small angles, their observed positions in the sky are always close to the ecliptic. The band of constellations through which the ecliptic passes defines the traditional ➤*zodiac*, though the effects of ➤*precession* and the precise definition of constellation boundaries mean that the ecliptic now passes through an additional constellation, Ophiuchus.

ecliptic coordinates   A celestial coordinate system, based on the plane of the ➤*ecliptic*. This system has applications in the study of planetary and solar system dynamics.

In the ecliptic system, the position of an object in the sky is defined by the coordinates ecliptic latitude (β, beta) and ecliptic longitude (λ, lambda). Latitude is measured in degrees north and south of the ecliptic; northerly latitudes are positive and southerly ones negative. Longitude is measured in degrees along the ecliptic. The zero point is the northern vernal ➤*equinox*, the point at which the ecliptic and the ➤*celestial equator* intersect. ➤*Precession* is making this zero point change slowly with time. ➤*obliquity of the ecliptic*.

ecliptic limits   The maximum angular distances the Sun or Moon can be at full or new Moon from the points where the Moon's orbit crosses the ➤*ecliptic* (the 'nodes') in order for an ➤*eclipse* to be possible.

If the Moon's orbit coincided with the ecliptic, a solar and lunar eclipse would be observed at each new and full Moon, respectively. Since the Moon's orbit is inclined at 5° to the ecliptic and the apparent diameters of the Sun and Moon are only half a degree, eclipses can occur only when the Earth, Sun and Moon are lined up in space, or very nearly so. From the point of view of an observer on Earth, this means that both the Sun and Moon must be within a certain angular distance of one of the nodes of the Moon's orbit. For a lunar eclipse to occur, the Moon must be within 24° of its node; the Sun must be within 37° of the node for a solar eclipse to take place.

E corona ➤*corona*.

Eddington limit   An upper limit on the ratio of the luminosity to mass of a stable star generating energy through the conversion of hydrogen to helium. Arthur S. Eddington (1881–1944) showed that the limit is 40,000 when units of the Sun's mass and luminosity are used. If the limit is exceeded, the outer layers of the star are blown away by ➤*radiation pressure* and a ➤*planetary nebula* is formed.

The Eddington limit also sets an important constraint on the rate at which a ➤*black hole* can grow by accreting matter.

Edgeworth–Kuiper Belt ➤*Kuiper Belt*.

effective area A parameter that expresses how effective a ➤*radio telescope* is at absorbing radiation of a particular frequency and from a particular direction, and at making the power available for measurement.

effective temperature A measure of the energy output of an object, such as a star, defined as the temperature of a ➤*black body* having the same total luminosity as the observed object. Effective temperature is often quoted as one of the physical characteristics of a star. Since the spectrum of a normal star is similar to that of a black body, effective temperature is a good indication of the actual temperature of its ➤*photosphere*.

Effelsberg The location 40 kilometres (25 miles) south-west of Bonn, Germany, of the observatory of the Max-Planck-Institut für Radioastronomie. The radio telescope located there is a 100-metre (330-foot) steerable dish, the largest of its type in the world.

Egeria Asteroid 13, diameter 244 km, discovered by A. de Gasparis in 1850.

EGG Abbreviation for 'evaporating gaseous globule'. An EGG is a condensed ball of gas surrounding a star in the process of formation; it is exposed when the less dense gas surrounding it is dispersed by the action of ultraviolet radiation.

Egg Nebula Popular name for a planetary nebula also known by the catalogue number CRL 2688. It is a very young planetary nebula and the central star is hidden behind a ring of dust. Numerous shells of gas ejected by the star, which was a red giant until a few hundred years ago, are illuminated by light from the central star.

Einstein Cross The four images of a ➤*quasar* formed by the galaxy G2237+0305 acting as a ➤*gravitational lens*. The resulting multiple image is in the form of a ring. ➤*Einstein ring*.

Einstein–de Sitter universe The simplest of the modern cosmological models, in which the universe has zero pressure and zero curvature (i.e. flat geometry), is of infinite extent, expands without limit and endures for all time. Proposed in 1932, it is a special case, with zero curvature, of the more general ➤*Friedmann universe*.

Einstein effect (Einstein shift) An alternative term for ➤*gravitational redshift*.

Einstein Observatory   An X-ray astronomy satellite launched by the USA on 13 November 1978. Officially designated HEAO-2 (2nd High Energy Astrophysical Observatory), it was informally renamed by the scientific staff working on it in celebration of the centenary of the birth of Albert Einstein.

It was the first X-ray observatory to be able to detect faint sources and to make images of complex or extended sources. Many important and significant observations were made until the supply of gas needed to control the telescope's position was exhausted in April 1981. ➤*X-ray astronomy*.

Einstein ring   A perfect circular image of a distant point source formed when a point mass along the line of sight acts as a ➤*gravitational lens*. This idealized case of lensing was described by Einstein and has been observed for the imaging of radio emission from the ➤*quasar* MG 1654+1348 by an intervening galaxy.

ejecta   Material that has been excavated during an impact or thrown out by volcanic activity and thus redistributed. Ejecta typically forms a circular 'blanket' of shattered rock fragments and solidified gas and liquid droplets around the impact or volcanic centre. Some impact ejecta can escape from the planet or satellite altogether.

Elara   A small satellite of Jupiter (number VII), discovered by C. Perrine in 1905. It is only about 80 kilometres (50 miles) across and belongs to a group of four satellites whose closely spaced orbits all lie between 11.1 and 11.7 million kilometres from Jupiter. (The others are Leda, Himalia and Lysithea.) ➤Table 6.

Electra   One of the brighter stars in the ➤*Pleiades*, also known as 17 Tauri.

electromagnetic radiation   A form of energy that propagates through a vacuum at a speed (c) of 3×10^8 metres per second. The name reflects the nature of the radiation, which consists of linked and rapidly varying electric and magnetic fields. Its character varies according to wavelength (λ).

Radio waves have the longest wavelengths, ranging from several metres to millimetres. The shortest radio waves are usually termed *microwaves*. These merge into the *infrared*, which ranges down to just under a micrometre. Visible *light* is a narrow band of wavelengths between about 700 and 400 nanometres (nm). *Ultraviolet* goes down to about 10 nm, then *X-rays* to 0.1 nm. The shortest waves of all are *gamma-rays*. The full *electromagnetic spectrum* is the whole range of radiation types, from the shortest wavelengths to the longest. (The ångström is also used as a unit of length for the measurement of wavelength. 1 Å = 0.1 nm.)

Electromagnetic radiation, in common with any wave, also has an associated frequency (ν). The link between frequency and wavelength is $\nu = c/\lambda$. Thus, as wavelength decreases, frequency increases.

The energy E associated with electromagnetic radiation increases in direct proportion to frequency, according to the relation $E = h\nu$, where h is Planck's constant. The energy is quantized in units of this size, which are termed ➤*photons*.

Electromagnetic radiation and its detection are crucial in the study of astronomy, which depends almost entirely on the receipt and analysis of such radiation from distant objects. Optical and radio astronomy can be undertaken from the ground because these wavelength bands pass relatively unobstructed through the atmosphere. Astronomical observations in other wavebands are largely carried out from orbiting spacecraft, though some are possible from high mountain sites and aircraft.

electromagnetic spectrum   ➤*electromagnetic radiation*.

electron density   The number of electrons per unit volume of space. In interstellar space the mean value is about 30,000 per cubic metre. In astrophysics the electron density is important in calculations concerned with the emission and propagation of ➤*electromagnetic radiation*.

electron temperature   For a population of electrons, the temperature required to account for their observed energy, if that energy is assumed to be entirely thermal.

electron volt (symbol eV)   A unit of energy, principally used in atomic and molecular physics. It is defined as the energy acquired by an electron when it is accelerated through a potential difference of 1 volt in a vacuum. 1 eV = 1.602×10^{-19} joules.

Electron volts are convenient units for measuring the energies of particles and electromagnetic radiation. The energies of X-rays are expressed in thousands of electron volts (keV). The wavelength associated with 1 keV is 0.124 nanometres.

Millions (MeV) and thousands of millions (GeV) of electron volts are also used as units for the highly energetic atomic particles. An electron with a kinetic energy of a few MeV is travelling at almost the velocity of light.

element (1) (chemical)   One of the basic materials of the universe, each of which is characterized by having a different number of protons in the atomic nucleus (the atomic number). Isotopes of the same element differ in the number of neutrons in the nucleus. There are 90 elements occurring naturally on Earth. Some, uranium for example, have no stable isotopes. Other radioactive elements can be created artificially. ➤*atom*, *nucleosynthesis*.

element (2) (orbital)   ➤*orbital elements*. Those most commonly used for the orbits of planets and comets around the Sun are ➤*semimajor axis*, ➤*perihelion distance*, ➤*eccentricity*, ➤*inclination*, ➤*argument of perihelion*, ➤*longitude of the ascending node* and ➤*period*.

ELFIR   Abbreviation for extremely luminous far-infrared sources. These are objects that are ten or a hundred times more luminous in infrared radiation than in visible light.

ellipse   A closed curve that is symmetrical about two perpendicular axes and has one axis longer than the other. The longer axis is the major axis; the other is called the minor axis.

The ellipse belongs to the family of curves known collectively as ►*conic sections* because they can all be generated by slicing through a solid cone in different ways. The other types of conic section are the circle, the parabola and the hyperbola. The shape of the orbit of a body moving under the influence of a central gravitational force is necessarily one of the conic sections. The ellipse is important in celestial mechanics since closed orbits are invariably elliptical. (The special case of the circle is rarely achieved naturally.) Planets have elliptical orbits around the Sun.

When a body is in an elliptical orbit under a gravitational force, the object providing the gravitational attraction lies at one focus of the ellipse. There are two foci located on the major axis, each the same distance (*c*) from the centre of the ellipse (see illustration). The more elongated the ellipse, the greater the

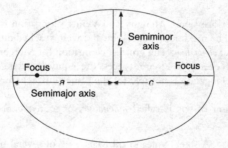

Ellipse.

value of *c* in relation to the *semimajor axis* (*a*). The ratio *c*/*a* defines the eccentricity (*e*) of the ellipse, which must be more than zero but less than unity (*e* = 0 for a circle; *e* = 1 for a parabola). The size of the *semiminor axis* (*b*) is linked to *a* and *e* by the formula $b = a \sqrt{(1-e^2)}$

The sum of the distances between any point on an ellipse and the two foci is a constant with the value 2*a*. This property means that an ellipse can be drawn with the help of a loop of string fixed between two points. A pencil moved so that the loop remains just taut will trace out an ellipse.

elliptical galaxy   A ►*galaxy* with an ellipsoidal shape and no spiral structure.

In most there is no evidence of interstellar matter nor sign of recent star formation. Elliptical galaxies differ only in mass and shape. Almost all their stars are older than about 10^{10} years, and much of their light comes from ►*red giant* stars. About 80 per cent of ►*normal galaxies* are elliptical.

ellipticity (oblateness)   A measure of the degree to which the shape of a planet, or other body, deviates from a perfect sphere. Planets and stars that are rotating tend to bulge at their equators to a degree that increases with speed of rotation and also depends on whether the body is solid or fluid. The shape taken up by the body is described as an oblate spheroid. A cross-section through the body that cuts through both poles is elliptical. The semiminor axis of the ellipse is the body's polar radius, R_p, and the semimajor axis of the ►*ellipse* is its equatorial radius, R_e. The ellipticity is defined as $(R_e - R_p)/R_e$.

Elnath (Alnath; Beta Tauri; β Tau)   The second-brightest star in the constellation Taurus, marking the tip of one of the bull's horns. Its name comes from the Arabic for 'the one butting with horns'. It also represented the right foot of neighbouring Auriga, and formerly shared the alternative designation Gamma Aurigae. Elnath is a ►*B star* of magnitude 1.7.

elongation   The angular distance between the Sun and a planet or the Moon as viewed from the Earth, i.e. the angle Sun–Earth–Moon/planet (see illustration).

For the inferior planets, Mercury and Venus, elongation is restricted to a limited range. The maximum values, east and west, reached during each orbit are called *greatest elongation*. The greatest elongation for Mercury lies between 18° and 28° according to circumstances; the equivalent range for Venus is 45°–47°. Any elongation is possible for the planets further from the Sun than the Earth.

An elongation of 90° is called *quadrature*, of 0° *conjunction* and of 180° *opposition*.

Elysium Mons   A shield volcano on Mars, one of several in the ►*Elysium Planitia* area.

Elysium Planitia   A large volcanic plain on Mars, more than 5,000 kilometres (3,000 miles) across.

emersion   The reappearance of a star, moon, planet or other body at the end of an ►*occultation* or ►*eclipse*.

emission line   A narrow wavelength range in a spectrum over which energy is emitted at a level above that of the surrounding continuum. Emission lines arise when transitions occur between different energy levels in the atoms or molecules of a gas, with a net release of electromagnetic energy. ►*atom, continuous spectrum*.

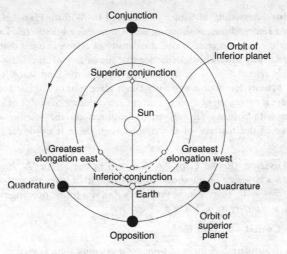

Elongation. The positions of the Sun and Earth in relation to inferior and superior planets at greatest elongation, conjunction, opposition and quadrature.

emission line galaxy Any ▶*galaxy* that displays ▶*emission lines* in its spectrum. This is a sign of unusual activity in the nucleus of the galaxy or of very rapid star formation. ▶*active galactic nucleus*.

emission line spectrum A ▶*spectrum* consisting solely of ▶*emission lines*, or containing emission lines in addition to a ▶*continuous spectrum* with or without ▶*absorption lines*.

emission line star Any star that exhibits emission lines in its spectrum in addition to or in place of the more typical absorption lines. The symbol 'e' in a spectral type (e.g. B5e) signifies the presence of emission lines.

▶*Wolf–Rayet stars*, the hottest of all, have totally emission line spectra. Emission lines occur in the spectra of certain ▶*O stars* and ▶*B stars* and also, at the cooler end of the range, in some ▶*M stars* and cool variable ▶*Mira stars*.

emission nebula A cloud of glowing gas in interstellar space. Interstellar clouds consist primarily of hydrogen, which can be excited or ionized by the action of the ultraviolet radiation from hot stars embedded within it. Energy is released by the recombination of ions and the collision of electrons with ionized atoms of heavier elements, such as oxygen and nitrogen, which are also present. The most intense radiation from hydrogen gives rise to the typical pinkish colour observed in ▶*ionized hydrogen* clouds (H II regions) such as the ▶*Orion Nebula*.

Enceladus   A satellite of Saturn, discovered by William Herschel in 1789. ➤*Voyager 2* returned images showing detail as fine as 2 kilometres. Large areas of the surface have no craters, and the density of craters in areas that do have them is relatively low. This is evidence that the surface of Enceladus has been completely remodelled by geological activity of some kind since it was first formed. Activity has almost certainly taken place within the last 100 million years; there is even a suggestion of current activity since the orbit of Enceladus coincides with Saturn's faint E ring. Eruptions on the satellite could be the source of the material of the ring, though there is no direct evidence. ➤Table 6.

Encke Division (Encke Gap)   A dark gap in the bright 'A' ring around ➤*Saturn*. A ➤*Voyager 2* image showed a narrow, rather wavy ringlet within the Encke Division. In 1990, a small satellite, ➤*Pan*, was found orbiting within the Division. ➤*planetary rings*, Table 7.

Encke's Comet   ➤*Comet Encke*.

English mounting   A form of ➤*equatorial mounting* for a telescope.

Engraved Hourglass (Hourglass Nebula)   Descriptive popular name for a planetary nebula with the catalogue designation MyCn18. Public attention was drawn to its spectacular form by a Hubble Space Telescope image published in 1996.

entrainment   A process by which jets streaming from ➤*radio galaxies* draw energy from the surrounding interstellar and intergalactic material. As material is sucked in, turbulent eddies are set up. These in turn create shocks which can heat the gas to extremely high temperatures. Pockets of the gas collect and cool; ultimately star formation takes place within them.

envelope   A gaseous region surrounding one or more stars or any other astronomical object.

Eos family   One of the ➤*Hirayama families* of asteroids. The members lie at a distance of 3.02 AU from the Sun and are intermediate in type between carbonaceous and silicaceous.

epact (1)   The Moon's age (i.e. stage in its cycle of phases) at the beginning of a calendar year.

epact (2)   The difference between a solar year and a lunar year of 12 lunar months. This period, of about 11 days, is known as the 'annual epact'.

epact (3)   The difference between a calendar month and a lunar month, known as the 'monthly epact'.

ephemeris (pl. ephemerides)   A table giving the celestial coordinates, magnitude and other data for astronomical bodies such as the Moon, Sun, planets and comets. The term is also used for a book containing a compendium of such tables and other astronomical data.

ephemeris time (ET)   The time used prior to 1984 in calculations involving gravitational theories of the solar system. In 1984 it was replaced by ➤*dynamical time*.

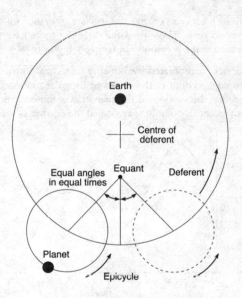

Epicycle.

epicycle   Motion in a circle, the centre of which is in turn moving around a larger circle called the ➤*deferent* (see illustration). Epicyclic motion was a fundamental aspect of the geocentric model of the solar system propounded by Ptolemy in the second century AD. To improve the accuracy of predictions based on the model, Ptolemy had the centre of the epicycle revolve with uniform angular motion, not about the centre of the deferent, but about a point called the *equant*, which he displaced to one side. By also placing the Earth to the other side of the centre of the deferent and choosing the radii of the epicycle and deferent appropriately, the Ptolemaic theory could be used to predict planetary positions to an accuracy of one degree.

Epimetheus   A small satellite of Saturn, discovered in 1980. It is co-orbital with Janus, the orbit lying just beyond the limit of the main ring system. The two satellites may be fragments of a larger object that was shattered by an impact. It is irregular, measuring 140 × 100 kilometres. ➤*co-orbital satellites*, Table 6.

epoch   A precise moment in time for which given celestial coordinates or orbital elements are strictly correct. It is necessary to define the epoch for such measurements because of the effects of ➤*precession* on celestial coordinate systems and gravitational ➤*perturbations* on the orbits of bodies such as comets and planets.

It has been the practice to change the *standard epoch* used for star charts and catalogues every 50 years. Those prepared since the 1970s have typically been for the epoch 2000.0; the previous standard epoch was 1950.0.

equant   A device incorporated by Ptolemy in his geocentric model of the solar system to take account of the observed irregular motion of the planets. He considered that the motion of the planets was uniform, not around the Earth itself, but around a point in space called the equant. ➤ *epicycle*.

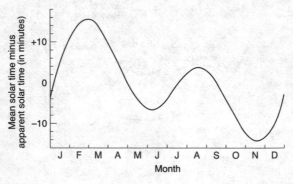

Equation of time.

equation of time   The difference between mean solar time, as shown on a uniformly running clock, and apparent solar time, as displayed on a sundial. It varies in a complicated way during the course of a year (see illustration), to a maximum of about 15 minutes. ➤*apparent solar time*.

equator   The great circle on the surface of a body defined by the plane that passes through the body's centre and is perpendicular to its rotation axis. In astronomy, if the context is clear, 'equator' is sometimes substituted for ➤*celestial equator*.

equatorial coordinates   A system of celestial coordinates based on the ➤*celestial equator*. Equatorial coordinates are those most commonly used when giving the position of an object on the celestial sphere for location purposes.

The coordinate equivalent to latitude is declination (Dec. or δ (delta)), which is measured in degrees north and south of the celestial equator. Northerly declinations are positive and southerly ones negative.

The other coordinate, right ascension (RA or α (alpha)), is the equivalent of longitude but is measured in hours, minutes and seconds of time, reflecting the rotation of the celestial sphere once in 24 hours of ➤*sidereal time*. Thus, a telescope pointing at a fixed direction in the sky will 'see' an angle of one hour of right ascension sweep by in one hour of sidereal time.

The zero point for right ascension is taken as the northern vernal ➤*equinox*. Because of ➤*precession*, this point is very slowly moving along the equator. Equatorial coordinates, therefore, have to be specified with reference to a particular ➤*epoch*.

equatorial head   Part of an ➤*equatorial mounting* for a telescope, including the telescope support and the mechanism for rotating the telescope, but not including the tripod or pillar.

equatorial mounting   A form of telescope mounting in which the instrument can rotate about a polar axis, parallel to the Earth's rotation axis, and a declination axis, perpendicular to the polar axis. Rotation about the two axes enables the two ➤*equatorial coordinates* to be set independently. Motion around the polar axis changes right ascension; motion about the other axis changes declination.

The equatorial mounting has a particular advantage. In order to compensate for the apparent motion of the sky produced by the Earth's rotation, the telescope needs to be motor-driven around only one of the axes, the polar axis. Once set correctly, the telescope continues to point at the required declination without any adjustment. For this reason, the design of the mounting of all telescopes of any significant size was for many years exclusively of the equatorial type. However, advanced computer control has made it possible to point and drive very large telescopes on the simpler ➤*altazimuth mounting* without difficulty. Equatorial mountings, though, remain popular and practical for many applications.

Different forms of equatorial mounting have been developed to give adequate support and freedom of movement for telescopes of different sizes and types. The main classes are the German, English, yoke, horseshoe and fork (see illustration on page 130). Because the polar axis must be parallel to the Earth's axis (i.e. it must point to the north celestial pole), a particular equatorial mount is suitable only for use at the latitude for which it is designed.

equinoctial colure   ➤*colure*.

equinox   Either of the two points at which the ➤*celestial equator* intersects the ➤*ecliptic*, and also the times when the Sun passes through either of these points. The Sun passes from south to north at the northern vernal (spring) equinox

equivalent width

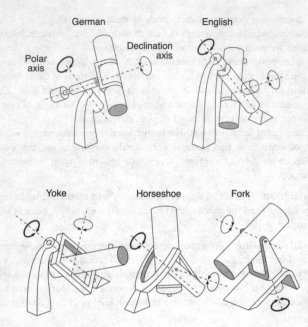

German **English**

Polar axis

Declination axis

Yoke **Horseshoe** **Fork**

Equatorial mountings.

and from north to south at the northern autumnal equinox. The approximate dates are 21 March and 23 September.

The position of the northern vernal equinox is also traditionally known as 'the first point of Aries' and is still often represented by the symbol, ♈, for Aries. However, the effects of ➤*precession* have gradually moved the point so that it now actually lies in the adjacent constellation of Pisces.

equivalent width   A measure of the strength of a ➤*spectral line*.

When a spectrum is displayed as a graph of light intensity against wavelength, absorption lines show up as bell-shaped dips in an otherwise smooth continuum level (see illustration). The equivalent width is calculated by measuring the area of the dip and dividing by the height of the continuum. The result is the wavelength range that the line would occupy if the amount of absorption were redistributed into a rectangle having the height of the continuum. This equivalent width is measured in units of wavelength, often milliångströms.

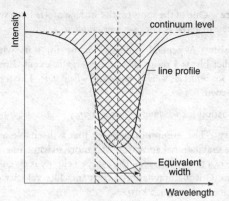

Equivalent width.

Equuleus (The Foal) The second-smallest constellation in the sky, next to Pegasus near the celestial equator. Though faint and obscure, it is one of the original 48 constellations listed by Ptolemy (*c.* AD 140). Its two brightest stars are fourth magnitude. ➤Table 4.

Erfle eyepiece A type of telescope ➤*eyepiece*, consisting of six lenses arranged as three doublets (see illustration), designed to give a particularly wide field of view – typically 68°.

erg A unit of energy equivalent to 10^{-7} joules.

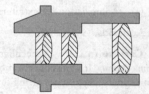

Erfle eyepiece.

Eridanus (The River Eridanus) A large, straggling constellation, spanning sixty degrees in declination in the southern hemisphere. It was included in Ptolemy's list of 48 constellations (*c.* AD 140), though its southernmost extension was added later. The most southerly point is marked by the first magnitude star ➤*Achernar*. There are eleven other stars brighter than fourth magnitude. ➤Table 4.

Eros Asteroid 433, size 7 × 16 × 35 km, discovered in 1898 by G. Witt. It was the first asteroid to be discovered with an orbit that brings it within the orbit of Mars. In 1931, when it came within 23 million kilometres of the Earth, measurements of its position were used in an attempt to determine the scale of the solar system more accurately than had been possible previously.

eruptive centre   One of the spots on the surface of ➤*Io* from which eruptions of gas and molten material occur. ➤*plume eruption.*

eruptive variable   A general collective name for stars whose brightness changes unpredictably as a result of some energetic event. Eruptive variables include ➤*novae,* ➤*supernovae,* ➤*dwarf novae* and ➤*flare stars.* The term 'cataclysmic variable' is also used.

ESA   Abbreviation for ➤*European Space Agency.*

escape velocity   The minimum velocity that will enable a small body to escape from the gravitational attraction of a more massive one. At a distance r from the centre of a body of mass m, the escape velocity is given by the formula $\sqrt{(2Gm/r)}$, where G is the gravitational constant. The velocity of escape from the Earth's surface is about 11.2 km/s.

Eskimo Nebula (NGC 2932)   A ➤*planetary nebula* in the constellation Gemini. The name comes from the non-uniform distribution of luminosity over the roughly circular shape, which suggests the features of a face, and a fainter outer ring, appearing rather like the fur fringe of an Eskimo hood.

ESO   Abbreviation for ➤*European Southern Observatory.*

ET (1)   Abbreviation for ➤*ephemeris time.*

ET (2)   Abbreviation for extraterrestrial. ➤*extraterrestrial intelligence.*

Eta Aquarids   ➤*Aquarids.*

Eta Carinae (η Car)   A peculiar star embedded in the heart of the ➤*Eta Carinae Nebula.* It is possibly the most massive, most luminous star known, but changes in its spectrum recurring on a 5½-year cycle suggest it may be a binary system of two 70-solar-mass stars. The total luminosity, including infrared, is five million times that of the Sun.

Eta Carinae has been observed to vary in brightness over the last 300 years. Halley recorded it as a fourth magnitude star in 1677. By 1843 it was the second-brightest star in the sky. Subsequently, it faded irregularly and has been below the limit of visibility to the naked eye for the last hundred years. It appears now to be increasing in brightness again.

The changes in apparent brightness seem to be caused by the presence of a small opaque dust cloud in which the star is now embedded. The visible nebulosity is known as the *Homunculus Nebula.* This is produced by an outflow of gas amounting to 0.07 solar masses a year, greater than any other known rate of stellar mass loss. It apparently started when the star was at maximum luminosity, but is so great that it cannot be sustained for long – possibly only a few hundred years. A star of such high mass is not stable, and it is thought that it could produce further dramatic changes and possibly a ➤*supernova.*

Eta Carinae Nebula (NGC 3372)   A large cloud of ►*ionized hydrogen* (H II region) in the southern Milky Way in the constellation Carina. The star ►*Eta Carinae* is embedded in it, near its centre.

The nebula has an apparent diameter of 3°, corresponding to a real size of 400 light years at a distance of 7,500 light years, making it much larger than the Orion Nebula. It is a region of star formation containing a number of young star clusters. ►*Keyhole Nebula*.

etalon   ►*Fabry–Perot interferometer.*

ether (aether)   A hypothetical medium through which electromagnetic waves were at one time thought to propagate. Einstein's ►*Special Relativity* theory accounted for the failure of physicists to detect the ether. The concept is not needed in modern physical theory. ►*electromagnetic radiation.*

ETI   Abbreviation for ►*extraterrestrial intelligence.*

E-type asteroid   A rare type of ►*asteroid* with high albedo. The composition may be similar to that of the ►*meteorites* known as enstatite achondrites.

Euclidean space   The space generally adopted for simpler cosmological models and the 'flat' space of everyday experience. Mathematically, the space has zero curvature and the angles of its triangles add up to 180°.

eucrite   A type of achondritic ►*meteorite* consisting of calcium-pyroxene and plagioclase. A close match between the reflectance spectrum of the asteroid ►*Vesta* and those of eucrites strongly suggests that Vesta is the parent body of these meteorites.

Eugenia   Asteroid 45, diameter 244 km, discovered in 1857 by H. Goldschmidt.

Eunomia   Asteroid 15, diameter 260 km, discovered in 1851 by A. de Gasparis.

Euphrosyne   Asteroid 31, diameter 248 km, discovered in 1854 by J. Ferguson.

Europa (1)   One of the four large Galilean moons of Jupiter (number II). The images returned by the ►*Voyager* spacecraft showed a bright reflective surface, criss-crossed by a complex network of dark lines. *Galileo* images revealed a tangled maze of both straight and curving grooves and bands. The relative absence of craters shows that the crust has been greatly modified since the satellite first formed. It is suspected that a thin outer icy crust overlies an ocean of liquid water or a mantle of rock and icy slush. Tidal forces induced by Jupiter could be sufficient to raise the temperature of the ice in Europa's mantle above freezing point. ►Table 6.

Europa (2)   Asteroid 52, diameter 312 km, discovered in 1858 by H. Goldschmidt.

European Southern Observatory (ESO)   A European research organization founded in 1962 to foster cooperation in astronomy and to provide European astronomers with a major modern observatory. The eight member-countries are Belgium, Denmark, France, Germany, Italy, the Netherlands, Sweden and Switzerland. Its headquarters are at Garching bei München in Germany, and its observatory is at ►*La Silla* in Chile.

European Space Agency (ESA)   A collaborative organization of twelve European countries (Austria, Belgium, Denmark, France, Germany, Ireland, Italy, the Netherlands, Spain, Sweden, Switzerland and the UK) for the design, development and deployment of satellites for scientific and commercial applications. ESA's largest establishment is its Research and Technology Centre (ESTEC) at Noordwijk in the Netherlands. Its launch vehicle is the Ariane rocket.

EUV   Abbreviation for extreme ultraviolet. The term is not precisely defined but is usually applied to the wavelength band between 10 and 100 nanometres. ►*electromagnetic radiation, ultraviolet astronomy, XUV*.

eV   Abbreviation for ►*electron volt*.

evection   A periodic perturbation in the motion of the Moon caused by the variation in the gravitational pull of the Sun as the Moon orbits the Earth in the course of a month. The period of variation is 31.8 days, and the maximum magnitude of the effect is 1.27° in the Moon's ecliptic longitude.

evening star   A term applied to the planet Venus (and occasionally to Mercury) when it is visible in the western sky at dusk and in the early evening.

event horizon   A surface surrounding a ►*black hole*, with the property that any light ray emitted inside the event horizon cannot escape because of the strength of the gravitational field. ►*Schwarzschild radius*.

Evershed effect   The radial flow of gas in the penumbra of a ►*sunspot*, discovered by John Evershed (1864–1956).

exit pupil   The image in an optical system (such as an eyepiece) of the stop defining the field of view. In practice the exit pupil is the optimum position for the eye. In a well-designed system intended for visual use, the exit pupil should be in a position and of a size for convenient and comfortable use.

exobiology   The practical or theoretical study of living organisms outside the Earth's environment.

EXOSAT   A European X-ray astronomy satellite launched in May 1983. Technical problems precluded the full intended programme of observations, but many useful observations were made of binary stars and galaxies until the mission ended in April 1986.

exosphere The outermost tenuous layers of a planetary atmosphere, which merge into the interplanetary medium. In this region, the density is so low that very few collisions occur between the atoms, and atoms moving rapidly are able to escape from the gravitational pull of the planet. The Earth's exosphere starts at a height of about 400 to 500 kilometres (250 to 300 miles).

expanding universe A model of the universe in which the fundamental length scale increases with the passage of time. The separation between clusters of galaxies would today be considered an appropriate length scale for this purpose.

The discovery that the redshifts of galaxies increase with distance (1929) and the detection of the ►*cosmic background radiation* (1964) are generally taken as evidence that the universe is expanding.

extinction The reduction in intensity of light as it passes through an absorbing or scattering medium, such as interstellar material or a planetary atmosphere. ►*atmospheric extinction, interstellar extinction*.

extragalactic Beyond the bounds of our own Galaxy, the Milky Way.

extrasolar planet A planet orbiting around a star other than the Sun. Techniques capable of detecting small cyclical changes in the velocities of stars by means of the ►*Doppler effect* led to the first substantial evidence for extrasolar planets around normal stars being presented in 1995 and 1996. Stars for which planets have been claimed include 51 Pegasi, 47 Ursae Majoris, Rho[1] Cancri, Tau Boötis, Upsilon Andromedae, 70 Virginis, HD 114762 and 16 Cygni B.

extraterrestrial intelligence (ETI) Forms of intelligent life, which may or may not exist elsewhere in the universe, other than life as we know it on Earth. Studies in this field focus on the search for radio signals, biologically significant molecules in space and possible planetary systems around stars.

Extreme Ultraviolet Explorer (EUVE) An orbiting spacecraft launched by NASA in 1992 with the object of mapping the entire sky at shorter ultraviolet wavelengths (7–76 nanometres) and carrying out high-sensitivity observations over a more restricted sky area. ►*ultraviolet astronomy*.

eyepiece A combination of small ►*lenses* mounted in a tube, used to magnify and focus the image formed by a telescope or other optical instrument. Eyepieces used for visual observation with a telescope are normally interchangeable, pushing or screwing into a draw-tube of standard size. The focal length of the eyepiece f_e determines the magnification in combination with the fixed focal length of the telescope, f_o: magnification $= f_o/f_e$.

Different types of eyepiece are used for different applications according to the magnification, image quality and field size required. ►*Huygens eyepiece, Ramsden eyepiece, Kellner eyepiece, orthoscopic eyepiece, Erfle eyepiece*.

eye relief The distance between the surface of the eyelens of an ►*eyepiece* and an observer's eye when the observer is positioned to see clearly the full ►*field of view*. In general, the higher the power of the eyepiece, the lower the eye relief.

F

Fabry – Perot interferometer   An optical instrument in which the phenomenon of interference produced by multiple reflection between two precisely parallel glass plates is exploited to achieve a high degree of resolution in the study of spectra. The pair of plates, with an adjustable air space between them, is called an *etalon*.

In transmission, the illuminated field of view appears as a series of bright concentric rings on a dark background, the angular diameters of which depend on the spacing of the etalon and the wavelength of the light. Inserted in front of the grating or prism in a ►*spectrograph*, it provides a means of revealing fine structure in a spectrum that could not be detected by the use of a diffraction grating alone.

facula (pl. faculae)   (1) A bright region of the Sun's photosphere. The appearance of faculae is linked to the subsequent emergence of ►*sunspots* in the same vicinity and ►*solar activity* in general.

facula (2)   A bright spot on a planetary surface, particularly on ►*Ganymede*.

fall   A ►*meteorite* recovered after it has been observed to fall.

falling star   A popular North American term for a ►*meteor*.

false colour image   A visual representation of an image in which the colours are not those that would be seen by a normal human eye in natural conditions. In astronomy, false colour is used to enhance the contrast of an image and so bring out details that would otherwise be difficult to see. False colour is also used to make visual representations of observations made in wavelength regions other than visible light.

False Cross   An ►*asterism* in the shape of a cross in the southern constellations ►*Carina* and ►*Vela*, made up of the stars ε (Epsilon) and ι (Iota) Carinae and δ (Delta) and κ (Kappa) Velorum. The name arises because of the possibility of confusion with the nearby constellation ►*Crux*.

Faraday rotation   The rotation of the plane of polarization of radio ►*synchrotron radiation*. The effect is caused by the presence of a magnetic field and free electrons in the region through which the radiation passes. ►*rotation measure*.

Far-Infrared and Sub-millimetric Space Telescope (FIRST)   A European Space Agency project for an orbiting observatory operating in the wavelength band from 100 microns to 1 mm. Launch is scheduled for 2007 and the operating lifetime is intended to be six years. The main scientific objectives are the study of the ►*interstellar medium*, star formation, the ►*cosmic background radiation* and the composition of comets.

farside   The hemisphere of the Moon that is permanently turned away from the Earth. ►*libration*.

Far Ultraviolet Spectroscopic Explorer (FUSE)   A NASA satellite for ultraviolet astronomy scheduled for launch in 1998.

F corona   One of the components of the solar ►*corona* caused by light scattered from dust particles in the vicinity of the Sun. The 'F' stands for Fraunhofer: the spectrum of the F corona is that of the Sun, including ►*Fraunhofer lines*. ►*zodiacal light*.

feedhorn   A horn-shaped antenna used in a radio telescope to convert radio energy into electrical signals, which then go to a receiver. The size of the horn has to match the wavelength of the radio signal it is designed for. Thus, a radio telescope capable of operating at several different wavelengths will require a set of feedhorns.

Fermi–Hart paradox   The argument that extraterrestrial life cannot exist because we see no evidence of it and have received no signals from extraterrestrials.

fibre optics   The use of fine glass fibres for transmitting light signals. The fibres used, which are normally less than 1 millimetre in diameter, are made of glass of a high refractive index, coated with a material of lower refractive index. Light signals fed into the fibre undergo successive internal reflections within the core fibre, which thus acts as a light guide. The method works even when the fibre is considerably curved. Images can be transmitted using bundles of very fine fibres in a fixed array. Such techniques have a variety of applications in astronomical instrumentation.

fibril   A characteristic feature observed in images made in ►*hydrogen alpha* light of active regions of the Sun. Fibrils show as dark streaks with a width of 725–2,200 kilometres and an average length of 11,000 kilometres. Individual fibrils have a lifetime of 10–20 minutes, though the overall fibril pattern of a region shows little change over several hours.

In the central parts of active regions, fibrils are arranged in patterns connecting spots and ►*plages* of opposite polarities. Regular spots are surrounded by a radial pattern of fibrils called the *superpenumbra*. They represent material flowing down into the spot at velocities around 20 km/s.

field equations   In ►*General Relativity*, the equations that describe the curvature of spacetime by connecting the ►*metric* to the distribution of matter and energy.

field galaxy A galaxy that appears in the same field of view as a cluster of galaxies but is not itself a member of the cluster. The alignment is purely coincidental and the field galaxy is nearer or further away than the cluster.

field of view The angular extent of the image formed by an optical instrument, such as a telescope. The field of view of a telescope becomes smaller the greater the magnification employed. Where an application requires a particularly wide field of view (e.g. for comet hunting or for sky survey photography) a special design of eyepiece or telescope is required.

field pattern In radio astronomy, a mathematical expression describing the amplitude and phase of the voltage at the terminals of an ►*antenna* as a function of the direction from which the radio signals are received.

field star A star that appears in the same field of view as a ►*star cluster* but is not itself a member of the cluster. The alignment is purely coincidental and the field star is nearer or further away than the cluster.

figuring The process of smoothing and shaping the surface of a lens or mirror to the high degree of accuracy required for use in a precision instrument, such as a telescope.

filament A solar ►*prominence* seen in the light of certain ►*spectral lines*, particularly ►*hydrogen alpha* and those of ionized calcium, projected against the bright disc of the Sun (i.e. viewed from above) and appearing as a dark streak across the photosphere.

filar micrometer ►*micrometer*.

filter An accessory used with an optical instrument or a detector of electromagnetic radiation to narrow down the wavelength band or reduce the total intensity passing into the instrument. Colour filters are used as an aid to improving contrast and bringing out particular features in the visual observation of planets, and also in multicolour ►*photometry*. A very narrow-band filter may be used to isolate, for example, the hydrogen alpha spectral line, particularly for making observations of the Sun. Filters are also used to help eliminate unwanted ►*light pollution* in certain types of observation.

filtergram A photograph of the Sun taken through a filter that transmits light over a very narrow wavelength range. ►*Lyot filter*.

find A ►*meteorite* discovered accidentally or fortuitously and recognized as such by its nature, as opposed to one actually seen to fall.

finder A small telescope attached to the tube of a larger instrument, used for locating the correct field of view. A finder usually incorporates a cross-wire and is aligned so that an object located on the cross-wire appears centrally in the smaller field of view of the main telescope.

fireball (1)   A particularly bright ➤*meteor*. There is no precise definition: magnitudes −3, −4 or −5 are variously quoted as the minimum for a meteor to be described as a fireball. The appearance of a fireball carries some likelihood that a ➤*meteorite* fall may occur.

fireball (2)   Alternative for ➤*primeval fireball*.

FIRST   Abbreviation for ➤*Far-Infrared and Sub-millimetric Space Telescope*.

first contact   In an ➤*eclipse* of the Sun, the point when the Moon's disc first begins to move across the Sun's ➤*photosphere*. In a lunar eclipse, first contact occurs when the Moon enters the full shadow (umbra) of the Earth. The term may also be used to describe the similar stage in the progress of a ➤*transit* or ➤*occultation*.

First Point of Aries (Υ)   One of the two points on the celestial sphere where the ➤*ecliptic* and the ➤*celestial equator* cross. It is used as the zero point for ➤*right ascension*. Because of the effects of ➤*precession*, the equator is slowly sliding around the ecliptic, taking about 25,800 years to make one revolution. Though the intersection point was in the constellation Aries several thousand years ago, it is now in Pisces, but the name has not been changed. The Sun is at the First Point of Aries on the northern vernal ➤*equinox*, and the expression 'vernal equinox' is also used to describe this place on the celestial sphere.

first quarter   The ➤*phase* of the Moon when half the visible disc is illuminated and the Moon is waxing towards full. First quarter occurs when the celestial ➤*longitude* of the Moon is 90° greater than the Sun's.

Fishes   English name for the constellation ➤*Pisces*.

FK   Abbreviation for the series of fundamental star catalogues published in Germany since 1907. The current version is FK 5, published in 1988. ➤*fundamental catalogue*.

Flaming Star Nebula   Popular name for IC 405, a star cluster surrounded by nebulosity in Auriga.

Flamsteed numbers   Identification numbers assigned to stars listed in *Historia coelestis Britannica* by John Flamsteed (1646−1719). In the official version of the catalogue, published posthumously in 1725, the numbers are not included explicitly. However, they do appear in a preliminary version published by Edmond Halley and Isaac Newton in 1712 without Flamsteed's approval. Few copies of this version survive because Flamsteed, greatly angered at the action of Halley and Newton, destroyed many of them.

The numbers were allocated by constellation and in order of right ascension. They were subsequently used by other cataloguers, including John Bevis (1750) and J.-J. de Lalande (1783). The Flamsteed numbers are still commonly used for stars that do not also have a Greek ➤*Bayer letter* designation.

flare   A phenomenon in the solar ➤*chromosphere* and ➤*corona*, caused by a sudden release of energy that heats and accelerates matter in the Sun's atmosphere. Flares are explosions lasting typically for a few minutes, during which matter reaches temperatures of hundreds of millions of degrees. Most of the radiation is emitted as X-rays but flares are more easily observed in visible light or radio waves. They are associated with active regions of the Sun. Charged particles ejected from the Sun by flares cause ➤*aurorae* when they reach the Earth a few days later.

flare star   A dwarf ➤*M star* that undergoes unpredictable outbursts lasting a few minutes, during which the brightness may increase by a magnitude or more. All flare stars have emission lines in their spectra and the flares are thought to be phenomena in the stellar ➤*chromosphere*, similar to ➤*flares,* but far more energetic. The nearest star to the Sun, Proxima Centauri, is a flare star. Flare stars are also known as UV Ceti stars.

flash spectrum   The ➤*emission line spectrum* of the solar ➤*chromosphere*, observed fleetingly at the onset and conclusion of a total solar eclipse.

flat   A plane mirror, optically figured to be flat with a high degree of accuracy.

flatness problem   The problem of explaining, in the standard ➤*Big Bang* models, why the universe as observed has a density so close to the ➤*critical density*. This means that it is expanding at a rate close to its escape velocity, and its geometry is very nearly that of ➤*Euclidean space*. The problem is resolved by ➤*inflationary universe* models.

flexus (pl. flexi)   A type of cusped linear feature on the surface of ➤*Europa*.

flocculi (sing. flocculus)   A fine mottled pattern seen on the Sun when the solar ➤*chromosphere* is imaged in monochromatic light, such as the light of singly ionized calcium.

Flora   Asteroid 8, diameter 162 km, discovered by J. R. Hind in 1847.

Flora group   A complex grouping of ➤*asteroids* near the inner edge of the asteroid belt at a distance of 2.2 AU from the Sun. The group is separated from the main belt by one of the ➤*Kirkwood gaps* and is not a true family with a common origin.

fluctus (pl. fluctus)   Terrain on the surface of ➤*Io* where flow of molten material has taken place.

flux collector   A telescope designed solely to collect radiation in order to measure its intensity or to carry out spectral analysis. No attempt is made to form an image, so a flux collector can have a more crudely figured reflecting surface than a conventional telescope.

flux density   The energy in a beam of radiation passing through unit area normal to the beam in one unit of time. In radio astronomy, flux density, measured in janskys (Jy), is the energy received from a radio source per unit area of detector per unit frequency interval.

Fly   English name for the constellation ➤*Musca*.

flyby   A space mission in which an exploratory probe carries out imaging and other experiments as its flight path passes close to the body under investigation, without landing or going into orbit.

Flying Fish   English name for the constellation ➤*Volans*.

Foal   English name for the constellation ➤*Equuleus*.

focal length (symbol F or f)   The distance between a lens or mirror and the point on the optical axis where parallel rays of light are brought to a focus.

focal plane   The surface on which fall the images of all points in the ➤*field of view* of an optical instrument such as a telescope. This surface may be flat and normal to the optical axis or, as in the Schmidt telescope, curved, in which case the term *focal surface* is preferable.

focal ratio (f-ratio)   The ratio of the focal length of a lens, mirror or compound optical system to its aperture. It is normally written using the prefix f: for example, $f/10$ means a focal ratio of 10.

following (f)   A term used to describe the position of a planetary feature or an object on the celestial sphere relative to another. Following features or objects come into view later than the reference feature or object; those coming into view earlier are described as ➤*preceding* (p).

Fomalhaut (Alpha Piscis Austrini; α PsA)   The brightest star in the constellation Piscis Austrinus, an ➤*A star* of apparent visual magnitude 1.2. The name, of Arabic origin, means 'the fish's mouth'. ➤Table 3.

Footprint Nebula   ➤*Minkowski's Footprint*.

forbidden lines   ➤*Spectral lines*, not normally observed under laboratory conditions because they have an intrinsically low probability of occurrence, or because they result from a transition between a metastable excited state and the ground state. Under typical conditions, an atom in a metastable state will lose energy through a collision before it is able to decay to the ground state. Under astrophysical conditions, however, where there are huge numbers of atoms and densities may be very low, it is possible for such 'forbidden' transitions to occur and to produce strong spectral lines.

fork mounting A particular form of ➤*equatorial mounting*.

Fornax (The Furnace) A small, inconspicuous constellation of the southern sky. It was introduced in the mid-eighteenth century by Nicolas L. de Lacaille with the longer name, Fornax Chemica – the chemical furnace. None of its stars are brighter than fourth magnitude. ➤Table 4.

Fornax A A strong radio source in the constellation Fornax associated with the spiral galaxy NGC 1316.

Forty-seven Tucanae (47 Tuc; NGC 104) The second-brightest ➤*globular cluster* in the sky (after Omega Centauri). To the naked eye it appears as a fifth magnitude star; this explains why it was originally designated 47 Tucanae, a form of name appropriate to a single star. It lies at a distance of 13,000 light years and is a relatively young globular cluster.

fossa (pl. fossae) A long, narrow, shallow depression on a planetary surface.

Foucault knife-edge test A technique for testing the accuracy with which the surface of a mirror or lens is figured. It was introduced by J. B. L. Foucault (1819–68) in 1851. The method utilizes the fact that a knife-edge moved slowly across the focal plane will cause the image of a beam of light to darken uniformly only if the imaging surface is perfectly spherical: any depressed or raised areas on the lens or mirror cause light and dark zones to appear in the image as the test is carried out.

Foucault's pendulum A long, free-swinging pendulum devised in 1851 by Foucault to demonstrate the rotation of the Earth. If undisturbed, the plane in which the pendulum swings slowly changes at a rate of 15 × sin(angle of latitude) degrees per hour of ➤*sidereal time*.

fourth contact In an ➤*eclipse* of the Sun, the point when the Moon's disc finally moves clear of the Sun's ➤*photosphere*. In a lunar eclipse, fourth contact occurs when the Moon leaves the full shadow (umbra) of the Earth. The term may also be used to describe the similar stage in the progress of a ➤*transit* or ➤*occultation*.

Fox English name for the constellation ➤*Vulpecula*.

Fra Mauro A lunar crater, 95 kilometres (60 miles) in diameter, close to the Apollo 14 landing site. It gives its name to the geological formation sampled by the Apollo 14 astronauts, this being part of the ejecta or debris created by the formation of the ➤*Imbrium Basin*. ➤*Apollo programme*.

Franklin Adams charts A photographic atlas of stars down to sixteenth magnitude, compiled by John Franklin Adams (1843–1912) and published in 1914.

ƒ-ratio   Abbreviation for ➤*focal ratio*.

Fraunhofer lines   The dark ➤*absorption lines* in the spectrum of the Sun and, by extension, in the spectrum of any star. Many of the stronger ones were first mapped by Joseph von Fraunhofer (1787–1826), who also labelled some of the most prominent with letters of the alphabet. Some of these identifying letters are still commonly used in physics and astronomy, notably the sodium D lines and the calcium H and K lines.

Fraunhofer's original (1817) designations of absorption lines in the solar spectrum

Letter	Wavelength (nm)	Chemical origin
A	759.37	Atmospheric O_2
B	686.72	Atmospheric O_2
C	656.28	Hydrogen α
D1	589.59	Neutral sodium
D2	589.00	Neutral sodium
D3	587.56	Neutral helium
E	526.96	Neutral iron
F	486.13	Hydrogen β
G	431.42	CH molecule
H	396.85	Ionized calcium
K	393.37	Ionized calcium

Note: Fraunhofer's original observations did not resolve the components of his D line.

Fred Lawrence Whipple Observatory   An observatory on Mount Hopkins in Arizona, operated by the ➤*Harvard–Smithsonian Center for Astrophysics*. The instruments located there, at an altitude of 2,600 metres (8,500 feet), are a 10-metre (32-foot) optical reflector for gamma-ray astronomy, 0.61-metre (24-inch) and 1.5-metre (60-inch) optical telescopes for planetary, stellar and extragalactic astronomy, automatic photoelectric telescopes, and the reconfigured ➤*Multiple Mirror Telescope* (MMT), operated jointly with the University of Arizona for optical and infrared observations. It is named in honour of a former Director of the ➤*Smithsonian Astrophysical Observatory* who is particularly well known and distinguished for his work on comets.

frequency (symbol ν or ƒ)   The number of times a repetitive phenomenon occurs in a unit of time. In the case of waves, the frequency is the number of waves passing a fixed point in a second. The basic unit of frequency is the hertz (Hz).

Friedmann universe   A model of the universe that can collapse in on itself. Alexander Friedmann (1888–1925)   discovered an error in Einstein's work on cosmology, from which he showed that ➤*expanding universes* and ➤*oscillating universes* were possible in ➤*General Relativity*. This work (1922 and 1924) was completely ignored until rediscovered and presented as the ➤*Lemaître universe*. The Friedmann universe can be closed if the density of matter is sufficient to reverse the expansion, and this has led to searches for the so-called ➤*missing mass*. ➤*open universe, closed universe, Robertson–Walker metric*.

F star   A star of ➤*spectral type* F. F stars on the ➤*main sequence* have surface temperatures in the range 6,000–7,400 K. Their spectra are characterized by strong absorption lines of ionized calcium (the ➤*H and K lines*), which are stronger than the hydrogen lines. There are also many medium-strength absorption lines due to iron and other heavier elements. ➤*Procyon* and ➤*Polaris* are examples of F stars.

F-type asteroid   A subclass of the ➤*C-type asteroids*, distinguished by weak or no ultraviolet absorption in their spectra.

full Moon   The ➤*phase* of the Moon when its celestial ➤*longitude* is 180° greater than the Sun's and its disc thus appears fully illuminated.

fundamental catalogue   A catalogue of star positions which have been determined absolutely and with the highest possible precision from a compilation of many observations. Absolute positions are determined by recording the times of transit across the meridian. Combination of the results from several observatories helps to reduce the effects of both systematic and random errors.

   A series of fundamental catalogues has been produced and published in Heidelberg, Germany. The current version, known as FK5, contains improved data on the 1,535 stars of magnitude 7 and brighter also listed in FK3 and FK4. Work is in progress on an 'Extension' to FK5 that will cover about a thousand stars in the magnitude range 5–7 selected from the FK4 Supplement, plus around two thousand others in the magnitude range 6.5–9.5.

fundamental epoch   The zero point of a system of time measurement, from which times may be measured forwards or backwards.

fundamental plane   A plane that forms the basis of a particular coordinate system.

FU Orionis star   A type of variable star characterized by a brightening over a period of the order of a year by as much as five or six magnitudes, followed by a long period (decades) of constancy or slow decline. These stars are ➤*supergiants* of ➤*spectral types* F and G, and are surrounded by dust and nebulosity.

Furnace   English name for the constellation ➤*Fornax*.

FUSE Abbreviation for ➤*Far Ultraviolet Spectroscopic Explorer.*

*f***-value** An alternative expression for ➤*oscillator strength.*

G

Gaia hypothesis   The notion that life on Earth regulates the composition of the lower atmosphere. Gaia was a Greek Earth goddess.

galactic centre   The central region of our Galaxy, which cannot be seen optically because it is obscured by dense concentrations of dust. Radio and infrared observations reveal a complex environment dominated by the radio source ►*Sagittarius A*. The compact radio source Sagittarius A* appears to mark the central point and is used as the origin for ►*galactic coordinates*.

Within ten light years of the centre there is a ring of gas and dust, rotating at about 110 km/s and surrounding a massive object, probably a black hole, of 3 million solar masses.

galactic cluster   An old term, no longer in common use, for an ►*open cluster* of stars.

galactic coordinates   A latitude and longitude coordinate system that takes the ►*galactic plane* as its equator and the ►*galactic centre* (RA 17h 42.4m, Dec. −28° 55′ as the zero point of longitude measurement. Galactic coordinates are used mainly in studies where the spatial distribution of objects within our Galaxy is of importance. For example, maps of the radio emission from hydrogen gas in the Milky Way are often plotted in this coordinate system. ►*galactic poles*.

galactic corona   A region around our Galaxy extending out to a radius of about 250,000 light years.

galactic halo   A spherical region around a spiral galaxy. The halo round our own Galaxy extends to a radius of about 50,000 light years. It is the region within which all the stars belonging to the galaxy are found, particularly the ancient globular clusters.

The halo contains very hot gas that emits X-rays. The stars outside the disc of the galaxy but within the halo are the oldest, and record the size of the galaxy before much of it collapsed to a disc.

galactic plane   The great circle on the sky that includes the galactic centre and the densest parts of the Milky Way. It is inclined at about 63° to the celestial equator.

galactic poles   The poles of the ►*galactic plane*, i.e. the points at galactic latitudes 90° north and south. The north galactic pole is in the constellation Coma Berenices at RA 12h 51.4m and Dec. 27° 7.7' (epoch 2000.0). The diametrically opposite south pole is in the constellation Sculptor.

galactic year The time taken for the Sun to complete one orbit around the galactic centre, roughly 220 million years.

Galatea A satellite of Neptune (1989 N3) discovered during the flyby of ➤*Voyager 2* in August 1989. ➤Table 6.

galaxy A family of stars, held together by their mutual gravitational attraction, and with a distinct identity separating it from other galaxies.

Galaxies cover a huge range of size and mass as well as exhibiting a variety of structures and properties. The smallest galaxies known are relatively nearby ➤*dwarf galaxies* containing only 100,000 stars, fewer than in a typical ➤*globular cluster*. At the other end of the scale, the most massive galaxy known, the giant elliptical M87, contains 3,000 billion solar masses, about 15 times more than our own ➤*Galaxy*.

Most galaxies can be categorized into a number of broad morphological types. ➤*Spiral galaxies* are disc-shaped, with a central bulge from which spiral arms appear to wind outwards. In *barred spirals*, a bar of stars extends out from the bulge and the arms appear to be attached to the ends of the bar. Spiral galaxies contain very luminous young stars and significant amounts of interstellar material concentrated in the arms.

Most of the conspicuous galaxies in the sky are spirals, but the most numerous type is the *elliptical galaxy*. Both the smallest and largest galaxies are of this kind. They are thought to consist entirely of old stars with relatively little interstellar material. The three-dimensional shape of galaxies in the elliptical category can be spheroidal or virtually spherical.

The third main group is that of ➤*irregular galaxies*, which are neither spiral nor elliptical. These account for up to a quarter of all known galaxies. At visible wavelengths, irregular galaxies show no particular circular symmetry and look chaotic.

A very small number of galaxies have unusual structure, often attributable to a gravitational interaction with another galaxy. Others emit exceptionally large amounts of energy and exhibit other evidence, such as variability, suggesting that unusual and violent processes are at work. Such active galaxies include ➤*Seyfert galaxies* and ➤*radio galaxies*. ➤*Hubble classification*.

Galaxy With a capital initial G, by convention, the galaxy or family of stars to which the Sun and solar system belong, visible in the night sky as the ➤*Milky Way*.

Our Galaxy is a spiral, possibly a mildly barred spiral galaxy, containing of the order of two hundred billion stars as well as much interstellar matter, both dark and luminous. It is disc-shaped with an almost spherical bulge at the centre.

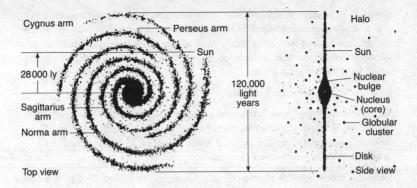

The Galaxy. Schematic views of the structure of our galaxy (excluding the outer corona), face on (left) and edge-on (right).

The disc is 100,000 light years across, but much of its content is concentrated in a thin layer only 2,000 light years thick towards its outer edges, though stars are distributed through a somewhat thicker disc. The central bulge has a radius of about 15,000 light years. Studies of the dynamics of stars and interstellar material suggest that the luminous material we can see accounts for as little as 10 per cent of the total mass of the Galaxy. The rest is so-called ➤*dark matter*, in a form not yet identified.

The spiral arms are concentrations of stars and interstellar material appearing to wind outwards from the edge of the bulge. Regions of star formation and ➤*ionized hydrogen* are concentrated in the arms. In the space between the arms, the average density of matter is a factor of two or three lower than within the arms. The Sun is located about 28,000 light years from the galactic centre, within the disc, near the inner edge of a spiral arm. The whole Galaxy is in rotation, but not as a rigid body, so it is constantly deforming. The Sun takes about 220 million years to complete a circuit, but stars nearer the centre take shorter times.

Centred on the nucleus is a sparsely populated, roughly spherical region, with a radius of at least 50,000 light years, known as the ➤*galactic halo*. The halo contains ➤*globular clusters* and, in general, the oldest stars in the Galaxy. There is very little luminous matter in the halo compared with the disc and central bulge, though gravitational studies suggest that the invisible component of the Galaxy's mass is probably distributed in a sphere around the Galaxy rather than concentrated in the disc. This dark matter is thought to extend up to 300,000 light years, far beyond the halo defined by visible objects, into what is sometimes termed the galactic corona.

The innermost nucleus, lying in the direction of the constellation Sagittarius, is concealed from direct optical observation by dense opaque dust. However, observations in the infrared and radio regions of the spectrum, and at gamma-ray and X-ray wavelengths, suggest that the core contains a tightly packed sphere of stars and a black hole. The likely mass of the black hole is disputed, some astronomers suggesting it could be as little as 100 solar masses, others saying it could be 3 million solar masses.

galaxy cluster ➤*cluster of galaxies.*

Galilean satellites The four largest moons of Jupiter, ➤*Io,* ➤*Europa,* ➤*Callisto* and ➤*Ganymede,* which were discovered telescopically by Galileo in 1610. They are easily observed with the aid of a small telescope or binoculars and their orbital motion around Jupiter can be readily followed.

Galilean telescope A simple telescope design, of the kind used by Galileo in the first astronomical telescopes, consisting of two lenses. A long-focus convex (converging) lens forms the objective and a single concave lens is used as the eyepiece. The resulting image is upright, and the system is still widely used in opera-glasses.

Galileo A NASA spacecraft for the exploration of ➤*Jupiter,* its rings and satellites. It was launched in October 1989 from the ➤*Space Shuttle* and arrived at Jupiter in December 1995. The flight plan included flybys of the asteroids ➤*Gaspra* and ➤*Ida* en route.

When *Galileo* was launched, more than twenty years had elapsed since the mission was first conceived. Initially, the launch had been planned for 1983. Following various problems, including the loss of the *Challenger* Space Shuttle, it was impossible to obtain sufficient power by means of the Shuttle launch for a direct journey to Jupiter. Engineers therefore designed a gravity-assisted flight plan, whereby the craft was accelerated by close passages by Venus and twice by the Earth during the first three years.

A major disappointment was the failure of the craft's high-gain communications antenna. Having only a low-gain antenna working limited the amount of data that could be transmitted back to Earth. However, *Galileo* otherwise worked well, despite some problems with its tape recorders. It entered orbit in the jovian system, returning high-resolution images of the Galilean moons.

A probe carried by *Galileo* separated from the craft and entered Jupiter's atmosphere on 7 December 1995. It parachuted down, returning data on the composition and physical state of the atmosphere for 57 minutes.

Galileo National Telescope ➤*Telescopio Nazionale Galileo.*

gamma-ray astronomy The study of gamma-rays from astronomical sources. Gamma rays are the most energetic form of ➤*electromagnetic radiation,* with wavelengths shorter than X-rays (i.e. less than about 0.1 nanometres). They are absorbed high in the Earth's atmosphere; only the most energetic can be detected at ground level and virtually all astronomical gamma-ray observations must be conducted from satellites. The detectors used are scintillation counters, spark chambers and solid-state detectors, and the angular resolution achieved is coarse by astronomical standards.

Detectors for ➤gamma-ray bursters have been carried on numerous spacecraft since 1969. Sky surveys have been carried out by the satellites SAS-2 and COS-B. SAS-2 was launched in 1972 and operated for seven months. COS-B was launched in 1975 and operated for over six years. A great advance in gamma-ray astronomy was achieved with the launch of the ➤Compton Gamma Ray Observatory by NASA in April 1991. Within months of launch, many new sources were being identified with greater positional accuracy than was possible previously.

Astronomical sources of gamma rays include solar ➤flares, ➤pulsars, ➤X-ray binary stars and ➤quasars, as well as the gamma-ray bursters. Known discrete sources of gamma-rays include the ➤Vela Pulsar, the ➤Crab Pulsar, ➤SS433 and ➤Geminga. The most intense diffuse gamma radiation comes from the galactic plane, generated by interactions between ➤cosmic rays and the interstellar gas. A gamma-ray spectrometer on HEAO-3 in 1979 observed lines produced by electron–positron annihilation from the direction of the ➤galactic centre

gamma-ray burster   An astronomical source of a transient burst of gamma-radiation and X-rays. The bursts are intense and short, lasting for between a few milliseconds and a few tens of seconds. Gamma-ray bursters were first discovered by chance in the late 1960s by military satellites designed for monitoring nuclear weapons tests and have since been observed by a variety of spacecraft carrying appropriate detectors. In 1979 a single burst, which seemed to come from the Large ➤ Magellanic Cloud, was detected simultaneously by nine satellites. Monitoring by the ➤Compton Gamma Ray Observatory (GRO) showed that bursts occur about twice a day, at random positions all over the sky. It has recorded several thousand.

Though the Compton GRO was able to determine the positions of the bursters with greater accuracy than was previously possible, the positions were still not accurate enough to allow optical identification. In 1997, however, the ➤BeppoSAX satellite, with the help of its narrow-field X-ray camera, was able to pinpoint the position of gamma-ray bursters precisely enough for them to be identified optically, and for radio emission to be detected. The first optical spectrum of a gamma-ray burster, obtained at the ➤Keck Observatories, showed it to be at a remote cosmological distance, about halfway to the edge of the observable universe. This implies that the energy output is immense. For a few seconds the burster emits more than a million times more energy than a whole galaxy. Though many theories have been advanced, the precise mechanism is not known. Some of the more favoured theories involve the merger of two ➤neutron stars.

Gamma Ray Observatory (GRO) ➤*Compton Gamma Ray Observatory*.

Ganymed Asteroid 1036, diameter 40 km, discovered in 1924 by W. Baade. It is a member of the ➤*Amor* group, and is possibly one of the largest asteroids that make close approaches to the Earth.

Ganymede One of the four Galilean moons of Jupiter (number III) and the largest natural satellite in the solar system.

The first high-resolution images of Ganymede were returned by ➤*Voyagers 1* and *2*. Images showing even finer detail were obtained by ➤*Galileo*. There are several different types of terrain, notably dark areas that are heavily cratered and a lighter grooved terrain that constitutes around 60 per cent of the surface photographed. *Galileo* images of the dark area suggest it is surface that has been changed by various episodes of shearing and furrowing. *Galileo* also revealed many small craters on the finely grooved areas.

One of the most significant discoveries made by *Galileo* was that Ganymede has a substantial magnetic field which is stronger at its surface than the fields of Mercury, Venus or Mars. Data from *Galileo*'s trajectory, combined with the magnetic, field, suggest that Ganymede must have a molten iron-rich core. Overall Ganymede's density is about twice that of water. It is likely that the core is surrounded by a rocky mantle, overlain by a thick layer of ice. ➤Table 6.

gardening The overturning of lunar ➤*regolith* (soil) by ➤*micrometeorite* bombardment.

Garnet Star An informal name, first used by William Herschel, for the strikingly red star μ (Mu) Cephei. It is a red ➤*supergiant* and a ➤*semiregular variable* ranging in magnitude between 3.6 and 5.1.

gaseous nebula A glowing cloud of gas in interstellar space, which may be either an ➤*emission nebula* or a ➤*reflection nebula*. The expression used to be employed more often, when ➤*galaxies* were referred to as 'extragalactic nebulae', in order to emphasize the distinction. The adjective 'gaseous' is now generally omitted since 'nebula' is normally used only for such interstellar clouds and not for galaxies.

Gaspra Asteroid 951, a member of the ➤*Flora group*, imaged by the ➤*Galileo* spacecraft, which passed Gaspra at a distance of about 16,000 kilometres on 29 October 1991. It is irregular in shape, measuring about 20 by 12 by 11 kilometres, with a cratered surface. The largest crater is 1.5 kilometres across. *Galileo* detected a magnetic field, suggesting that Gaspra has a metallic composition.

Gassendi A lunar crater, 100 kilometres (60 miles) in diameter, on the northern border of the Mare Humorum. Clefts cross the crater floor and there are multiple peaks. Gassendi has been particularly associated with the search for ➤*transient lunar phenomena*.

gas tail ➤*ion tail.*

gegenschein ➤*zodiacal light.*

GEM Abbreviation for Giotto Extended Mission. ➤*Giotto.*

Geminga A powerful gamma-ray source in the constellation Gemini discovered in 1972 by the orbiting observatory SAS (➤*Small Astronomy Satellite*) 2. Weak X-rays from Geminga were detected by the ➤*Einstein Observatory* and its optical counterpart appears as a twenty-fifth magnitude star. Geminga is thus a very unusual object in that it emits almost all its energy as gamma rays; the X-ray emission is a thousand times weaker and its luminosity in visible light a thousand times weaker again. It is thought to be relatively close, probably within 700 light years.

Observations made by ➤*ROSAT* confirmed the X-ray emission and showed it to be pulsating with a period of about a quarter of a second. Gamma-ray pulsations have also been detected by the ➤*Compton Gamma Ray Observatory*. Geminga thus appears to be a gamma-ray and X-ray ➤*pulsar*. Why the bulk of its radiant energy is emitted in such a high-energy form is not known.

Gemini (1) (The Twins) One of the twelve constellations of the zodiac in the list drawn up by Ptolemy (*c.* AD 140). The two brightest stars in Gemini, both first magnitude, bear the names of the twins Castor and Pollux of classical mythology. Pollux, though the brighter of the pair, was given the designation β (Beta) by Johann Bayer. ➤*Table 4.*

Gemini (2) A series of manned, orbiting spacecraft, launched by the USA in the 1960s. They were an important part of the development of manned spaceflight technology in preparation for the ➤*Apollo programme* of Moon landings. *Gemini 3* in 1965 was the first American flight with a crew of more than one astronaut, and *Gemini 8* in March 1966 achieved the first successful docking in space. The last of the series was *Gemini 12* in November 1966. Many of the astronauts who later took part in the Apollo Moon landings also flew in the Gemini programme.

Geminids A major annual ➤*meteor shower*, the radiant of which lies near the star Castor in the constellation Gemini. The shower peaks around 13 December, and the normal limits are 7–16 December. The meteor stream has an unusual orbit with a perihelion distance of only 0.14 AU. IRAS, the ➤*Infrared Astronomical Satellite*, discovered a cometary nucleus in 1983, designated as asteroid 3200 Phaethon, which appears to be the parent body for this stream.

Gemini 8-meter Telescopes Two 8-metre telescopes for optical and infrared astronomy, resulting from an international collaboration between the USA, the UK, Canada, Chile, Brazil and Argentina. One is sited in the northern hemisphere, at the ➤*Mauna Kea Observatories* in Hawaii, the other in the southern hemisphere on Cerro Pachón in Chile, near the ➤*Cerro Tololo Inter-American Observatory*. The siting of the two telescopes ensures complete sky coverage between them. The Hawaii telescope is being completed during 1998, and its more southerly twin in 2000.

Gemma   An alternative name for the star ➤*Alphekka*.

General Relativity   A theory of gravitation, published in its final form in 1916. It was developed by Albert Einstein (1878–1955) from his earlier (1905) ➤*Special Relativity* theory.

One of the fundamental postulates of the general theory is that, over a limited region of ➤*spacetime*, it is impossible for observers to tell whether they are undergoing uniformly accelerated motion or are in a gravitational field. This is known as the *principle of equivalence*. Einstein showed that it was not necessary to think of gravity as a force acting at a distance. Instead, he described gravity in terms of its local effects on space and time, i.e. as the curved geometry of spacetime, which is determined by the distribution of matter and energy.

A good three-dimensional analogy to help understand the meaning of curved, four-dimensional spacetime is geometry on the surface of a sphere. For example, two travellers who set out from different places on the equator and travel due north will eventually find that their paths cross, even though they started out travelling parallel to each other. This contrasts with what happens on a flat surface, where parallel lines never cross. The two travellers might say they had been pulled together by some force (gravity, for example), but their experience is more effectively explained in terms of geometry.

In regions where the gravitational field is weak, General Relativity approximates to the theory set out by Isaac Newton. For a strong gravitational field, General Relativity gives the best description yet devised. There are other theories of gravity but none has met with the total success enjoyed by General Relativity.

Several areas of astronomy have proved to be testing grounds for the theory. The perihelion of Mercury's elliptical orbit around the Sun advances by 43 arc seconds per century more than is predicted by Newton's gravitational theory, but General Relativity explains it exactly. The light from stars deviates from a straight line if it passes very close to the Sun, and this has been observed at solar eclipses. The motion of pulsars in binary systems is readily accounted for by General Relativity. The most frequent application of General Relativity is in cosmology, since gravity is the dominant factor in all attempts to make mathematical models of the universe.

geocentric model   A model of the solar system that places a stationary Earth at the centre of the motion of the Sun, Moon and planets. This was the universally accepted view of the cosmos until Copernicus (1473–1543) suggested that a solar system with the Sun in the middle would provide a more elegant explanation of observed planetary motions. Predictions of the movements of the planets in the geocentric system were made on the basis of a complex theory of epicycles set out by the Greek astronomer and mathematician Ptolemy (*c.* AD 100–170). ➤*heliocentric model.*

geocorona The outermost part of the Earth's atmosphere, consisting of a halo of hydrogen gas that extends over a region extending to about 15 Earth radii.

geodesic The shortest path between two points in ➤*spacetime* and therefore the path followed by photons.

geodesy The measurement of the precise figure of the surface of the Earth and the Earth's gravitational field.

Geographos Asteroid 1620, diameter 2 km, first discovered in 1951 by R. Minkowski and A. Wilson, and recovered in 1969 when it made a close approach to the Earth. It is a member of the ➤*Apollo* group.

geoid A surface defined by mean sea level in the open ocean and, on land, the surface that would be taken up by water in an imaginary network of frictionless channels connected to the sea.

geomagnetic field The magnetic field in the vicinity of the Earth. To a first approximation, the Earth's magnetic field is like that of a bar magnet (dipole) currently displaced 451 kilometres from the centre of the Earth towards the Pacific Ocean and tilted at 11° to the rotation axis. The strength and shape of the geomagnetic field varies gradually over a timescale of years.

The intensity of the geomagnetic field is denoted by a vector quantity F or B, and is measured in gauss (G), tesla (T) or gamma (γ). (1 tesla = 10,000 gauss; 1 gamma = 1 nanotesla = 10^{-5} gauss.) The direction of the field at any point can be described by two angles: I, the dip angle or inclination, the angle between the horizontal and the field, taken as positive when downwards; D, the declination, the azimuth from the northward horizontal direction, measured towards either east or west.

geomagnetic storm A large decrease, typically of a few hours' duration, in the horizontal component of the Earth's magnetic field. The cause is the arrival in the vicinity of the Earth of electrically charged particles (plasma) from the Sun, generated usually by a solar ➤*flare.* Auroral activity and disruption of radio communications are common during such storms.

geometric albedo The ratio between the brightness of a planetary body, as viewed from the direction of the Sun, and the brightness of a hypothetical white, diffusely reflecting sphere of the same size and at the same distance. ➤ *albedo.*

geospace The region of space around the Earth incorporating Earth's ➤*magnetosphere* and ➤*ionosphere*.

geostationary orbit ➤*geosynchronous orbit*.

geosynchronous orbit An orbit around the Earth in which a satellite has a period exactly equal to the Earth's sidereal rotation period of 23 hours 56 minutes 4.1 seconds. If such an orbit is circular and in the plane of the Earth's equator, the satellite appears to be practically stationary in the sky and the orbit is described as *geostationary*. A geostationary orbit is at an altitude of 35,900 kilometres.

A satellite in a geosynchronous orbit inclined to the Earth's equatorial plane appears to trace out a figure-of-eight shape over the course of a day.

German mounting A particular form of ➤*equatorial mounting*.

Ghost of Jupiter Popular name for NGC 3242, a ➤*planetary nebula* in Hydra.

Giacobinids Alternative name for the ➤*meteor shower* also known as the ➤*Draconids*. ➤*Comet Giacobini–Zinner*.

Giacobini–Zinner, Comet ➤*Comet Giacobini–Zinner*.

giant branch The region on a ➤*Hertzsprung–Russell diagram* occupied by points corresponding to stars that have evolved into ➤*red giants*. ➤*stellar evolution*.

Giant Metrewave Radio Telescope (GMRT) A ➤*radio telescope* located near Poona in India. It consists of thirty 45-metre (146-foot) dishes arranged in an array extending for 25 kilometres (16 miles) and is the most powerful telescope for ➤*radio astronomy* at metre wavelengths.

giant molecular cloud ➤*molecular cloud*.

giant planet A term used to describe the planets Jupiter, Saturn, Uranus and Neptune to contrast them with the smaller, rocky 'terrestrial' planets.

giant star A member of a broad category of stars that have luminosities between 10 and 1,000 times that of the Sun, and radii typically between 10 and 100 times the Sun's.

A star becomes a giant when the hydrogen fuel available for nuclear fusion reactions in its core is depleted and the adjustment to the new energy balance causes the outer layers to expand greatly. The surface temperature drops, but the total luminosity rises because of the great increase in surface area. Examples are Capella, Aldebaran and Arcturus.

Massive hot stars, which are very large in comparison with the Sun even before they reach a late stage of evolution, are sometimes also referred to as giants. ➤*Hertzsprung–Russell diagram, red giant, stellar evolution*.

gibbous An adjective used to describe the ➤*phase* of illumination of a body shining by reflected sunlight, such as the Moon, when it is between half and full.

Ginga   A Japanese X-ray astronomy satellite launched on 5 February 1987, originally called Astro-C.

Giotto   A European Space Agency probe, which encountered ➤*Halley's Comet* in March 1986.

Up to two seconds before closest approach, a distance of 605 kilometres from the nucleus, all experiments worked perfectly. At that moment, a collision with a small particle caused the craft to wobble so that the communications antenna was not directed towards Earth. Communications were re-established after the encounter.

The multicolour camera returned images, including close-ups of the nucleus, until the moment when contact was lost; it was subsequently destroyed by impacts. The other instruments on board included a dust impact detector and an ion mass spectrometer, which observed that the gas in the coma is 80 per cent water (by mass). When the dust and gas are considered together, the composition was found to be (by mass) 45 per cent water, 28 per cent stony material and 27 per cent organic material.

ESA named the mission after the artist Giotto di Bondone, who is thought to have used the 1301 appearance of Halley's Comet as a model for the star of Bethlehem in his fresco, *The Adoration of the Magi*, painted in 1303 in the Scrovegni chapel in Padua.

In 1992, *Giotto* was successfully reactivated after two years' dormancy and seven years in space for an encounter with Comet 26P/Grigg–Skjellerup, a project known as the *Giotto Extended Mission* (GEM).

Giraffe   English name for the constellation ➤*Camelopardalis*.

glitch   A sudden change in the rotation rate of a ➤*pulsar*. These are particularly prominent in the ➤*Vela Pulsar* and the ➤*Crab Pulsar* but many others also show them. In the Vela pulsar the jumps amount to 200 nanoseconds, which is twenty times larger than the steady decrease in period. Glitches are thought to be caused by ➤*starquakes*.

Global Oscillation Network Group (GONG)   A project under the auspices of the ➤*National Solar Observatory* of the USA for the continuous monitoring of solar oscillations by a network of six observing stations spaced around the world at approximately 60° intervals in longitude.

The instrument used, called a Fourier tachometer, is based on the ➤*Michelson interferometer*. It measures small ➤*Doppler shifts* of the spectral line of nickel at 676.8 nanometres, which is isolated by means of a ➤*Lyot filter*. The whole disc of the Sun is focused at the same time on to an electronic detector composed of a matrix of 256 by 256 pixels. The Doppler shift at each pixel can be measured independently as a function of time. ➤*helioseismology*.

globular cluster   A roughly spherical, densely packed cluster of hundreds of thousands or even millions of stars. The brightest globular cluster in the sky is ►*Omega Centauri* (ω Cen), which has a diameter of 620 light years. It is also one of the oldest globular clusters known, believed to be 15 billion years old. The globular clusters in our Galaxy contain some of its oldest stars. They are distributed within a spherical halo around the Galaxy, in contrast with ►*open clusters*, which are found only in the disc, and seem to move in highly elliptical orbits around the centre of the Galaxy.

The stars in globular clusters contain low abundances of the elements heavier than helium. This is consistent with their having been formed from the original material of the Galaxy, before the composition of the interstellar medium had been enriched by elements made only inside stars.

Globular clusters have also been recognized in other galaxies.

globule   A small, almost spherical cloud of dark opaque gas and dust which shows up against a brighter background such as star clouds or a luminous nebula. It is thought that globules represent an early stage in the star formation process. The name of the Dutch-American astronomer, Bart Bok (1906–83), is associated with small globules, known as *Bok globules*, which may be only a few thousand astronomical units across.

GMAT   Abbreviation for ►*Greenwich Mean Astronomical Time*.

GMC   Abbreviation for giant ►*molecular cloud*.

GMRT   Abbreviation for ►*Giant Metrewave Radio Telescope*.

GMST   Abbreviation for ►*Greenwich Mean Solar Time*.

GMT   Abbreviation for ►*Greenwich Mean Time*.

gnomon   A rod or plate mounted vertically to form a shadow-stick, such as that used on a sundial. The altitude of the Sun may be calculated from the height of the rod and the length of the shadow. The direction of the shadow gives the ►*apparent solar time*.

Goat   English name for the constellation ►*Capricornus*.

Goddard Space Flight Center   A large NASA establishment, located at Greenbelt, Maryland, 16 kilometres (10 miles) north-east of Washington, DC. Established in 1959, its many thousands of staff are engaged in basic astronomical research and the design, development and management of near-Earth orbiting spacecraft. It is also responsible for the Wallops Flight Facility in Virginia, a special centre for sub-orbital flights and the NASA Balloon Program, and for the Goddard Institute for Space Studies in New York City.

Goldstone   The location in southern California of a 70-metre (230-foot) radio dish that was the first antenna of the NASA/JPL ►*Deep Space Network*, coming into operation in 1966. It is also used for radio astronomy work, particularly for ➤*very-long-baseline interferometry* in conjunction with other radio telescopes.

GONG Abbreviation for ➤*Global Oscillation Network Group.*

Gossamer Ring The outermost of Jupiter's three known rings. ➤*planetary rings*, Table 7.

Gould's Belt A formation of many of the brightest, most conspicuous stars in the sky, which appear to lie in a band tilted at 16° to the plane of the ➤*Milky Way.*

The belt includes the bright stars of Orion and Taurus in the northern hemisphere and of Lupus and Centaurus in the southern hemisphere. Its existence was first noted in 1847 by Sir John Herschel and later studied by B. A. Gould (1879).

Gould's Belt is thought to be a band-like structure of young stars in the form of a spur off the nearest spiral arm of the ➤*Galaxy.*

graben A channel or rille on the surface of a planet caused by vertical faulting.

GRANAT A Russian space observatory for gamma-ray and X-ray astronomy, launched in December 1989. The main instrument it carried was the French coded-aperture gamma-ray telescope SIGMA. It also carried Russian and Danish high-energy X-ray experiments.

Grand Unified Theory (GUT) An attempt to describe the strong and weak nuclear forces and electromagnetic force in a single unified theory. Partial success has been achieved by physicists such as Steven Weinberg, Abdus Salam, Sheldon Glashow and Howard Georgi. Weinberg and Salam united the weak nuclear and electromagnetic forces. ➤*Superstring theory* tries to unify gravity with these grand unified theories to produce a 'theory of everything' (TOE).

granulation A cellular pattern observed in high-resolution images of the Sun's ➤*photosphere*, caused by the convective movement of hot gases rising from hotter layers at greater depths. ➤*supergranulation.*

granule A bright convective cell up to 1,000 kilometres across in the solar photosphere. ➤*granulation.*

graticule Cross-wires, or a grid of fine lines, used for measuring positions and separations, particularly such markers set in the focal plane of a telescope eyepiece.

grating spectrograph A ➤*spectrograph* in which the light is dispersed into a spectrum by means of a ➤*diffraction grating.*

Graving Tool English name for the constellation ➤*Caelum.*

gravitation The attractive force that appears to act between all masses. According to the theory formulated by Isaac Newton, the force between two masses is proportional to their product divided by the square of the distance between them. In ➤*General Relativity*, gravitation is viewed as curvature in the geometry of ➤*spacetime*.

Gravitation, one of the four fundamental interactions in physics, finds particular application in astronomy because very large masses (stars and galaxies) are commonplace and because it is the only force that needs to be considered in models of the universe. ➤*cosmology*.

gravitational collapse The sudden collapse of a massive star when its central temperature falls and the internal pressure pushing outwards is no longer sufficient to balance the inward gravitational force. The gravitational collapse of a massive evolved star is very sudden and catastrophic, perhaps taking less than a second. The enormous energy released triggers a ➤*supernova* explosion, and the core of the collapsed star may become a ➤*neutron star*, ➤*pulsar* or ➤*black hole*.

gravitational instability The tendency for a small perturbation in the density or equilibrium of a system to result in further disruption of the initial state, owing to the attractive nature of the gravitational force. In gas clouds, for example, a slight compression that raises the gravitational force locally causes more matter to be sucked in, which, in turn, raises the local field further. Instabilities of this sort, perhaps started by compression in the spiral arms of galaxies, are probably the means by which star formation is initiated in giant ➤*molecular clouds*.

gravitational lens A massive object, such as a galaxy, which has the effect of distorting the appearance and/or magnifying the image of more distant objects that are in the same line of sight, or very close to it. The paths of the light rays from the distant light source are bent in the gravitational field of the object acting as a gravitational lens in a manner similar to the bending of light rays by refraction when they pass through a glass lens. A number of examples of the phenomenon are known, in both visible light and radio maps. These include double and multiple images of ➤*quasars* and clusters of galaxies in which the images of many members are distorted into concentric arcs. The natural magnification produced by a gravitational lens makes it possible to obtain detailed spectra of some remote objects that would otherwise have been too faint.

On a smaller scale, the phenomenon of ➤*microlensing* may be observed when a dark stellar-sized object crosses the line of sight to a more distant star. ➤*Einstein ring*, *General Relativity*.

gravitational radiation ➤*gravitational waves*.

gravitational redshift A reddening of the light from a massive object caused by photons losing energy as they travel away from a region of high gravity. For the Sun, the gravitational redshift is 0.000 002; for the surface of a ➤*neutron star* the theoretical value is about 1. Theorists have suggested that the large ➤*redshifts* of ➤*quasars* could be gravitational in origin rather than caused solely by the ➤*Doppler effect*.

gravitational waves Ripples in the structure of ➤*spacetime*, which may occur singly or as continuous radiation. They travel at the speed of light.

According to ➤*General Relativity* theory, massive objects subject to accelerations or changes in shape will emit gravitational waves. Gravitational radiation is emitted most intensely in regions of spacetime where gravity is so strong that General Relativity must be applied and where the velocities are close to that of light. In practice, this means that collapsing stellar cores or interactions of bulk matter with black holes are the most likely sources of gravitational radiation. Rotating neutron stars and binary star systems also emit gravitational radiation, and this mechanism of energy transfer plays an important role in the evolution of close binary systems.

Support for the existence of gravitational waves comes from observations of the only known ➤*binary star* believed to consist of two neutron stars (though pulsed radiation is seen from only one of them). A slight decrease in the orbital period is accounted for precisely if energy is being carried away by gravitational radiation at the rate predicted by General Relativity.

Several experiments have attempted to detect gravitational radiation, but none has yet succeeded. Major technical problems remain in achieving the degree of sensitivity required in a detector.

gravity ➤*gravitation*.

gravity assist The technique of using the gravitational field of a planet to change the speed and direction of a spacecraft without consuming any fuel.

grazing occultation An ➤*occultation* in which the occulted object appears just to touch the limb of the occulting body.

Great Attractor An aggregation of galaxies, containing perhaps 5×10^{16} solar masses of matter, roughly 150–350 million light years from our Galaxy in the direction of the constellations Hydra and Centaurus. Measurements of the velocities of galaxies, out to about 300 million light years, show deviations of up to 500 km/s from the pure ➤*Hubble flow*, and our Galaxy also moves relative to the ➤*cosmic background radiation*. The gravitational pull of the Great Attractor is thought to be partly responsible for these motions.

Great Bear English name for the constellation ➤*Ursa Major*.

great circle Any circle on the surface of a sphere that divides the sphere into two equal hemispheres.

Great Dark Spot   An oval feature on the planet ➤*Neptune*, discovered on the images returned by the ➤*Voyager 2* spacecraft in 1989. It was a storm system in Neptune's cloud layers, similar to the ➤*Great Red Spot* on Jupiter, but not so long lived: it had disappeared in 1994 when the next observations of Neptune with sufficient resolution were made by the Hubble Space Telescope. Its longest dimension was about the same as the Earth's diameter (approximately 12,000 kilometres or 7,500 miles), making it about half the size of the Great Red Spot.

Greater Dog   English name for the constellation ➤*Canis Major*.

Great Nebula in Orion   ➤*Orion Nebula*.

Great Red Spot   A large red oval feature on ➤*Jupiter*, 24,000 kilometres (15,000 miles) long and 11,000 kilometres (7,000 miles) wide. It was first reported by Robert Hooke in 1664 and has been observed ever since, during which time it has apparently varied to some extent in size and colour. The reason for its existence remains uncertain but it rotates like a giant ➤*anticyclone*, with a westerly wind on its northern edge and an easterly wind to the south.

Great Rift   An apparent rift in the ➤*Milky Way* where it passes through the constellations Cygnus and Aquila, due to the presence of large dark ➤*absorption nebulae*.

Green Bank   The location, in West Virginia, of a radio astronomy observatory forming part of the facilities of the ➤*National Radio Astronomy Observatory* of the USA. The 92-metre (300-foot) dish built in 1962 collapsed in 1988 and was a total loss. Its successor, the 100-metre (325-foot) Green Bank Telescope, is due for completion in 1998. It is the largest fully steerable dish in the world. An unusual off-axis feed arm means that the dish is not obstructed. The 43-metre (140-foot) dish at Green Bank, completed in 1965, is the largest equatorially mounted telescope in the world. There is also a radio interferometer, consisting of three 26-metre (85-foot) dishes, two of which can be moved along a track 1.6 km (1 mile) long.

Green Bank Telescope   ➤*Green Bank*.

green flash   A phenomenon that may be observed at the moment of sunset over a clear horizon, especially over the sea. Refraction by the Earth's atmosphere makes the last fragment of the Sun to sink below the horizon appear to break free and flash momentarily green before disappearing. The phenomenon is widely documented and has been recorded photographically.

greenhouse effect   The internal heating effect in a planetary atmosphere, caused by its opacity to infrared radiation. The name arises because the mechanism is essentially the way a greenhouse works, with glass playing the same role as the atmosphere.

The primary source of heat for a planet's surface and atmosphere is energy radiated from the Sun in the visible and infrared regions of the spectrum. Longer-wavelength infrared radiation emitted by the warm planetary surface becomes trapped in the atmosphere, causing the equilibrium temperature of the atmosphere (and hence the planetary surface) to be higher than it would otherwise be. On the Earth, the increase in temperature amounts to about 33 K. On Venus, a 'runaway' greenhouse effect raises the temperature by 500 K. Mars is warmed by a modest 5 K.

How much heating the greenhouse effect causes depends on how opaque the atmosphere is to infrared radiation. Carbon dioxide is one of the main sources of the opacity, but water vapour and rarer gases also play a part.

Concern has mounted that global warming of the Earth will result from increased concentrations of carbon dioxide and other so-called *greenhouse gases* released by human activities, particularly the burning of fossil fuels such as coal and oil.

Greenwich Mean Astronomical Time (GMAT)   A time system formerly used for astronomical purposes, which was ➤*Greenwich Mean Time* with the day changing at noon rather than midnight. GMAT is converted to ➤*Universal Time* by the addition of 12 hours.

Greenwich Mean Solar Time (GMST)   ➤*Mean solar time* on the Greenwich meridian.

Greenwich Mean Time (GMT)   ➤*Mean solar time* on the Greenwich meridian.

Greenwich Observatory   ➤*Royal Greenwich Observatory*.

Greenwich sidereal date   The number of ➤*sidereal days* that have elapsed at Greenwich since the beginning of the Greenwich sidereal day that was in progress at ➤*Julian date* o.o. The whole-number part of the date constitutes the Greenwich sidereal day number; the fractional part is the Greenwich Sidereal Time.

Greenwich Sidereal Time (GST)   The ➤*sidereal time* on the Greenwich meridian. ➤*Greenwich sidereal date*.

Gregorian calendar   The civil calendar now in use in most countries, introduced by Pope Gregory XIII in 1582 to replace the ➤*Julian calendar*.

A workable civil calendar needs to be organized such that the seasons remain in step with the months of the year. This is a problem because the time taken by the Earth to orbit the Sun is not a whole number of days. The introduction of an extra day each fourth or leap year makes a first-order correction, but further adjustments are necessary if the calendar is to stay synchronized with the seasons over centuries.

In the Gregorian system, years exactly divisible by four are leap years, except that century years must be exactly divisible by 400 to be leap years. Thus 2000 is a leap year, but 1900 and 2100 are not. Averaged over 400 years, the rule gives an average year length of 365.2425, which is close to the true length of the ►*tropical year,* 365.2422 days.

The Gregorian calendar came into effect in Roman Catholic countries in October 1582, when the seasons were brought back into step by eliminating 10 days from the calendar. Thursday 4 October was followed by Friday 15 October. Also, on the introduction of the Gregorian system, the new year began on 1 January for the first time, instead of 25 March. Britain and its colonies did not introduce the Gregorian calendar until September 1752, by which time an 11-day correction was needed.

Gregorian telescope   A type of reflecting telescope proposed by James Gregory in 1663. The primary mirror is a paraboloid with a central hole and the secondary is ellipsoidal. In practice, Gregory was unable to obtain mirrors figured accurately enough to construct his telescope before Newton made the first working reflector to a simpler design with a flat secondary. Subsequently, the Gregorian design was rendered obsolete by the ►*Cassegrain telescope.* ►*Newtonian telescope.*

Griffith Observatory   A public observatory owned and administered by the city of Los Angeles for educational purposes. It was established in accordance with the will of Colonel Griffith J. Griffith, and formally dedicated in 1935. The facilities include a planetarium, a science exhibition and a 30-centimetre (12-inch) telescope for free public viewing.

Grimaldi   A large lunar crater, 222 kilometres (138 miles) in diameter, situated near the western limb of the Moon on the border of the Oceanus Procellarum.

GRO   Abbreviation for Gamma Ray Observatory. ►*Compton Gamma Ray Observatory.*

Groombridge Catalogue   A catalogue of the positions of 4,243 stars in the region of the north celestial pole, by S. Groombridge, published in London in 1838.

Grus (The Crane)   A small southern constellation, introduced probably by sixteenth-century navigators and included by Johann Bayer in his atlas ►*Uranometria,* published in 1603. It contains four stars brighter than fourth magnitude. Delta Gruis (δ Gru) is a double star that can be 'split' by the unaided eye. ►Table 4.

GST   Abbreviation for ►*Greenwich Sidereal Time.*

G star   A star of ►*spectral type* G. G stars on the ►*main sequence* have temperatures in the range 4,900–6,000 K and are yellow in colour. Many absorption lines of neutral and ionized metals feature in the spectrum, and there are some molecular bands. The Sun is a typical dwarf G star; ➤*Capella* is an example of a giant G star.

G-type asteroid   A subclass of the ➤*C-type asteroids*, distinguished by strong ultraviolet absorption in the spectrum.

Guardians   The two stars β (Beta) and γ (Gamma) in the constellation ➤*Ursa Minor*.

guest star   An expression used by Chinese astronomers in historical times to denote the appearance of a ➤*nova*, ➤*supernova* or ➤*comet*.

guide star   A star on which the manual or automatic guidance system of a telescope can be locked in order to ensure that some fainter object being observed is correctly followed as the Earth rotates.

Guide Star Catalog   A catalogue of 18,819,291 celestial objects created as the database for operations of the ➤*Hubble Space Telescope*. Including 15 million stars and over three million galaxies, this is the largest catalogue of celestial objects ever produced.

guide telescope   A telescope on the same mounting as one being used for photography or with instruments, provided for the purpose of guiding the telescope accurately during the course of an observation.

Gum Nebula   A large, circular ➤*emission nebula* in the southern constellations Vela and Puppis, discovered by an Australian astronomer, Colin Gum. The nebula is 30° across, which is equivalent to a diameter of 800 light years at its distance of 1,300 light years. It is thought to be the result of ionization of the interstellar medium caused by a ➤*supernova* that exploded perhaps a million years ago. This would mean that there is no longer a source of energy, and the nebula is now gradually fading as ionized hydrogen recombines and ceases to be luminous.

GUT   Abbreviation for ➤*Grand Unified Theory*.

H

H I region (H^0 region)   An interstellar cloud of ►*neutral hydrogen*.

H II region (H^+ region)   An interstellar cloud of ►*ionized hydrogen*.

HA   Abbreviation for ►*hour angle*.

Hadar (Beta Centauri; β Cen; Agena)   The second-brightest star in the constellation Centaurus. It is a giant ►*B star* of magnitude 0.6. ►Table 3.

Hadley Rille   A sinuous channel on the Moon, running across Palus Putredinis. It is close to the landing site of the Apollo 15 mission and is believed to be a collapsed lava tube. ►*Apollo programme, rille*.

hadron era   The period from 10^{-35} to 10^{-6} seconds after the primordial ► *Big Bang*. During this interval, the behaviour of the universe was dominated by the strong interaction, the force governing the interactions between unstable elementary particles known as hadrons, which are more massive than protons and neutrons.

Hakucho   Name given to the satellite ►*Corsa-B*.

halation   The formation of an unwanted halo of scattered light, for example in a defective optical system or on a photographic film, by internal reflection of light from a bright object.

HALCA   A Japanese radio astronomy satellite launched on 11 February 1997. It was the first satellite for mapping celestial radio sources. Before launch, HALCA was known as *MUSES-B*. ►*very-long-baseline interferometry*.

Hale Observatories   A name used between 1970 and 1980 for a group of observatories comprising the ►*Palomar Observatory*, the ►*Mount Wilson Observatory*, the ►*Big Bear Solar Observatory* and the ►*Las Campanas Observatory*.

Hale Telescope   The 5-metre (200-inch) reflecting telescope at the ►*Palomar Observatory*. Work on the telescope began in the 1930s following the award of a grant to the California Institute of Technology from the Rockefeller Foundation. Completion was delayed by World War II. It was officially opened in 1948 and dedicated to the memory of George Ellery Hale (1868–1938), who had been the driving force behind the initiation of the project.

Halley's Comet (Comet 1P/Halley)   The most famous of all periodic comets.

It travels in an elongated elliptical orbit around the Sun, returning to the Earth's vicinity in the inner solar system every 76 years. From historical records it has been shown that Halley's Comet has been observed for over 2,200 years.

Edmond Halley (1656–1742), in whose honour the comet is named, was not the comet's discoverer but the first person to realize the connection between the comet he saw in 1682 and certain other recorded appearances of comets separated by intervals of 76 years. He calculated the orbits of a number of comets, using Isaac Newton's newly published theory of gravitation. Noticing the similarity between the orbits of the comets seen in 1531, 1607 and 1682, he went on to predict a return in 1758–9, which was duly observed after his death.

The orbit of Halley's Comet has a perihelion distance of 0.59 AU, between the orbits of Mercury and Venus. At its most distant, it travels beyond the orbit of Neptune. The orbit is inclined to the main plane of the solar system at an angle of 162° and the comet travels around its orbit in the direction opposite to the motion of the planets.

The return of 1986 was very unfavourable for observation from Earth but several countries launched space probes to investigate the comet, with considerable success. The closest approach was made by the European ➤*Giotto* probe, which passed within 605 kilometres (375 miles) of the nucleus on 14 March 1986. The Soviet probes *Vega 1* and *Vega 2* surveyed the nucleus from distances of 8,890 and 8,030 kilometres (5,550 and 5,020 miles) on 6 and 9 March 1986, and information they gathered was used for last-minute corrections to the course of *Giotto*. Two small Japanese probes were also deployed.

The results demonstrated conclusively the existence of a solid nucleus, probably made of ice and dust. It has an irregular elongated shape, reminiscent of a potato, measuring 16 × 8 kilometres (9 × 5 miles). It is also dark, reflecting only 4 per cent of incident sunlight. The nucleus rotates slowly – once in 7.1 days, with a 3.7-day precession. On the side facing the Sun, temperatures as high as 350 K were measured, enough to melt ice, and escaping jets of material were observed.

Two meteor showers, the Eta ➤*Aquarids* and the ➤*Orionids*, are associated with Halley's Comet.

halo Any roughly circular or spherical distribution of light or matter around another object. ➤*galactic halo*.

halo star A star belonging to the population forming the ➤*galactic halo*.

H alpha ➤*hydrogen alpha*.

Hamal (Alpha Arietis; α Ari) The brightest star in the constellation Aries. It is a giant ➤*K star* of magnitude 2.0. The name is derived from the Arabic for 'sheep'. ➤Table 3.

h and χ (chi) Persei (Double Cluster in Perseus; NGC 869 and 884)   A pair of open star clusters in the constellation Perseus. They are visible to the naked eye as faint hazy patches. Their names, of a type used for individual stars, were given before their nature was known. The two clusters are very similar in appearance and are less than one degree apart in the sky. They are 7,100 light years away and estimated to be only 50 light years apart. ➤*open cluster*.

H and K lines   The strongest lines in the visible spectrum of ionized calcium, lying in the violet at wavelengths of 393.4 and 396.8 nanometres. They are conspicuous features in the spectra of many stars, including the Sun. The designations H and K were given by Fraunhofer and are still commonly used. ➤*Fraunhofer lines*.

Hare   English name for the constellation ➤*Lepus*.

Harmonice mundi (*The Harmony of the World*)   A book by Johannes Kepler (1571–1630) published in 1619. It included the third of ➤*Kepler's laws* of planetary motion. Kepler himself believed his greatest achievement was the discovery of musical harmonies for the planets reflecting the divine harmony of the universe, and named the work accordingly.

Haro galaxy   A member of a class of ➤*galaxies* characterized by their blue colour and narrow emission line spectra.

Hartmann test   A test for the optical quality of a mirror. The mirror is covered with a screen in which there is a pattern of regularly spaced holes. The quality of the mirror can be judged from the image produced in the focal plane.

Harvard classification   ➤*Henry Draper Catalogue*.

Harvard College Observatory   The observatory of Harvard College, established in 1839. In 1847 it was equipped with a 0.38-metre (15-inch) refracting telescope, which is still in its original building, the Sears Tower on Observatory Hill. In 1973, the Harvard–Smithsonian Center for Astrophysics was formed by combining the resources of the ➤*Smithsonian Astrophysical Observatory* and Harvard College Observatory under one director, George Field.

Harvard–Smithsonian Center for Astrophysics   ➤*Harvard College Observatory*.

harvest Moon   The full Moon nearest the time of the autumnal ➤*equinox*. At this time of the year the inclination of the Moon's orbit to the horizon is low and the Moon rises at approximately the same time each evening for a short period.

Hat Creek Observatory   A radio astronomy observatory in California operated by the radio astronomy laboratory of the University of California, Berkeley.

Haute Provence Observatory   An observatory in southern France, 100 km (62 miles) north of Marseilles, at an altitude of 650 metres (2,100 feet) in the foothills of the Alps. It was established in 1937 as a national facility for French astronomers and is owned by the Centre National de la Recherche Scientifique (CNRS). The main instrument is a 1.93-metre (76-inch) reflector which has been operated since 1958. A 1.52-metre (60-inch) telescope, operating since 1967, is equipped only with a ➤*coudé focus* and a modern ➤*spectrograph* introduced in 1989. There are also 1.2-metre (47-inch) and 80-centimetre (31-inch) telescopes, and a 60/90-centimetre (24/36-inch) ➤*Schmidt camera.*

Hawking effect   The evaporation of mini ➤*black holes*. In general, black holes are an end-point for matter. However, Stephen Hawking showed that quantum physics allows mini black holes (less than 10^{-5} grams) to evaporate matter and antimatter in equal quantities, possibly leaving a naked singularity behind.

Hayashi track   The evolutionary track on the ➤*Hertzsprung–Russell diagram* of a ➤*protostar*, before it becomes a main-sequence star. The track shows the changes in luminosity and surface temperature that occur in the early phases of a star's life. The exact position of the track on the diagram depends on the star's mass.

Hayashi tracks are named after the Japanese astrophysicist, Chushiro Hayashi, who pioneered the theoretical work in the 1960s.

Haystack Radio Telescope   A 37-metre (120-foot) radio dish located at the Haystack Observatory in Massachusetts, north-west of Boston. It is a facility of the Massachusetts Institute of Technology, and the astronomy research is conducted under the auspices of a consortium of thirteen educational establishments. It is equipped for use in the wavelength range 2.6 mm–13 cm, the surface is accurate to half a millimetre. The antenna is fully steerable and of a ➤*Cassegrain* configuration. Among its discoveries are a number of ➤*interstellar molecules*. It is also used for ➤*very-long-baseline interferometry*. When first built in the 1960s, it was used mainly for radar studies of the Moon and nearer planets.

HD   A prefix used for star catalogue numbers in the ➤*Henry Draper Catalogue*.

head   The nucleus and coma of a ➤*comet*, excluding the tail.

head–tail galaxy   A ➤*radio galaxy* with radio emission streaming away to one side of the corresponding optical galaxy, giving a shape resembling a tadpole.

HEAO   Abbreviation for ➤*High Energy Astrophysical Observatory*.

Heavenly Twins   English name for the constellation ➤*Gemini*.

Hebe   Asteroid 6, diameter 204 km, discovered in 1847 by K. L. Hencke.

Heinrich Hertz Submillimeter Telescope   A 10-metre (33-foot) telescope at the ➤*Mount Graham International Observatory* operating in the submillimetre waveband between 0.3 and 1 mm. First observations were made in 1994. It is operated jointly by the University of Arizona and the Max-Planck-Institut für Radioastronomie in Bonn, Germany.

Hektor   Asteroid 624, the largest of the ➤*Trojan asteroids*, discovered by A. Kopff in 1907. Its brightness varies by a factor of three as it rotates in a period of just under 7 hours. Measurements of the variation indicate that Hektor is cylindrical in shape, 150 km wide by 300 km long. It has been suggested that Hektor may in fact be two asteroids in contact or a close binary.

Helene   A small satellite of Saturn discovered in 1980. It is an irregular object, measuring 36 × 30 kilometres. ➤Table 6.

heliacal rising   The rising of a bright star just before the Sun. In practice, the date of heliacal rising would be when the star was first observable in the eastern dawn sky. The heliacal rising of Sirius was used by the ancient Egyptians as a herald of the flooding of the River Nile.

heliocentric model   A model of the solar system that places the Sun at the centre with the planets in orbit around it. Although such a system had been suggested as early as *c.* 200 BC by Aristarchus of Samos, it offered no particular advantage at the time for predicting the positions of the planets and the idea of a moving Earth was philosophically unacceptable. The ➤*geocentric model*, refined by Ptolemy (*c.* AD 100−170), was in general use until the work of Copernicus (1473−1543). By this time, the idea that the Earth was the centre of the created universe was strongly rooted in religious dogma.

In his book *De revolutionibus,* Copernicus argued the advantages of considering the solar system as Sun-centred. However, the idea did not gain general acceptance until the observational work of Galileo (1564−1642) and Kepler (1571−1630), whose results made better sense in the context of a heliocentric system.

In the Copernican system, the planetary orbits were assumed to be circular. This meant that the theory was no more successful on a practical level than the Ptolemaic theory in predicting planetary positions, though it was more elegant and provided a natural explanation for the ➤*retrograde* motion of the planets. Kepler's discovery that the planetary orbits are elliptical resolved this problem and the first telescopic observations by Galileo revealed phenomena, such as the phases of Venus, that could be explained only on the basis of a heliocentric model.

heliocentric parallax   ➤*annual parallax*.

heliographic latitude   Angular distance on the Sun's surface north (positive) or south (negative) of the Sun's equator. The solar equator, which intersects the ecliptic at an angle of 7° 15′, changes its apparent location on the solar disc as the Earth orbits the Sun. Tables are published in handbooks for observers giving the heliographic latitude of the centre of the Sun's disc, from which other latitudes may be calculated.

heliographic longitude   Longitude measured for points on the Sun. There is no fixed zero point on the Sun so heliographic longitude is measured from a nominal reference great circle: the solar meridian that passed through the ascending node of the solar equator on the ecliptic on 1 January 1854 at 12.00 UT. Relative to this, longitude is calculated by assuming a uniform sidereal rotation rate of 25.38 days. Tables published in observing handbooks are used by solar observers to determine the position of the solar reference meridian for a given date and time.

heliometer   An obsolete form of refracting telescope in which the objective lens was cut into two parts which could be moved relative to each other. Heliometers were formerly used to measure small angular separations.

heliopause   ➤*heliosphere.*

helioseismology   The study of the interior of the Sun by the analysis of its natural modes of oscillation, which are observed spectroscopically as ➤*Doppler shifts* in the ➤*absorption line spectrum.* ➤*Global Oscillation Network Group.*

heliosphere   The spherical volume of space extending to between 50 and 100 AU from the Sun, bounded by the zone where the ➤*solar wind* merges with the ➤*interstellar medium.* This boundary region is called the *heliopause.*

heliostat   A movable flat mirror used to reflect sunlight into a fixed solar telescope. Solar telescopes are large structures with long focal lengths but they need to point only at a small part of the sky. The heliostat is controlled so as to move synchronously with the motion of the Sun across the sky. It is a simple device but produces an image that slowly rotates during the course of a day, so the more complex ➤*coelostat* is sometimes preferred.

helium flash   An explosive event in the interior of a low-mass star (less than two solar masses) marking the start of helium burning when all the hydrogen available for nuclear fusion in the core has been exhausted. ➤*stellar evolution.*

helium problem   The problem of explaining why the observed abundance of helium in the universe is about 25 per cent by mass. This is part of the much larger problem of accounting for the observed distribution of all the elements heavier than hydrogen. The amount of helium is much too large for it all to have been synthesized in stars, although that is where all of the heavier elements were probably made. The problem was solved by Gamow in 1946, who proposed a hot ➤*Big Bang* model of the universe in which helium nuclei are manufactured at the start of the ➤*radiation era.*

helium star   A ►*B star* in the spectrum of which the helium lines are abnormally strong.

Helix Galaxy   A popular name for the ►*spiral galaxy* NGC 2685 in the constellation Leo.

Helix Nebula (NGC 7293)   A large, ring-shaped ►*planetary nebula* in the constellation Aquarius. Its apparent diameter is a quarter of a degree (half the size of the full Moon) and, at a distance of about 500 light years, it is the nearest planetary nebula.

Hellas Planitia   An almost circular impact basin on the surface of Mars. Hellas Planitia is 1,800 kilometres (1,100 miles) in diameter and is a feature that has long been recognized and mapped, being conspicuous by its colour, which is lighter than surrounding areas. It was formerly known simply as 'Hellas'.

hemispherical albedo   The fraction of incident light scattered by a surface as a function of the angle of incidence. ►*albedo*.

Henry Draper Catalogue (HD Catalogue)   A catalogue of stellar spectra, compiled at Harvard College Observatory. The work was made possible through funds donated by the widow of the pioneering astrophysicist Henry Draper (1837–82), and the catalogue was named as a memorial to him. Under the direction of Edward C. Pickering (1846–1919), Annie Jump Cannon (1863–1941) classified most of the 225,300 stars in the nine-volume catalogue between 1911 and 1915, though the first volume was not ready for publication until 1918 and the ninth did not appear until 1924.

The Harvard spectral classification system adopted the sequence of classes still in use today – O, B, A, F, G, K and M. This apparently random order resulted from earlier work in which classes were originally named in alphabetical order. ►*spectral type*.

Herbig–Haro object (HH object)   One of a number of peculiar nebulous objects associated with newly forming stars. The first three were discovered on images of the nebula NGC 1999 in Orion in 1946/7 by the American astronomer George Herbig and the Mexican Guillamero Haro. Many more similar objects have since been identified.

Herbig–Haro objects are thought to result from the interaction between a strong ►*bipolar outflow* from a protostar and the interstellar gas, which is heated and compressed. They are typically between 500 and 4,000 AU in size and have masses in the range 0.5–30 Earth masses, making them among the least massive objects to have been detected outside the solar system. Their velocities are high and, in many cases, their motion can be traced back to ►*T Tauri stars* or infrared-emitting objects, suggesting that they have been ejected by young stars.

►*Hubble's Variable Nebula* is an example of an HH object.

Hercules   A large constellation of the northern sky, included by Ptolemy in his list of 48 (*c.* AD 140). It is named after the hero of classical mythology. There are no first magnitude stars. The brightest ►*globular cluster* in the northern hemisphere, M13, lies in Hercules. ►Table 4.

Hercules A   The strongest radio source in the constellation Hercules, associated with an elliptical galaxy. Two long jets extend for half a million light years into space from the faint nucleus.

Hercules X-1   An X-ray ►*pulsar* in the constellation Hercules, consisting of a rotating neutron star which is accreting matter from its companion in a binary system. The rotation period of the neutron star is 1.2 seconds and the orbital period of the system 1.7 days.

Herdsman   English name for the constellation ►*Boötes*.

Hermes   Asteroid 1937 UB, discovered by K. Reinmuth in 1937 when it passed within 800,000 km of the Earth, in what was then the closest approach of an asteroid ever recorded. It reached eighth magnitude and travelled across the sky at a rate of 5° per hour. It was observed for a only a few days and was subsequently lost.

Herschel   The largest impact crater on ►*Mimas*. Its diameter is 130 kilometres (80 miles), one-third the diameter of Mimas.

Herschelian telescope   A type of ►*reflecting telescope* designed by William Herschel (1738–1822), in which the paraboloid primary mirror is tilted so that its focus lies outside the main tube of the telescope and can be accessed without obstructing the incoming light. The system has the disadvantage that distortions are introduced, and it was soon superseded by other types of reflector.

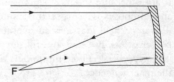

Herschelian telescope. The tilted primary mirror brings light to a focus at F, to one side of the telescope tube.

Herschel Telescope   ►*William Herschel Telescope*.

Herschel wedge   A device for safely reducing the intensity of light transmitted through a telescope during observation of the Sun. It consists of a thin prism with unsilvered faces. The first face is inclined at 45° to the optical axis and reflects 5 per cent of the incident radiation. A baffle is used to intercept the reflected beam from the second surface. The remaining radiation, which passes through, should be safely absorbed in a heat trap. An infrared rejection filter should also be used in conjunction with a Herschel wedge.

Hertzsprung gap   A region on the ►*Hertzsprung–Russell diagram*, between

173

the giant branch and the main sequence, where very few stars are represented. The gap reflects a general absence of giant stars of ►*spectral types* F and G, because the stage in the evolution of a star when it is observable as a giant of this type is very short. Thus at any one time, relatively few such stars exist.

Hertzsprung—Russell diagram   (HR diagram) A graph displaying, for any sample of stars, the relationship between their spectral type and luminosity (see illustration).

Colour, temperature or some other comparable quantity may be substituted for spectral type as the quantity plotted along the horizontal axis. Temperature conventionally decreases towards the right. Either ►*magnitude* or luminosity relative to the Sun are frequently used for the vertical scale. The resulting graph may also be called a *colour—magnitude diagram* or *colour—luminosity diagram* according to the actual quantities used.

What is now known as a Hertzsprung—Russell diagram was first plotted by Henry Norris Russell in 1913. It was later recognized that Ejnar Hertzsprung had independently put forward similar ideas at around the same time.

Any star whose spectral type and luminosity are known may be plotted as a single point on the HR diagram, but the diagram acquires particular significance when plotted for a related group of stars, such as a star cluster. For any sample of stars, the points are not distributed randomly: most lie on a band running diagonally from the upper left to the lower right, the so-called *main sequence*. It arises because the most significant factor determining a star's spectral type and luminosity is its mass – the main sequence is in effect a mass sequence. The idea once held that it is an evolutionary sequence is known to be wrong. Nevertheless, hot stars are still often called 'early-type stars' and cooler ones 'late-type stars'. These misnomers are a legacy from the misinterpretation of the main sequence.

The effects of evolution in fact move stars away from the main sequence, which represents stars burning hydrogen in thermonuclear reactions in their cores. When the central hydrogen is exhausted, a sequence of internal changes leads to a great expansion of the star, coupled with a decrease in surface temperature. Such evolved stars are found in the giant and supergiant branches lying above the main sequence. The highly evolved ►*white dwarfs* form a group well below the main sequence.

The HR diagram for a star cluster immediately makes clear how many stars there are at each stage of evolution. This, coupled with the theoretical knowledge of how evolution rate increases with stellar mass, gives an important key to the ages of clusters. Plotting apparent magnitude on the vertical axis rather than absolute magnitude for a cluster provides a method of measuring a cluster's distance.

HR diagrams are also useful for displaying the sequence of changes in colour and luminosity that take place in an individual star in the course of its

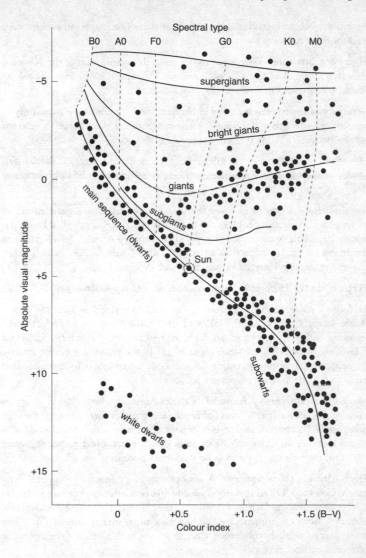

Hertzsprung–Russell diagram. A schematic diagram showing the main sequence and areas occupied by stars at different stages of evolution. Absolute magnitude is plotted against colour index in this example. The dashed lines link stars of different luminosities with the same spectral types but different colour indices.

evolution – before, on and after the main sequence. The result is an evolutionary track. ➤*stellar evolution*.

Herzberg Institute   The astrophysics research organization of the National Research Council of Canada. ➤*Dominion Astrophysical Observatory, Dominion Radio Astrophysical Observatory*.

heterodyne spectrometer   An instrument used in microwave astronomy to measure the strength of the ➤*cosmic background radiation* by switching a detector rapidly between a stable reference source and the sky.

Hevelius   A lunar crater, 118 kilometres (73 miles) in diameter, on the western border of the Oceanus Procellarum. There is a system of clefts on the crater floor.

hexahedrite   A type of iron ➤*meteorite* containing less than 6 per cent nickel by weight. Hexahedrites contain the form of iron–nickel alloy called kamacite, which has cubic symmetry. The polished surfaces of such meteorites are featureless except for numerous striations, called *Neumann lines*, which occur in some examples. They are caused by shock deformation.

HH object (H–H object)   Abbreviation for ➤*Herbig–Haro object*.

Hidalgo   Asteroid 944, diameter 40–60 km, discovered in 1920 by Walter Baade. Its highly elliptical orbit takes it between the main ➤*asteroid belt* at 2 AU from the Sun to a distance of 9.7 AU, beyond the orbit of Saturn. The orbit is inclined at the relatively steep angle of 42° to the plane of the solar system. Its unique orbit among asteroids has led some astronomers to speculate that Hidalgo may be a 'dead' comet nucleus.

hierarchical universe   A model of the universe in which clumping occurs on all scales, like a fractal: stars form galaxies, galaxies form clusters, these in turn form superclusters, and so on. Present observations indicate that galaxies clump together on scales up to 150 million light years. Over distances greater than this, the universe appears to be uniformly populated on average.

High Altitude Observatory   A solar physics observatory and research institute in Colorado, USA. It was founded in 1940 under the auspices of ➤*Harvard College Observatory* and is now a division of the National Center for Atmospheric Research. Solar experiments are also located at other sites and on satellites. The research is particularly concerned with variations in the Sun and their possible terrestrial consequences.

High Energy Astrophysical Observatory   Name of three NASA orbiting observatories launched between 1977 and 1979, two of which were for X-ray astronomy and the third for gamma-ray astronomy.

HEAO-1, launched in September 1977, conducted an all-sky survey for

X-ray sources which resulted in the most complete catalogue up to that time. HEAO-2 became known as the ►*Einstein Observatory*. HEAO-3 was launched in September 1979 and carried a gamma-ray spectrometer and two cosmic ray instruments.

highlands, lunar   A general term applied to the regions on the lunar surface other than the ►*mare* areas. The highlands are characterized by a high crater density and a lighter colour in comparison with the maria.

high-velocity cloud   A cloud of ►*neutral hydrogen* in our Galaxy moving at a speed (up to 460 km/s) higher than would be expected from the rate of rotation of the Galaxy alone (220 km/s). High-velocity clouds occur mainly in the northern hemisphere, and it is suggested that they could be clouds of intergalactic hydrogen falling towards the Milky Way.

high-velocity star   A star travelling at an exceptionally high velocity (i.e. more than about 65 km/s) relative to the Sun. High-velocity stars are very old stars that do not share the motion of the Sun and most stars in the solar neighbourhood in circular orbits around the centre of the ►*Galaxy*. Rather, they travel in elliptical orbits, which often take them well outside the plane of the Galaxy. Although their orbital velocities in the Galaxy may be no faster than the Sun's, their different paths result in the high relative velocities.

Hilda asteroids   A group of asteroids at the outer edge of the main ►*asteroid belt*, at a distance of 4.0 AU from the Sun. They are named after asteroid 153 Hilda, diameter 180 km (110 miles), discovered by J. Palisa in 1875. Their orbital periods are commensurable with that of Jupiter in the ratio 3:2, and they are separated from the rest of the asteroid belt by a ►*Kirkwood gap*. ►*commensurability*.

Hill sphere   The approximately spherical region within which a planet, rather than the Sun, dominates the motion of particles through its stronger gravitational field.

Himalia   A satellite of Jupiter (number VI), 180 kilometres in diameter, discovered in 1904 by Charles Perrine. It belongs to a group of four satellites whose closely spaced orbits all lie between 11.1 and 11.7 million kilometres from Jupiter. (The others are Leda, Lysithea and Elara.) ►Table 6.

Hind's Nebula (NGC 1554/5)   A variable ►*reflection nebula* surrounding the star T Tauri. ►*T Tauri star*.

Hipparcos   A European Space Agency satellite designed to carry out astrometric surveys to an unprecedented degree of accuracy. The launch in 1989 was only partially successful, in that the satellite went into a highly elliptical orbit instead of the intended geostationary orbit. However, it was still possible for the scientific objectives to be met. The cumulative effect of radiation damage forced the termination of observations on 15 August 1993.

The name 'Hipparcos' is an acronym for High Precision Parallax Collecting Satellite, chosen also for its similarity to the name of the Greek astronomer, Hipparchus (also spelt Hipparchos), who measured the Moon's parallax and made an accurate star map, which led to the discovery of the ➤*precession* of the equinoxes.

The basic instrument was an all-reflective Schmidt telescope with a 0.29-metre (11.5-inch) primary mirror and the programme required the advance compilation of an *Input Catalogue*, including specially made ground-based observations.

The satellite observations resulted in the Hipparcos catalogue, containing the positions, parallaxes and proper motions of 118,000 stars with an accuracy of 2 milliarcseconds, together with the Tycho catalogue which lists these data for over a million stars with less precision but with systematic errors of only about one milliarcsecond. Hipparcos more than doubled the number of known variable stars and discovered many thousands of new double and multiple star systems.

Hirayama family   A concentration of asteroids having similar orbits and thus located near each other in space. The existence of such groupings was first noted by Kiyotsugo Hirayama in 1918. More than a hundred have been identified. In many cases the members are of similar or related types, strongly suggesting that they were formed from the break-up of one parent body; some well-known examples are the Eos, Koronis and Themis families. About half of all asteroids are thought to belong to Hirayama families.

Hoba meteorite   The largest known ➤*meteorite* in the world. It is of the iron type and weighs about 55,000 kilograms. It is still at its impact site in Namibia, where it was discovered in 1928. A layer of rusty weathered material surrounds the meteorite; when this weathering is accounted for, the original mass must have exceeded 73,000 kilograms.

Hobby*Eberly Telescope (HET)   A large telescope at the ➤*McDonald Observatory* in Texas, designed specifically for spectroscopy. A partnership between the University of Texas at Austin and several other universities in the USA and Germany, it began full operation in 1997. It has an 11-metre (36-foot) segmented mirror permanently tipped at a 35-degree angle to the zenith, mounted on a structure that can rotate in azimuth to point in any direction. The telescope tracks its targets with the aid of a movable secondary mirror. Though the tilt of the main mirror is fixed, the telescope will nevertheless be able to observe objects in about 70 per cent of the sky accessible from the site.

homogeneity   The property of being the same everywhere in space. Cosmologists assume that the universe is homogeneous on the largest scales.

Homunculus Nebula   A small nebula surrounding the star ➤*Eta Carinae*.

Hooker Telescope   The 2.54-metre (100-inch) reflecting telescope at the ➤*Mount Wilson Observatory*, near Pasadena in California. It was completed in 1917, having been financed by a gift from John D. Hooker. Until the opening of the 5-metre (200-inch) ➤*Hale Telescope* in 1948, it was the largest telescope in the world. It was temporarily closed in 1985 but was subsequently renovated and brought back into use in the early 1990s.

horizon   The great circle defined by points 90° away from a ground-based observer's zenith, which marks the boundary between the visible and invisible halves of the celestial sphere for that observer.

The term 'horizon' is also used for any boundary in spacetime between events that can in principle be observed and those that cannot. ➤*event horizon*, *particle horizon*.

horizontal branch   In the ➤*Hertzsprung–Russell diagram* for a ➤*globular cluster*, a short horizontal strip to the left of the giant branch. It represents low-mass stars that have lost mass during the giant phase of their evolution.

horizontal coordinates   A coordinate system in which points on the celestial sphere are identified by the two coordinates altitude and azimuth. Altitude is the angular distance above the horizon, and azimuth the angular distance, measured eastwards, along the horizon from the north point. The altitude and azimuth of a celestial object vary according to the latitude and longitude of the observer and the time of observation.

horizontal parallax   ➤*diurnal parallax*.

horn antenna   A radio telescope in the shape of a horn, specifically designed to collect microwaves and to observe the ➤*cosmic background radiation*. Its design ensures that very little radiation is picked up from outside the main beam of the antenna, so that the noise level is very low.

Horologium (The Clock)   A faint and inconspicuous constellation of the southern sky introduced by Nicolas L. de Lacaille in the mid-eighteenth century. ➤Table 4.

Horsehead Nebula (NGC 2024)   A dark dust nebula, in the shape of a horse's head, which protrudes into a bright emission nebula, IC 434, in the constellation Orion.

Horseshoe Nebula   An alternative name for the ➤*Omega Nebula*.

hot dark matter   ➤*dark matter*.

hour (1)   The period of time equal to one twenty-fourth part of a day.

hour (2)   The unit in which ➤*right ascension* is measured, equivalent to 15 degrees of arc.

hour angle (HA)   For a celestial object, the length of ►*sidereal time* that has elapsed since it last made a transit of the meridian. Since an object reaches its maximum altitude when it is on the meridian, and is then best placed for observation, the hour angle is a measure of how far an object is from its optimal position for observation. One hour of sidereal time corresponds to the apparent rotation of the sky through 15°. By convention, it is positive in the westward direction.

hour circle   Any great circle on the celestial sphere passing through the north and south celestial poles. An hour circle is a line of constant ►*right ascension*.

Hourglass Nebula (1)   A bright, luminous nebula within M8, the ►*Lagoon Nebula*. It was first noted by John Herschel, and its name is derived from its shape.

Hourglass Nebula (2)   Alternative name for the ►*Engraved Hourglass*.

howardite   A stony ►*meteorite* of the ►*achondrite* type containing fragments of different types of rocks that have been subsequently fused together.

HR   A prefix, standing for Harvard revised photometry, used to identify star catalogue numbers in the ►*Bright Star Catalogue* drawn up originally by Pickering in 1908 and subsequently revised and republished on a number of occasions.

HR diagram (H–R diagram)   Abbreviation for ►*Hertzsprung–Russell diagram*.

HST   Abbreviation for ►*Hubble Space Telescope*.

Hubble classification   A method devised by Edwin Hubble (1889–1953) of classifying ►*galaxies* according to their shape (see illustration). The scheme classifies elliptical galaxies on a scale from E0 for a circular disc, through E1, E2, and so on, to E7 in order of increasing elongation. Spirals are designated as Sa, Sb or Sc in order of increasing openness of the arms and decreasing size of the nuclear bulge in relation to the overall size of the galaxy. There is a parallel sequence for barred spirals, which are designated SBa, SBb or SBc. Galaxies that are neither elliptical nor spiral in form are designated Ir for 'irregular'. Hubble suggested in 1925 that a transitional type, S0, was the 'missing link' in an evolutionary chain from E0 through to the open spirals Sc and SBc. This progression is no longer accepted as an evolutionary sequence, but Hubble's classification continues to be widely used as a simple way of describing the shapes of galaxies.

Hubble constant (symbol H_0)   The constant of proportionality in the ►*Hubble's law*. It describes the rate at which the universe expands with time. The value is not easy to determine, on account of uncertainties in the extragalactic distance scale, but increasingly precise measurements by many different

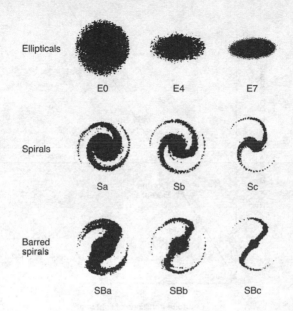

Ellipticals

E0   E4   E7

Spirals

Sa   Sb   Sc

Barred
spirals

SBa   SBb   SBc

Hubble classification.

researchers using a variety of techniques point to a value between 60 and 80 kilometres per second per megaparsec. In an evolving universe, its value changes with time, rather than being a true constant. For that reason, some prefer to refer to it as the Hubble 'parameter'. Its inverse is the ►*Hubble time.*

Hubble Deep Field An image made in December 1995 with the ►*Hubble Space Telescope* of a small area of sky in the constellation Ursa Major. Among the 1,500 galaxies recorded, some were as dim as thirtieth magnitude, the faintest ever seen. It was estimated that some would be ten billion light years away. The image was assembled from 342 separate exposures, effectively amounting to a continuous exposure 10 days long.

Hubble diagram In its original form, a graph on which the redshifts of a sample of galaxies are plotted against their apparent magnitudes (see over). The term is now applied more generally to any plot of redshift or recession velocity against a distance indicator for a carefully selected sample of galaxies.

This diagram gave the first strong evidence (1929) that the universe is expanding. It continues to be a crucial tool in observational cosmology because the shape of the diagram at high redshifts is dependent on the geometry of the universe. ►*Hubble's law, Hubble constant, expanding universe.*

Hubble flow

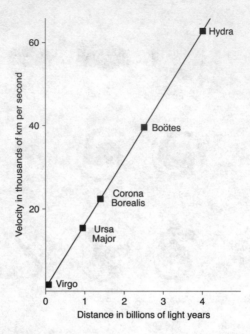

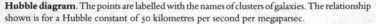

Hubble diagram. The points are labelled with the names of clusters of galaxies. The relationship shown is for a Hubble constant of 50 kilometres per second per megaparsec.

Hubble flow   The uniform motion of galaxies, due solely to the expansion of the universe. In practice, random and systematic velocities of a few hundred kilometres per second are superimposed on the idealized flow. ➤*Great Attractor*.

Hubble–Sandage variables   Superluminous variable stars noted by Edwin Hubble and Allan Sandage in their studies of the galaxies M31 and M33. They have been identified with the class of variables called ➤*P Cygni* stars, or S Doradus stars.

Hubble's law   The recession velocities of distant galaxies are directly proportional to their distance from us. The law is an immediate consequence of the uniform expansion of the universe. The relationship was discovered from observations of the redshifts of galaxies and was first announced by Edwin Hubble (1889–1953) in 1929. During the 1930s, he went on to more detailed work on the form of the relationship.

Hubble Space Telescope (HST)   An orbiting observatory built and operated jointly by NASA and ESA.

After a delay of six years beyond the originally scheduled launch date, it was successfully placed in orbit on 25 April 1990 from the ➤*Space Shuttle*. However, after extensive testing in the first few weeks, it became apparent that a fault in the optical figuring of the main mirror was introducing ➤*spherical aberration* and that sharp focusing would be impossible until corrective measures were carried out by a maintenance mission to the telescope. This was a grave disappointment, eliminating the possibility of carrying out the most eagerly awaited observations in the first phase of operation. During the first servicing mission in December 1993, a Space Shuttle crew successfully installed a unit known as COSTAR, which corrected the faulty optics, and they replaced the solar arrays.

The concept of the HST was to escape from the limitations imposed by the atmosphere on the quality of image that might otherwise be obtained. It was designed with a 15-year lifetime as an observatory that could be maintained and upgraded in orbit. It is a ➤*Ritchey–Chrétien telescope* with a main mirror 2.4 metres (94.5 inches) in diameter made of ultra-low-expansion glass. The compact, folded optical system is contained in a tube 13 metres (43 feet) long. The wavelength range of operation is from 110 to 1,100 nanometres.

The major elements of the telescope are the optical telescope assembly, the support system module, the fine guidance system (which can be used for astrometric measurements) and a suite of scientific instruments. The pointing accuracy is 0.007 arc seconds. Power is provided from two solar arrays and communication is via the ➤*Tracking and Data Relay Satellite System*.

The five original science instruments, operating in the optical and ultraviolet part of the spectrum, were a faint object camera (FOC), a wide-field and planetary camera (WF/PC), the Goddard high-resolution spectrograph (GHRS), a faint object spectrograph (FOS) and a high speed photometer. However, during the 1993 servicing mission, the high-speed photometer was removed to accommodate the COSTAR corrective unit. The WF/PC was also replaced by an improved version, WFPC-2.

A second servicing mission took place in February 1997. Various items of worn-out and outdated hardware, including a fine guidance sensor, were replaced, but the principal change was the substitution of two new instruments in place of the GHRS and FOS. These were the Near Infrared Camera and Multi-Object Spectrometer (NICMOS) and the Space Telescope Imaging Spectrograph (STIS).

NICMOS was the first HST instrument designed for infrared observation, giving the telescope an important new capability, and the first needing to be cooled. Its sensitive infrared detectors are maintained at 58 K inside an insulated vessel containing frozen nitrogen, which had been expected to have a lifetime of up to five years. However, a technical problem that came to light after installation is likely to reduce significantly the planned lifetime. The instrument contains three cameras, each with different resolution. It also operates as a spectrograph, a coronagraph and a polarimeter.

STIS is a powerful imaging spectrograph covering the ultraviolet, visible and near infrared parts of the spectrum. It is capable of recording simultaneously up to 512 spectra at different locations over an extended object, such as a galaxy. Its performance includes all the major capabilities of both the GHRS and FOS.

The third servicing mission is scheduled for December 1999, and operation of the telescope is expected to continue to 2010, beyond the 15 years initially envisaged.

The HST has been used to make observations of virtually every kind of celestial object, from planets in the solar system to the most remote galaxies detectable. The science operations are conducted from the Space Telescope Science Institute (STScI) in Baltimore, Maryland.

Hubble's Variable Nebula (NGC 2261)   A luminous triangular-shaped nebula in the constellation Monoceros. From photographs taken between 1900 and 1916, Edwin Hubble discovered that the nebula varied in shape and brightness. An irregularly variable star, R Monocerotis, is embedded in the nebula. R Mon is a strong source of infrared radiation and is probably a very young star surrounded by a circumstellar disc and ejecting a ▸*bipolar outflow*. The nebula is now thought to be an example of a ▸*Herbig–Haro object*. Light from the star reflected from interstellar dust is also seen.

Hubble time   A measure of the length of time for which the universe has been expanding, the presently accepted value is between 13,000 and 15,000 million years. It is defined as the reciprocal of the ▸*Hubble constant*, and it represents the time required for the universe to double its present size. The actual age of the universe is less than the Hubble time.

Humason, Comet   ▸*Comet Humason*.

Humboldt   A large lunar crater, 207 kilometres (128 miles) in diameter, on the extreme south-eastern limb of the Moon.

Hungaria   Asteroid 434, diameter 11.4 km, discovered by M. Wolf in 1898.

Hungaria group   A group of asteroids at the inner edge of the asteroid belt, 1.95 AU from the Sun, with orbits inclined at 24° to the plane of the solar system. The group is separated from the main belt by a ▸*Kirkwood gap*, and is not a true family with a common origin.

Hunter   English name for the constellation ▸*Orion*.

Hunting Dogs   English name for the constellation ▸*Canes Venatici*.

Huygens eyepiece   One of the simplest designs of telescopic ▸*eyepiece*, consisting of two planoconvex lenses (see illustration). It works well on long-focus refractors, but image distortion by spherical aberration becomes apparent at short focal ratios.

Huygens probe ➤*Cassini mission.*

Hyades An open star cluster in the constellation Taurus. Its members appear to be scattered around the star Aldebaran (which does not itself

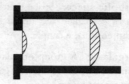

Huygens eyepiece.

belong to the cluster but is in the foreground), over an area 8° in diameter. It is the nearest star cluster, lying at a distance of about 150 light years. Because it appears so scattered, it was not listed in either the ➤*Messier Catalogue* or the ➤*New General Catalogue.* ➤*open cluster.*

Hydra (Sea Monster) The largest constellation in the sky by area, but a difficult one to identify since it contains only one moderately bright star, the second magnitude Alphard. It was included by Ptolemy in his list of 48 constellations (*c.* AD 140). ➤Table 4.

Hydra A The brightest radio source in the constellation Hydra. It is associated with a large elliptical galaxy at the centre of a small cluster of galaxies about one billion light years away.

hydrogen alpha (H alpha; Hα) The most prominent line in the visible part of the spectrum of hydrogen. It is the leading line of the Balmer series, with a wavelength of 656.28 nanometres, and is red in colour. It arises from transitions between the second and third energy levels in the hydrogen atom. ➤*Balmer lines*

hydrostatic equilibrium The condition of stability that exists when gravitational forces are exactly balanced by counteracting gas and radiation pressure.

hydroxyl ➤*OH source, maser.*

Hydrus (The Water Snake) An inconspicuous southern constellation, introduced by Johann Bayer in his 1603 star atlas. Its three brightest stars are third magnitude. ➤Table 4.

Hygeia Asteroid 10, diameter 430 km, discovered by A. de Gasparis in 1849. It is the fourth-largest asteroid.

hyperbolic Having the shape of a hyperbola, one of the family of curves known as ➤*conic sections.* The hyperbola is one of the possible shapes for the orbit of a body under the influence of a central gravitational force. The hyperbolic shape results if the body has enough energy to be able to escape from the gravitational force. A body travelling fast enough for this to occur is sometimes described as having a hyperbolic velocity.

hyperbolic space Space that has negative curvature and within which the angles of a triangle add up to less than 180°. The surface of a saddle is a two-dimensional example.

hypergalaxy A group of galaxies consisting of a dominating spiral galaxy surrounded by a cloud of dwarf galaxies. Our own ➤*Galaxy* and the ➤*Andromeda Galaxy* are examples.

Hyperion A satellite of Saturn, discovered in 1848 by W. C. Bond. It is an elongated irregular body measuring approximately 350 × 200 kilometres. There are large craters and a curving scarp-like feature, apparently 300 kilometres long. The evidence suggests that it may be a remnant of a larger body shattered by an impact. ➤Table 6.

I

Iapetus   A satellite of Saturn, discovered by Giovanni Cassini in 1671. The *Voyager* probes confirmed a hypothesis, proposed by Cassini after he noticed variations in Iapetus's brightness, that one hemisphere is very much darker than the other. The satellite always keeps the same face towards Saturn and, as it orbits the planet, its dark and light hemispheres alternately face the Earth. The bright surface is cratered and probably ice-covered; the dark part is blanketed with a material ten times darker whose nature and origin are unknown. The density of 1.2 times that of water suggests that Iapetus contains a high proportion of ice, possibly including frozen methane and ammonia. ➤Table 6.

IAU   Abbreviation for ➤*International Astronomical Union*.

IC   Abbreviation for ➤*Index Catalogue*.

Icarus   Asteroid 1566, diameter 1.4 km, discovered in 1949 by W. Baade. A member of the ➤*Apollo* group, its highly elliptical orbit takes it closer to the Sun than Mercury.

ICE   Abbreviation for ➤*International Cometary Explorer*.

ice dwarf   A planetary body characterized by relatively small size and a composition consisting of a mixture of ices and rock. Examples include ➤ *Pluto*, planetary moons such as ➤ *Triton*, and objects populating the ➤*Kuiper Belt*.

ICM   Abbreviation for ➤*intracluster medium*.

Ida   Asteroid 243, a member of the ➤*Koronis family*, measuring 58 km by 23 km. Close-up images of Ida were obtained on 28 August 1993 by the ➤*Galileo* spacecraft en route to Jupiter. *Galileo* discovered that Ida has a small satellite, subsequently named Dactyl, which measures about 1.6 km by 1.2 km. Observations of Dactyl's orbital motion made it possible to determine that Ida's density lies in the range 2.2–2.9 g/cm³. The two bodies do not have an identical composition, suggesting that the system may result from the collision and break-up of larger bodies that created the Koronis family. The surfaces of both bodies are heavily cratered.

IDP   Abbreviation for ➤*interplanetary dust* particle.

igneous rock   A rock formed directly by cooling and solidification from a molten state.

IGY   Abbreviation for ►*International Geophysical Year.*

Ikeya–Seki, Comet   ►*Comet Ikeya–Seki.*

image intensifier (image tube)   An electronic device for amplifying the brightness of an image. Amplification is achieved by the multistage production of photoelectrons. A photocathode is coated on the inside of the entry window. The electrons produced when a light signal strikes the window are accelerated through a potential of about 40 kV. Focusing is by the magnetic field of a surrounding solenoid combined with the electrostatic fields of ring electrodes; alternatively, the focusing may be entirely electrostatic. Further amplification is introduced by the use of more than one photocathode stage. The image finally formed on a phosphor screen is many times brighter than the original. The final detector may be a television camera or a ►*charge-coupled device.* A drawback of the system is the unavoidable distortion in the image; nevertheless, such devices have found wide application in astronomy.

Image Photon Counting System   An electronic imaging light detector sensitive enough to record and count the arrival of single photons.

image tube   An alternative name for an ►*image intensifier.*

Imbrium Basin   The largest and youngest of the very large circular impact features on the Moon. It was subsequently flooded by lava to form the dark area known as the Mare Imbrium, which is 1,300 kilometres (800 miles) in diameter. The Imbrium Basin is surrounded by three concentric rings of mountains, though only the outer one is defined with any clarity, by the Carpathian, Apennine and Caucasus mountains. The Alps form part of the second ring.

IMF   Abbreviation for ►*initial mass function.*

immersion   The disappearance of a star, planet, moon or other body at the beginning of an ►*occultation* or ►*eclipse.*

inclination (symbol *i*)   One of the principal ►*orbital elements* used to define an orbit, giving the angle between the orbital plane and a reference plane. For the orbits of planets and comets around the Sun, for example, the reference plane is the ►*ecliptic.* For satellite orbits, it is the plane of the equator of the parent planet.
    The term inclination is also used for the angle between a body's rotation axis and a reference plane, normally the body's orbital plane.

Index Catalogue (IC) A supplement to the *New General Catalogue* (NGC) of nebulae and star clusters, compiled and published by J. L. E. Dreyer in two parts, in 1895 and 1908.

Indian   English name for the constellation ►*Indus.*

Indus (The Indian)   An inconspicuous southern constellation, representing an American native Indian. It was introduced in the 1603 star atlas of Johann Bayer and contains no stars brighter than the third magnitude. ➤Table 4.

inequality   An irregularity in orbital motion.

inferior conjunction   The position of either of the planets Mercury or Venus when it lies directly between the Earth and the Sun. Because of the relative tilts of the planetary orbits, only rarely does Mercury or Venus actually ➤*transit* the face of the Sun. Normally, at inferior conjunctions, the planets pass in the sky to the north or south of the Sun.

inferior culmination   The crossing of the meridian by a ➤*circumpolar star* when at its lowest point in the sky.

inferior planet   Either of the two planets Mercury or Venus, whose orbits around the Sun lie inside that of the Earth.

inflationary universe   A class of ➤*Big Bang* models of the universe that include a finite period of accelerated expansion in their early histories. Such an event would have released enormous energy, stored until then in the vacuum of spacetime. The horizon of the universe expanded, temporarily, much faster than the speed of light. This theory is able to account satisfactorily for the present vast extent of the universe and its uniformity.

infrared array   A two-dimensional infrared imaging device, consisting of an array of small, individual electronic detectors, each of which records a pixel in the image.

Infrared Astronomical Satellite (IRAS)   A highly successful orbiting infrared telescope which operated from launch on the night of 25 January 1983 until the supply of coolant ran out on 23 November 1983. It was a collaborative mission between NASA, which designed and built the telescope, the Netherlands Aerospace Agency, which was responsible for the basic satellite, and the UK, which was responsible for the day-to-day tracking and data reception.

IRAS was a Ritchey–Chrétien telescope with mirrors made of beryllium rather than glass in order to withstand the low operating temperature. The diameter of the primary was 57 centimetres (22.5 inches). The telescope was cooled to 2 K by means of liquid helium. The detector consisted of an array of 62 elements, and filters were used for operation in wavebands centred on 12, 25, 60 and 100 micrometres. The different wavebands were used to distinguish between sources at different temperatures. The satellite orbit was oriented in the north–south direction and devised to rotate by about 1° per day so that it was always along the ➤*terminator* with the telescope pointing away from the Sun.

During the ten-month mission, 96 per cent of the sky was scanned twice so that an overall map of the infrared sky could be plotted. In addition, there was a spectrometer and a mapping facility for individual sources. A quarter of a million individual sources were detected, including stars, galaxies, dense interstellar dust clouds and some unidentified objects. Five comets were discovered. The first and brightest, Comet IRAS−Araki−Alcock, discovered in May 1983, passed within five million kilometres (3 million miles) of the Earth − the closest approach of any comet for 200 years.

infrared astronomy The study of infrared radiation from astronomical sources. Infrared is ▶*electromagnetic radiation* with wavelengths in a range between the visible red spectrum and radio waves. The definition is not precise but the wavelength range from about 0.1 to about 100 micrometres is normally considered to be the infrared. It is invisible to the human eye and is absorbed almost completely in the lower layers of the Earth's atmosphere, primarily by water vapour. For this reason, infrared astronomy observations have to be conducted from the highest mountain sites, or from aircraft or satellites.

The first infrared observation was made accidentally by William Herschel in 1800 when a thermometer he placed just to one side of the red end of a visible solar spectrum recorded a rise in temperature. Infrared images predominantly show the distribution of heat. All warm objects radiate infrared, so infrared telescopes must be cooled to a few degrees above absolute zero to prevent their being blinded by their own emission.

The development of systematic infrared astronomy began in the 1960s, when suitable detectors became available. The first infrared survey of the sky was carried out by Gerry Neugebauer and Robert Leighton of the California Institute of Technology (Caltech). They published a list of 5,612 sources in 1969. In 1968, Eric Becklin and Neugebauer announced that the infrared emission from the galactic centre at a wavelength of 2.2 micrometres is more than a thousand times stronger than might have been anticipated from radio observations. Infrared astronomy has made important advances with the development since the 1980s of two-dimensional arrays of infrared detectors, capable of making a complete image in a single exposure.

Infrared astronomy received an enormous boost from the successful operation of IRAS, the ▶*Infrared Astronomical Satellite*, in 1983. Its successor, the ▶*Infrared Space Observatory* (ISO) was launched in November 1995. The best ground-based site for infrared astronomy is the ▶*Mauna Kea Observatori*es in Hawaii. Three infrared telescopes started operation there in 1979: the ▶*United Kingdom Infrared Telescope* (UKIRT), NASA's ▶*Infrared Telescope Facility* (IRTF) and the ▶*Canada−France−Hawaii Telescope* (CFHT), which also functions as an optical telescope. The telescopes of the ▶*Keck Observatory* also have infrared capability.

Infrared radiation is detected from stars and galaxies, and from dust clouds within the solar system and in the interstellar medium. Strong infrared emission is particularly characteristic of dust that has been heated by shorter-wavelength visible and ultraviolet radiation from stars. Protostars in process of formation and evolved red giant stars are surrounded by shells of dust, giving rise to infrared emission. Unlike visible light, infrared radiation passes relatively unimpeded through dust clouds. So, for example, the ►*galactic centre*, which is largely obscured by dust in the visible spectrum, can be explored by means of infrared and radio astronomy. The way in which infrared radiation is scattered from the surfaces of objects in the solar system provides important clues to their composition. Infrared observations are also important for remote objects with large ►*redshift*.

infrared cirrus Wispy cloud-like structures seen above and below the plane of the Galaxy in the infrared maps of the sky produced by the ►*Infrared Astronomical Satellite*. They are thought to be caused by dust – probably micrometre-sized graphite and silicate particles – associated with relatively local ►*neutral hydrogen* clouds. The estimated temperature of the dust is 35 K. ►*infrared astronomy*.

infrared galaxy A galaxy that emits most of its energy (typically more than 90 per cent) in the infrared region of the spectrum. Such galaxies are thought to have unusually high rates of star formation and, for that reason, they are also described as ►*starburst galaxies*.

Infrared Space Observatory (ISO) An orbiting infrared telescope launched by the ►*European Space Agency* on 17 November 1995. It was equipped to make observations in the waveband between 2.5 and 200 micrometres with a sensitivity much greater than that of its predecessor, the ►*Infrared Astronomical Satellite*, IRAS. The instruments on ISO were a camera, an imaging photopolarimeter and two spectrometers to cover shorter and longer wavelengths.

infrared star A star that is a source of infrared radiation. ►*infrared astronomy*, *Infrared Astronomical Satellite*.

infrared telescope A telescope for astronomical observations in the infrared. ►*infrared astronomy*, *Infrared Astronomical Satellite*.

Infrared Telescope Facility (IRTF) A NASA infrared telescope located at the ►*Mauna Kea Observatories* in Hawaii, where it has been in operation since 1979 as a national facility for the USA. The main mirror is 3 metres (120 inches) in diameter.

initial mass function A mathematical expression describing the relative number of objects found in different ranges of mass for a cluster of stars or galaxies (or any specified volume of space) at the time of its formation.

inner planets The planets Mercury, Venus, Earth and Mars collectively. They are also known as the *terrestrial planets*.

insolation   The amount of radiative energy received from the Sun per unit area per unit time.

instability strip   A region of the ➤*Hertzsprung—Russell diagram* occupied by pulsating ➤*variable stars*. The strip delineates combinations of luminosity and temperature corresponding to unstable states that result in regular pulsations of a star's outer layers.

Institut de Radio Astronomie Millimétrique (IRAM)   A collaborative project between France, Germany and Spain for studies in ➤*millimetre-wave astronomy*. The institute operates a 30-metre (98.5-foot) dish in the Sierra Nevada, Spain, and a four-dish interferometer located in France, south of Grenoble.

INTEGRAL   Acronym for ➤*International Gamma Ray Laboratory*.

intensity interferometer   ➤*Brown—Twiss stellar interferometer*.

interacting binary   A ➤*binary star* system in which ➤*mass transfer* between the components takes place. ➤*contact binary, dwarf nova, mass transfer, Roche lobe*.

interacting galaxies   Galaxies that are close enough for their mutual gravitational attraction to produce distortions of shape and structure. Most galaxies are in clusters, and gravitational or tidal interactions between pairs are not uncommon. They often result in the production of long wisps or filaments, forming bridges between them. Computer simulations have been used to demonstrate that interactions can indeed produce the distortions observed.

Interamnia   Asteroid 704, diameter 338 km, discovered in 1910 by V. Cerulli. It is the sixth-largest asteroid known.

interference filter   A filter that uses the phenomenon of interference in a thin film to transmit light selectively over a narrow wavelength band. ➤*interferometer*.

interferometer   In astronomy, an instrument in which the electromagnetic radiation from a celestial object is collected along two or more different path lengths and then brought together to form an interference pattern.

   When any electromagnetic wave is sampled it has amplitude (how strong the wave is) and phase (the point the wave pattern has reached between peaks and troughs). If two light beams from the same point source have travelled slightly different paths and are then brought together, an interference pattern is formed. Where the two sets of waves are in phase, the pattern is bright, and where they cancel it is dark.

   Astronomical interferometers exploit this basic physical principle as a way of increasing ➤*resolving power*. A star image, for example, is a disc and not a point source. With a pair of mirrors there will be a critical separation at which the interference pattern disappears because the bright fringes from one side of the disc coincide with dark fringes from the other side. The first successful astronomical application of this principle was the ➤*Michelson stellar interferometer*. In basic radio interferometry, telescopes are used in pairs and the resulting interference patterns analysed by computer.

By using more than two elements (e.g. mirrors or antennas) in an interferometer, it is possible to produce high-resolution maps or images, a technique often known as ➤*aperture synthesis*.

Interferometry has been an essential observational tool in radio astronomy for decades. More recently it has become feasible to extend the technique fully into the infrared and optical wavebands. A number of dedicated instruments for optical interferometry have been constructed, such as the Cambridge Optical Aperture Synthesis Telescope in the UK and the Navy Prototype Optical Interferometer in the USA. In addition, several projects for the construction of very large telescopes have been specifically designed so that optical interferometry can be carried out. These include the ➤*Keck Observatories*, the ➤*Very Large Telescope* and the ➤*Large Binocular Telescope*. ➤*radio interferometer, very-long-baseline interferometry*.

intergalactic medium ➤*intracluster medium*.

intermediate polar ➤*polar*.

International Astronomical Union (IAU) An organization formed in 1919 for fostering international cooperation in astronomy. It is composed of member countries (represented by national academies or similar institutions, not governments) and around 8,000 individual members. Along with similar organizations for other branches of science, it belongs to the International Council of Scientific Unions (ICSU), which has its headquarters in Paris.

The history of the IAU may be traced back to the international cooperation established for the ➤*Carte du Ciel* project. From 1887 the Permanent Commission on the Photographic Carte du Ciel extended its interest into other areas of astronomy and may be regarded as the parent organization of the IAU.

The IAU is recognized as the international authority on astronomical matters requiring cooperation and standardization, such as the official naming of astronomical bodies and features on them. The Central Bureau for Astronomical Telegrams and the Minor Planets Center located at the ➤*Smithsonian Astrophysical Observatory* operate under its auspices. The IAU is also concerned with the promotion of astronomy in developing countries. Its General Assembly meets every three years, and specialist symposia and colloquia are sponsored regularly. Work is organized through Commissions and Working Groups.

International Atomic Time (TAI) A continuous timescale resulting from analyses by the Bureau International des Poids et Mesures of atomic time standards in many countries. ➤*atomic clock*.

International Cometary Explorer (ICE)   A spacecraft, originally known as ISEE-3 (International Sun–Earth Explorer) when it was launched in 1978, which was diverted and reactivated in order to pass through the tail of ►*Comet Giacobini–Zinner* in September 1985, and to observe ►*Halley's Comet* from 28 million kilometres (17 million miles) in March 1986.

International Dark Sky Association   An organization based in the USA, established to combat ►*light pollution*.

International Gamma Ray Laboratory (INTEGRAL)   A European/ Russian orbiting observatory for spectroscopy and accurate imaging of gamma-ray sources. Launch by a Russian Proton launcher is planned for 2001.

International Geophysical Year (IGY)   The period from 1 July 1957 to 31 December 1958 during which an internationally coordinated programme of geophysical research was organized to coincide with a period of maximum ►*solar activity*.

International Sun–Earth Explorer (ISEE)   Name of three spacecraft built by NASA and ESA to study the influence of the Sun on the Earth's space environment and magnetosphere. ISEE-1 and 2 were launched in 1977. ISEE-3, launched in 1978, later became the ►*International Cometary Explorer* (ICE).

International Ultraviolet Explorer (IUE)   An astronomical telescope with a 45-centimetre (18-inch) primary mirror, designed to work in the ultraviolet region of the spectrum, which was launched into Earth orbit in 1978. A joint NASA–ESA–UK project, it continued to observe successfully for 18 years, finally ceasing operations in September 1996.

International Years of the Quiet Sun (IQSY)   The period 1964–5 when, following the success of the earlier ►*International Geophysical Year*, an international programme of geophysical research was coordinated to coincide with a period of minimum solar activity.

interplanetary dust   Micrometeoroids present in the space between the planets. The particles are thought to originate in collisions between asteroids in the ►*asteroid belt* and from the gradual break-up of comets. Fragments from comets initially form ►*meteor streams*, but will disperse over long periods of time. The zodiacal dust cloud extends out from the centre of the solar system for at least 600 million kilometres (373 million miles). Its presence is observed as the ►*zodiacal light*: sunlight scattered by dust particles a few tens of micrometres in size. Its very low density is equivalent to one grain in a cube hundreds of metres across at the Earth's distance from the Sun. Near the Sun, the particles are eroded by collisions, evaporation and the action of the ►*solar wind*.

The ►*Infrared Astronomical Satellite* (IRAS) located three dust bands between

Mars and Jupiter in the asteroid belt. The brightest is centred on the ecliptic, but there are fainter bands 10° to the north and south of the ecliptic. These are thought to consist of dust particles in orbits tilted at about 10° to the ecliptic. ➤*Poynting–Robertson effect*.

interplanetary medium The medium between the planets in the solar system composed of ➤*interplanetary dust*, electrically charged particles from the Sun and neutral gas from the ➤*interstellar medium*.

The charged particles consist of electrons, protons and helium nuclei (alpha particles) streaming outwards from the Sun and forming the ➤*solar wind*. Atoms of neutral hydrogen and helium gas are also present, replenished from the interstellar medium in the Sun's vicinity. Under the influence of solar ionizing radiation, these atoms have a lifetime against ionization of only twenty days at the Earth's distance from the Sun.

interplanetary scintillation Fluctuations in the signal received from a distant radio source observed along a line of sight close to the Sun. The scintillation is caused by irregularities in the ➤*solar wind*.

interstellar dust Small particles in the ➤*interstellar medium*. Interstellar dust particles range in size between 0.005 and 1 micrometre and are generally mixed in with gas in the interstellar medium. Though accounting for less than one per cent of the mass of the typical interstellar medium, the dust absorbs far more light and emits far more infrared radiation than the gas. It causes both ➤*interstellar extinction* and ➤*interstellar reddening*. Starlight scattered from dust particles creates ➤*reflection nebulae*.

The absorption of energy from starlight by dust raises its temperature to a few tens of degrees above absolute zero. At such temperatures, the dust emits ➤*thermal radiation* peaking in intensity in the infrared. Dust heated to temperatures higher than about 1,500 K is destroyed.

It is unlikely that all interstellar dust is composed of the same material. Graphite (a common form of carbon) and silicates of iron, aluminium, calcium and magnesium are thought to be among the commonest, though the broad spectral features produced by the dust are difficult to identify with certainty. ➤*Polarization* effects suggest that at least some of the particles are not spherical.

Most of the dust is thought to originate in the outflow of material from cool ➤*red giant* stars. As the gas cools with increasing distance from the star, solid materials condense out. Infrared emission detected from such stars shows that they are indeed surrounded by dust shells. Grains of material may also condense within ➤*molecular clouds*.

interstellar extinction The dimming of light from distant stars by absorption and ➤*scattering* by interstellar dust. The effect decreases with increasing wavelength. Extinction is less effective for red light than for blue, resulting in the phenomenon of ➤*interstellar reddening*. The blue light from a star near the centre of the Galaxy is reduced in brightness by 25 magnitudes by the interstellar material along our line of sight. At infrared and radio wavelengths, which are longer than those of visible light, the interstellar medium is increasingly transparent. In the ultraviolet, extinction continues to increase towards shorter wavelengths; it has been studied down to 100 nanometres.

interstellar medium (ISM) The diffuse material in the space within galaxies between individual stars, which are typically separated by several light years. In our own Galaxy, the mass of material in the interstellar medium is estimated to be at least one-tenth that in the stars. It is concentrated in the central region and in four spiral arms. In general, spiral galaxies have substantial amounts of interstellar material and elliptical galaxies little or none.

There is a continuous interaction between stars and the interstellar medium, which is not uniform but consists of a number of diverse components: dark clouds of gas and dust, regions of ►*ionized hydrogen* and ►*neutral hydrogen*, ►*molecular clouds*, ►*globules*, a very hot dilute gas and high-energy ►*cosmic ray* particles.

Interstellar clouds are the sites of star formation but are also enriched by material ejected by ►*supernovae* and other processes of mass loss from stars. Over distance scales of thousands of light years, the structure of the interstellar medium is probably dominated by the coalescing of supernova remnants. The thick shells surrounding them ultimately cool and condense into small clouds. Such clouds can interact, either coalescing or fragmenting on collision. ►*astration*. ►*Local Bubble*.

interstellar molecules Molecules present in the interstellar medium, especially in ►*molecular clouds*. They can survive only when shielded from the destructive effect of ultraviolet radiation from stars and so are found in dense interstellar or circumstellar clouds. Prior to 1963, CH (methylidyne), CH$^+$ and CN (cyanogen) were the only interstellar molecules known, having spectra in the visible range. Radio emission at 18 centimetres wavelength was recognized as being from hydroxyl (OH) in 1963. Over 90 different molecules have been identified since 1968, primarily by means of their characteristic spectra at millimetre wavelengths. Most are simple organic molecules.

interstellar reddening The apparent reddening of light from distant stars by interstellar dust, which causes ►*scattering*. The extent to which light is scattered and absorbed in the interstellar medium depends strongly on wavelength: blue light is dimmed more than red light. As a result, the colours of stars viewed through interstellar material are altered and appear redder. The degree of reddening increases with the quantity of intervening matter. A similar effect in the Earth's atmosphere causes the reddening of the Sun when it is close to the horizon.

intracluster medium (ICM) The material between the galaxies in a ►*cluster* *of galaxies*. The ICM contains several components. The presence of tenuous hot gas is revealed through its X-ray emission. It typically consists of only one atom in 1,000 cm^3 but accounts for an estimated 10 per cent of the total cluster mass. Diffuse radio emission from clusters is probably ➤*synchrotron radiation* produced by high-energy particles in the ICM. Stars that have been torn from galaxies by the gravitational interactions between them are also found in intergalactic space. These observed components, and the galaxies themselves, still account for only 20 per cent of the mass of a typical cluster judged by applying gravitational theory to the motion of member galaxies. Though some of this unseen mass is almost certainly associated with galaxies, the behaviour of clusters suggests that there is also unseen matter in the ICM, but its nature is unknown. ➤*dark matter*.

invariable plane The plane that includes the centre of mass of the solar system and is at a right angle to the ➤*angular momentum* vector (i.e. the spin axis) of the solar system. This plane is the fundamental reference plane in computations of the dynamics of the solar system.

inverse Compton effect A collision between a photon and a high-energy electron in which some of the electron's energy is transferred to the photon. ➤*Compton effect*.

inverse square law A relationship between two physical quantities in which one declines in proportion to the reciprocal of the square of the other. Gravity is an example where the inverse square law operates. Mathematically, the attractive force F between two masses m and M can be expressed as $F = GmM/r^2$, where r is the distance between the masses and G is Newton's gravitational constant.

Io One of the four Galilean moons of Jupiter (number I), the nearest to the planet and arguably the most remarkable. Its surface is brightly coloured – much of it a greenish yellow dappled with patches of orange and white. Eight active eruptive centres were identified in the images returned during the encounter of the ➤*Voyager 1* spacecraft, six of which were still active when ➤*Voyager 2* flew by four months later. Continuous monitoring from ground-based observatories, and comparisons between *Voyager* and *Galileo* images confirm the high level of continuous eruptive activity on Io. The eruptive centres appear as dark spots. Many are surrounded by roughly circular 'haloes' of ejected material, and lava flows are also visible. The coloured crust is made of sulphur and solid sulphur dioxide. No impact craters are seen; any that were formed in Io's early history have long since been covered by erupted material.

Io is the only body in the solar system, apart from the Earth, definitely observed to be volcanically active, though Triton and Enceladus show evidence of likely activity, and Venus might also be active. Io's activity was predicted on the basis of the strong tidal effects Jupiter has on the interior of Io. The satellite is surrounded by a thin atmosphere of sulphur dioxide and a ring of electrically charged particles – a *plasma torus* surrounds Jupiter, enclosing the orbit of Io. Data from *Galileo* indicates that Io has a substantial metallic, electrically conducting core. ➤Table 6.

ion   An atom that has gained or lost one or more electrons and so is not electrically neutral.

ionization   The process in which electrons are removed from ➤*atoms* or molecules through collisions between particles or by absorbing a ➤*photon*. The electrically charged particles are positive *ions*.

ionization front   The transition region in an interstellar cloud between ➤*neutral hydrogen* and ➤*ionized hydrogen*. Ionization fronts are found near ➤*O stars* and ➤*B stars*, which radiate strongly in the ultraviolet. The ultraviolet photons ionize the medium near such stars.

ionized hydrogen   Hydrogen in which electrons have become separated from their parent protons. (A neutral hydrogen atom consists of a single proton, forming the nucleus, and one electron.)

Hydrogen clouds in interstellar space become ionized largely through the absorption of ultraviolet photons, which have enough energy to detach the single electrons from the atoms. Ionized hydrogen is the main constituent of H II (or H$^+$) regions, discrete hot clouds that are roughly spherical and up to 600 light years across. Ionization is produced by intense ultraviolet radiation from young O and B stars embedded in the clouds.

H II regions are strong sources of radio waves, emitted by the free electrons, and of ➤*recombination lines*. The ➤*Orion Nebula* is a giant H II region and one of the nearest.

Ionized hydrogen is also present in ➤*supernova remnants* and the shells of ➤*planetary nebulae*.

ionosphere   An ionized layer in a planetary atmosphere where free electrons and ions with thermal energies exist under the control of the gravitational and magnetic fields of the planet. The Earth's ionosphere lies between heights of about 50 and 600 kilometres (30 and 350 miles), though the extent varies considerably with time, season and ➤*solar activity*. It is created by the effect of ultraviolet and X-radiation from the Sun. Four different layers with different characteristics are known as the D, E, F_1 and F_2 layers in order of increasing height. The D region, between 50 and 90 kilometres, has low electron density. The E and F_1 regions, between 90 and 230 kilometres, form the main part of the ionosphere.

ion tail (Type I tail)   One of the two distinct types of tail developed by ➤*comets* as they near the Sun. The ion tail, also known as the *gas tail* or *plasma tail*, consists of ionized atoms and molecules that are emitting light by the mechanism of resonance fluorescence. In this process, energy is absorbed from sunlight then re-emitted later. The ion tail lies in the plane of the comet's orbit. It is nearly straight but curves away from the direct radial line to the Sun by a few degrees. The ion tail is 'blown' away from the comet by the ➤*solar wind* and the effects of its magnetic field. ➤*disconnection event*.

IPCS   Abbreviation for ➤*Image Photon Counting System*.

IQSY   Abbreviation for ➤*International Years of the Quiet Sun*.

IR   Abbreviation for infrared. ➤*electromagnetic radiation, infrared astronomy*.

IRAM   Abbreviation for ➤*Institut de Radio Astronomie Millimétrique*.

IRAS   Abbreviation for ➤*Infrared Astronomical Satellite*.

Iris   Asteroid 7, diameter 208 km, discovered by J. R. Hind in 1847.

iron meteorite   A type of ➤*meteorite* composed almost entirely of iron and nickel.

irradiation (1)   Exposure to electromagnetic radiation or energetic particles.

irradiation (2)   The spreading of a photographic image by scattering of light within the emulsion.

irradiation (3)   An optical illusion that makes bright objects seen against a dark background appear larger than they really are.

irregular galaxy   Any galaxy that is not obviously an ➤*elliptical galaxy* or a ➤*spiral galaxy*. About a quarter of known galaxies are irregular. Many appear to be undergoing star formation, and are dominated by regions of luminous gas and bright young stars. Radio observations of the hydrogen gas in irregular galaxies often reveal the underlying symmetry of a rotating disc of gas; in this respect, and in their star content, they resemble spiral galaxies.

irregular satellite   A natural planetary satellite that is in a ➤*retrograde* orbit, or an orbit of high inclination to the equatorial plane and/or large eccentricity.

irregular variable   A pulsating variable star that changes in brightness slowly and in an irregular way. Such stars belong mainly to ➤*spectral types* K, M, C and S.

IRTF   Abbreviation for ➤*Infrared Telescope Facility*.

Isaac Newton Telescope (INT)   A 2.5-metre (98-inch) reflecting telescope in the Isaac Newton Group at the Observatorio del Roque de los Muchachos, La Palma, Canary Islands. Observing time is shared between the collaborating countries, which are the UK, Spain and the Netherlands. The telescope was originally installed at Herstmonceux, the UK base of the Royal Greenwich

Observatory until 1990. It was rebuilt and provided with a new primary mirror on removal to La Palma, where it came into operation in 1984.

ISAS   The Japanese national space agency.

ISEE   Abbreviation for ►*International Sun–Earth Explorer*.

Ishtar Terra   One of the major upland areas on the planet ► *Venus*, comparable in size with the continent of Australia. It includes the highest mountain peaks on Venus, Maxwell Montes.

island universe   An obsolete term, formerly used to mean a giant galaxy. It is associated particularly with the philosopher Immanuel Kant (1724–1804), who introduced it in 1755. It was used until the 1920s in popular writing.

ISM   Abbreviation for ►*interstellar medium*.

ISO   Abbreviation for ►*Infrared Space Observatory*.

isophote   A line connecting points of equal light intensity on a map showing the distribution of brightness over, for example, an area of sky.

isotropy   The property of being the same in every direction. Liquid water is isotropic, whereas a snowflake, which has sixfold symmetry, is not. The universe on the largest scales is believed to be isotropic. It is observed to be expanding isotropically to a high degree of precision, and the ►*cosmic background radiation* is equally isotropic.

IUE   Abbreviation for ►*International Ultraviolet Explorer*.

Ivar   Asteroid 1627, diameter 6.2 km, discovered in 1929 by Ejnar Hertzsprung. It is a member of the ►*Amor* group.

J

Jacobus Kapteyn Telescope (JKT) A 1.0-metre (40-inch) reflecting telescope in the Isaac Newton Group at the ➤*Observatorio del Roque de los Muchachos*, La Palma, Canary Islands. Observing time is shared by the collaborating countries, which are the UK, Eire, Spain and the Netherlands. It is a specialist telescope for photometry and wide-field photography, and came into operation in 1984.

James Clerk Maxwell Telescope A submillimetre-wave telescope located at the ➤*Mauna Kea Observatories* on the island of Hawaii. It is managed by the Joint Astronomy Center, Hilo, Hawaii, on behalf of the UK, the Netherlands and Canada, which are the sponsoring countries.

It is a ➤*Cassegrain* design, with a paraboloidal reflector 15 metres (49 feet) in diameter. The dish is constructed from 276 lightweight panels supported on a mild steel frame, designed to deform uniformly and so maintain the paraboloid shape when the orientation is changed. The mount is altazimuth. The telescope is inside a rotating protective enclosure; during observations, the viewing aperture is covered with a membrane of woven Teflon to protect the surface from wind and heating by the Sun. Only 10 per cent of the submillimetre radiation is absorbed by the membrane.

The telescope's capabilities were extended in 1996 by the provision of a newly developed instrument, which is a combined imaging camera and photometer. Known as SCUBA (Submillimetre Common-User Bolometer Array), it consists of two arrays of detectors cooled to below 0.1 K. One array is optimized for use at 850 microns and the other at 450 microns.

jansky (symbol Jy) The unit of ➤*flux density* used in radio astronomy. One jansky is 10^{-26} watts per square metre per hertz. The unit is named after Karl Jansky, who discovered radio emission from the Milky Way in 1932.

Janus A small satellite of Saturn, discovered by Audouin Dollfus in 1966 when the ring system was edge-on as seen from Earth. Janus orbits just beyond the outer edge of the ring system and is co-orbital with Epimetheus. The two may be fragments of a larger object that was shattered by an impact. Janus is irregular in shape, measuring 220 × 160 kilometres. ➤*co-orbital satellite*, Table 6.

JD Abbreviation for ➤*Julian date*.

Jeans criterion   An idealized set of initial conditions for the steady contraction under gravity of a density perturbation in an interstellar cloud.

Jeans length   The minimum wavelength that a density perturbation in a gas cloud must have if it is to grow in response to gravitational forces. The concept is central to theories of star formation in interstellar clouds.

Jeans mass   The minimum mass for a density perturbation in an interstellar cloud that can contract under its own gravity.

jet   A narrow beam of matter or radiation emerging, for example, from an ➤*active galactic nucleus* or from an ➤*accretion disc*.

Jet Propulsion Laboratory (JPL)   An institution in Pasadena, California, operated by the California Institute of Technology (Caltech) in support of programmes of the US ➤*National Aeronautics and Space Administration* (NASA) and of other agencies. It is distinguished as the primary facility in the USA for the development and operation of interplanetary space probes. Activities managed by JPL have included the ➤*Viking*, ➤*Voyager*, ➤*Galileo* and ➤*Magellan* projects. Although often referred to as a NASA centre, JPL employees are not US government civil servants, unlike, for example, those who work at the Goddard Space Flight Center and similar establishments that are formally part of NASA.

Jewel Box (NGC 4755)   An open star cluster in the constellation Crux. Its brightest member is the sixth magnitude blue supergiant star Kappa Crucis. There are a number of blue and red supergiants, resulting in an impressive appearance in a small telescope; this contrast is said to be why John Herschel gave the cluster the name 'Jewel Box'. The cluster's distance is 7,800 light years.

JILA   Abbreviation for ➤*Joint Institute for Laboratory Astrophysics*.

JKT   Abbreviation for ➤*Jacobus Kapteyn Telescope*.

Jodrell Bank   The location in Cheshire of the ➤*Nuffield Radio Astronomy Laboratories* of the University of Manchester.

Johnson Space Center   ➤*Lyndon B. Johnson Space Center*.

Joint Institute for Laboratory Astrophysics (JILA)   An establishment operated jointly since 1962 by the University of Colorado and the US National Institute of Standards and Technology (formerly the National Bureau of Standards) as a centre for advanced research and teaching on subjects such as atomic interactions, spectroscopy, gravitational physics, radiative transfer, stellar interiors and much more. Both theoretical and experimental work is carried out. JILA is located on the main campus of the University of Colorado at Boulder.

jovian   Pertaining to the planet Jupiter.

jovian planets   A collective name for the planets Jupiter, Saturn, Uranus and Neptune, which share the general property of being gaseous giants in contrast with the rocky ➤*terrestrial planets*.

JPL   Abbreviation for ➤*Jet Propulsion Laboratory*.

JSC   Abbreviation for ➤*Lyndon B. Johnson Space Center*.

J star   ➤*carbon star*.

Julian calendar   A calendar instituted for use in the Roman Empire by Julius Caesar from 46 BC. There were twelve months in a year and three years of 365 days followed by one of 366 days, giving an average of 365.25 days. Since this is 11 minutes 14 seconds longer than the ➤*tropical year*, which governs when the seasons occur, the seasons gradually moved in relation to the civil year. Because of this, the ➤*Gregorian calendar* was introduced from AD 1582.

Julian date (JD)   The interval of time in days since noon at Greenwich on 1 January 4713 BC.

Julian year   A period of 365.25 days. Since 1984, standard ➤*epochs* have been defined in terms of the Julian years, whereas previously the ➤*Besselian year* was used.

Juliet   A small satellite of Uranus, 80 kilometres in diameter, discovered by ➤*Voyager 2* in 1986. ➤Table 6.

Juno   Asteroid 3, diameter 248 km, discovered by Karl L. Harding in 1804.

Jupiter   The largest planet in the solar system and the fifth in order from the Sun. After Venus, it is the second-brightest planet as seen from Earth.

Jupiter is ten times the size of the Earth and one-tenth of the Sun's diameter. Its mass is 0.1 per cent that of the Sun and its composition (by number of molecules) is very similar to the Sun's: 90 per cent hydrogen (in its molecular form in Jupiter) and 10 per cent helium. Of trace gases, the most significant are water vapour, methane and ammonia. There is no solid surface beneath the cloud layer. Instead, a gradual transition from gas to liquid takes place as the pressure increases with depth below the outermost layers, followed by an abrupt change to a metallic liquid, in which the atoms are stripped of their electrons. At the very centre there may be a small core of rock and perhaps ice.

A source of internal energy, heat generated when Jupiter formed by gravitational collapse, causes the planet to radiate between 1.5 times and twice as much heat as it absorbs from the Sun.

Observed visually, the disc of Jupiter is seen to be crossed by alternating light *zones* and dark *belts*. Results from four space probes that passed by Jupiter

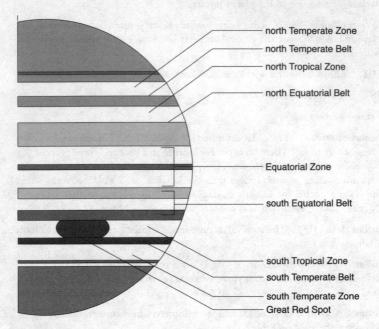

- north Temperate Zone
- north Temperate Belt
- north Tropical Zone
- north Equatorial Belt
- Equatorial Zone
- south Equatorial Belt
- south Tropical Zone
- south Temperate Belt
- south Temperate Zone
- Great Red Spot

Jupiter. The names used for the light zones and dark belts running parallel to the equator in Jupiter's atmosphere.

between 1973 and 1981 (*Pioneers 10* and *11*, *Voyagers 1* and *2*), and from the ➤*Galileo* mission have revealed the full complexity of the flow patterns within these bands. There are five or six in each hemisphere, correlating with wind currents.

White or coloured ovals appear as relatively long-lived features. The best-known and most conspicuous is the *Great Red Spot*, which has been observed for around 300 years. The origin of this feature, which is as wide as the Earth, is uncertain; one popular theory is that it is essentially a huge ➤*anticyclone*.

The coloured clouds are in the highest layers of Jupiter in a region with a depth of only 0.1–0.3 per cent of the total radius. The origin of their coloration remains a mystery, though it seems certain that it must have to do with trace constituents of the atmosphere, and is evidence of complex chemistry. Cloud colour correlates with altitude: blue features are the deepest, followed by brown, then white, with red being the highest.

A probe released by the *Galileo* spacecraft in 1995 parachuted through Jupiter's upper atmosphere and returned data on the composition and physical conditions. Ground-based observations of the entry site indicated that it may have been a relatively cloud-free spot, explaining why hardly any evidence was found for the expected three layers of cloud consisting of ammonia crystals at the highest level, ammonium hydrosulphide in the middle, with water and ice crystals below. Winds up to 530 km/hour (330 mph) were even faster than anticipated. The abundance of helium was only about half that expected. A likely explanation is the concentration of helium towards the centre of the planet. The probe also discovered an intense radiation belt.

The existence of a faint ring around Jupiter was first suggested by results from *Pioneer 11* in 1974 and confirmed by direct *Voyager* images. The main part lies between 1.72 and 1.81 Jupiter radii from the centre of the planet. The nature of the ring is such that many of the particles must have dimensions measured in micrometres. A constant source of replenishment is required, which may be a population of orbiting boulder-sized objects, constantly bombarded by high-velocity particles.

There are sixteen known natural satellites orbiting Jupiter. They fall into four distinct groups. The four small inner satellites (Metis, Adrastea, Amalthea and Thebe) and the four large Galilean satellites (Io, Europa, Ganymede and Callisto) are in circular orbits in the equatorial plane. The third group (Leda, Himalia, Lysithea and Elara) are small satellites in circular orbits, inclined at angles between 25° and 29° to the equatorial plane and at distances between 11 and 12 million kilometres from Jupiter. The outermost group (Ananke, Carme, Pasiphae and Sinope) are small satellites in ►*retrograde* orbits that are relatively eccentric ellipses, inclined substantially to the equatorial plane. These orbits all lie between 21 and 24 million kilometres from Jupiter. The four Galilean satellites and their movements in orbit are easily visible with a small telescope or binoculars.

Radio emission from Jupiter was discovered in 1955. It was the first indication of the presence of the strong magnetic field, which is 4,000 times stronger than the Earth's. The ►*magnetosphere* is consequently 100 times larger. The radio emission is caused by the spiralling of electrons around the field lines. Trapped electrons near the planet give rise to ►*synchrotron radiation* at decimetre wavelengths. Decametric radiation, observed only from certain regions of the planet, is associated with the interaction between Jupiter's ionosphere and Io, whose orbit lies within a huge *plasma torus*: this interaction also creates aurorae. Radiation at kilometre wavelengths was discovered by the *Voyager* probes, originating at high latitudes near the planet and in the plasma torus. ►Tables 5, 6 and 7.

K

kamacite   An alloy of iron and nickel that occurs as platelets or single crystals in iron ►*meteorites* and also as grains in ►*chondrites* and ►*achondrites*.

Kamiokande   ►*Super-Kamiokande*.

KAO   Abbreviation for ►*Kuiper Airborne Observatory*.

Kappa Crucis (κ Cru)   The most prominent star in the open cluster NGC 3324, popularly known as the ►*Jewel Box*.

Kapteyn's Star   An eighth magnitude ►*M star* (HD 33793), notable for its relatively large ►*proper motion* (8.7 arc seconds per year) – the second largest known after ►*Barnard's Star* – and high ►*radial velocity* (245 km/s). At a distance of 12.7 light years, it is among the stars nearest to the solar system. Its properties were discovered in 1897 by the Dutch astronomer J. C. Kapteyn (1851–1922), and it lies in the southern constellation Pictor.

Kapteyn Telescope   ►*Jacobus Kapteyn Telescope*.

Karl Schwarzschild Observatory   A German observatory, located at Tautenburg near Jena, which was founded in 1960 as part of the Academy of Sciences of the former German Democratic Republic. It is equipped with a Zeiss 2-metre (78-inch) reflecting telescope which can be used as a photographic ►*Schmidt camera* or at the ►*Cassegrain* or ►*coudé focus*.

Kaus Australis (Epsilon Sagittarii; ε Sgr)   The brightest star in the constellation Sagittarius. With Kaus Meridionalis (Delta) and Kaus Borealis (Lambda), it marks the Archer's bow. It is a ► *B star* of magnitude 1.9. ►Table 3.

K corona   ►*corona*.

Keck Observatories   Two 10-metre (400-inch) reflecting telescopes belonging jointly to the California Institute of Technology (Caltech) and the University of California. They are located at the ►*Mauna Kea Observatories* on the island of Hawaii and have been funded by the W. M. Keck Foundation. The first telescope was completed in 1992 and the second in 1996.

The primary mirrors of these ►*Ritchey–Chrétien telescopes* are of a unique pioneering design, each consisting of 36 individual hexagonal segments. The precise figure of the mirrors is maintained by a combination of specially designed passive supports and active control, monitored by computer. By this means, it

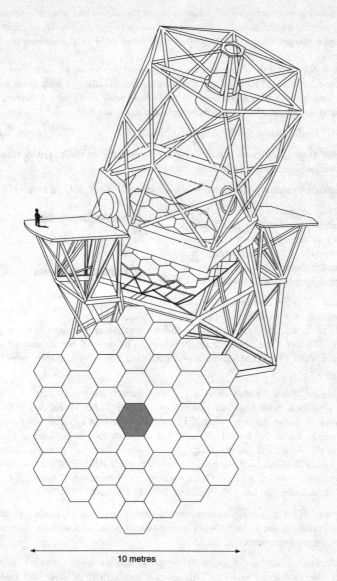

10 metres

Keck Observatories. The 10-metre mirrors are made of 36 hexagonal segments. The structure of the telescopes is shown behind, with a human figure for scale.

was possible to construct such large-aperture telescopes and assemble them at a remote mountain-top site. The use of ►*adaptive optics* makes it possible to produce images with a resolution of 0.04 arcseconds at a wavelength of 2 microns.

It will be possible to use the two telescopes together as an ►*interferometer*. Because Keck I and Keck II are about 85 metres (nearly 280 feet) apart, they will have a resolution equivalent to a telescope with an 85-metre mirror, or about 0.005 arc seconds at a wavelength of 2 microns.

Keel English name for the constellation ►*Carina*.

Keeler Gap A narrow gap towards the outer edge of the bright A ring of ►*Saturn*. ►*planetary rings*, Table 7.

Keenan's System The galaxies NGC 5216 and 5218 in Ursa Major.

Kellner eyepiece A type of telescopic ►*eyepiece* with a planoconvex field lens and an achromatic doublet as an eyelens (see illustration). It is generally regarded as a good all-purpose eyepiece and is widely used.

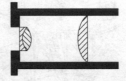

Kellner eyepiece.

kelvin The SI unit in which temperature is measured. It is defined formally as the fraction 1/273.16 of the thermodynamic temperature of the triple point of water. One kelvin corresponds to an interval of one degree on the Celsius scale. The kelvin temperature scale starts from absolute zero, which is −273.16°C.

Kennedy Space Center (KSC) The primary space launch facility of the US ►*National Aeronautics and Space Administration* (NASA), located at Cape Canaveral on the Atlantic coast of Florida. It is the sole location from which the ►*Space Shuttle* has been launched, and is one of the few facilities at which a Shuttle has landed. In the past, all the manned launches in the ►*Mercury*, ►*Gemini*, ►*Apollo* and ►*Skylab* programmes took place at KSC. Many unmanned expendable vehicles, such as Delta rockets, are also launched from KSC. It was named after the late US President John F. Kennedy.

Kepler A lunar crater, 32 kilometres (20 miles) in diameter, in the Oceanus Procellarum. It has terraced walls and a central peak, and is the centre of a large, bright ray system.

Keplerian telescope A simple telescope in which convex lenses are used as both objective and eyepiece. This gives a larger field of view and higher magnification than the ►*Galilean telescope*, but the image is inverted.

Kepler's equation ➤*anomaly*.

Kepler's laws Three fundamental statements about planetary motion, derived empirically by Johannes Kepler (1571–1630) on the basis of detailed observations of the planets made by Tycho Brahe (1546–1601):

    1. The orbit of each planet is an ellipse with the Sun at one of the foci.

    2. Each planet orbits the Sun such that the radius vector connecting the planet and the Sun sweeps out equal areas in equal times.

    3. The squares of the sidereal periods of any two planets are proportional to the cubes of their mean distances from the Sun.

The first two were published in 1609 in *Astronomia Nova* and the third in 1619 in ➤*Harmonice mundi*. The physical basis for the laws was not understood until Isaac Newton (1642–1727) formulated his law of gravity.

Kepler's Star A ➤*supernova* in the constellation Ophiuchus in October 1604. It was observed and its position determined by Johannes Kepler. It reached a maximum magnitude of about −2.5 and the light curve shows it to have been a Type I supernova. The remnant is a source of radio emission, and a faint optical remnant has also been identified.

Kerr metric A solution discovered by Roy Kerr in 1963 of Einstein's equations for the spacetime of a rotating ➤*black hole*. The Kerr metric shows that about half the mass of a black hole can theoretically be extracted as rotational energy. This is of significant importance in models of ➤*quasars* and ➤*active galactic nuclei* that invoke rotating black holes as the 'central engine'. ➤*metric*.

Keyhole Nebula (NGC 3372) A dark dust nebula near the centre of the ➤*Eta Carinae Nebula*. The name comes from the shape. The 'eye' of the keyhole is a bubble expanding at 40 km/s.

Keystone An ➤*asterism* formed by the four stars ε (Epsilon), ζ (Zeta), η (Eta) and π (Pi) in the constellation Hercules.

Kids The group of three stars ε (Epsilon), ζ (Zeta) and η (Eta) in the constellation Auriga.

kiloparsec (symbol kpc) A unit of distance equal to one thousand ➤*parsecs*, which is 3,261.61 light years.

kinematics The mathematical analysis of movement, without reference to mass or force.

Kirkwood gaps Spaces in the radial distribution of asteroids due to ➤*commensurabilities* and resonances of orbital period with that of Jupiter. There are notable gaps corresponding to orbital period ratios of 4:1, 3:1, 5:2, 7:3 and 2:1. Any asteroids formerly in such orbits would be perturbed by regular gravitational interactions with Jupiter. The explanation was first given by Daniel Kirkwood in 1857. However, beyond 3.5 AU from the Sun, resonances correspond not to gaps but to isolated groups of asteroids at ratios of 3:2, 4:3 and 1:1. The reason for this is not fully understood.

Kitt Peak A mountain-top observatory site near Tucson, Arizona, which is home to one of the largest collections of astronomical research instruments in the world. These include the ➤*Kitt Peak National Observatory*, facilities of the US ➤*National Solar Observatory*, and the ➤*WIYN Telescope* operated by the US ➤*National Optical Astronomy Observatories*. A number of universities and other research organizations also lease space on Kitt Peak for astronomical work.

Kitt Peak National Observatory (KPNO) A facility of the US ➤*National Optical Astronomy Observatories* located at ➤*Kitt Peak* in Arizona. The largest telescope is the 4-metre (160-inch) ➤*Mayall Telescope*. The other instruments include a 2.1-metre (82-inch) telescope and the Burrell ➤*Schmidt camera*.

Kleinmann–Low Nebula An extended source of infrared radiation in the ➤*Orion Nebula*. It is a region of star formation in the central dark part of the nebula, lying behind the optically luminous gas.

K line ➤*Fraunhofer lines*.

Kohoutek, Comet ➤*Comet Kohoutek*.

Koronis family One of the ➤*Hirayama families* of asteroids, at a mean distance of 2.88 AU from the Sun. The members are very similar silicaceous types and are presumed to come from the break-up of a single parent body about 90 km (56 miles) in diameter. The largest member is 208 Lacrimosa, which is about 45 km (28 miles) in diameter. The family is named after 158 Koronis, diameter 35 km (22 miles), discovered in 1876.

KPNO Abbreviation for ➤*Kitt Peak National Observatory*.

KREEP An acronym given to a group of basaltic rocks found on the Moon, which are particularly rich in potassium (symbol K), rare earth elements (REE), and phosphorus (P), though the composition varies widely between different types. Samples of KREEP were returned from the Oceanus Procellarum by the Apollo 12 mission. These were rich also in uranium, thorium and zirconium.

Kreutz group ➤*sungrazer*.

Krüger 60 A faint binary star in the constellation Cepheus. The two members of the system, magnitudes 10 and 11, orbit each other in a period of 44 years. Since the orbit is seen 'face on' and the two stars are easily resolved, the relative motion shows clearly in photographic sequences taken over a period of decades. Both are dwarf ➤*M stars* and the fainter component is a ➤*flare star*. At a distance of 13 light years, it is one of the nearest stars to the solar system.

KSC Abbreviation for ➤*Kennedy Space Center.*

K star A star of ➤*spectral type* K. K stars have surface temperatures in the range 3,500 – 4,900 K and are orange in colour. Lines of neutral and ionized calcium are prominent in the spectrum; there are also numerous lines of neutral metals and molecular bands, particularly at the cooler end of the range. ➤*Arcturus* and ➤*Aldebaran* are examples of K stars.

Kuiper Airborne Observatory (KAO) A 0.915-metre (36-inch) Cassegrain reflecting telescope mounted in a Lockheed C141 Starlifter jet transport aircraft, which was operated between 1975 and 1996 by NASA as a national facility in the USA. It was based at NASA's Ames Research Center, Mountain View, California. Important discoveries made with the observatory include the ring system around the planet Uranus. Its replacement, ➤*SOFIA*, is due to begin operations in 2001.

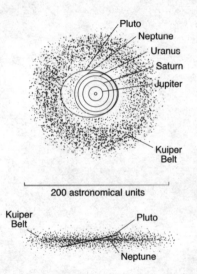

Kuiper Belt. The location and extent of the Kuiper Belt in relation to the orbits of the outer planets, seen from above the plane of the solar system (top) and along the plane (bottom).

Kuiper Belt A population of small icy bodies, similar in size to ➤*asteroids*, occupying a ring-shaped region in the plane of the solar system extending from the orbit of Neptune (30 AU from the Sun) out to possibly 100 or 150 AU. This population, members of which are variously described as 'Kuiper Belt objects', 'trans-Neptunian objects' or 'ice dwarfs', is believed to be the source of ➤*short-period comets*. The name of Gerard Kuiper, a distinguished

Dutch-American planetary scientist, became associated with the belt because, in 1951, he predicted its existence from theoretical work on the origin of the planetary system. However, an Irish writer and theorist, Kenneth E. Edgeworth, had published similar arguments in 1943 and 1949. In recognition of his contribution, the belt is also sometimes termed the Edgeworth–Kuiper Belt.

The first observational evidence for the existence of the Kuiper Belt was the discovery in 1992 of the faint object known as 1992 QB1 in a near-circular orbit about 50 AU from the Sun. About 50 further objects in similar orbits were found in the following few years. It has been suggested that ➤ *Pluto* is the largest member of the Kuiper Belt.

L

labes (pl. labes)   A landslide. A descriptive term used in naming planetary features.

labyrinthus (pl. labyrinthı)   A complex system of intersecting valleys on the surface of a planet.

Lacerta (The Lizard)   A small inconspicuous constellation between Cygnus and Andromeda. It was introduced by Johannes Hevelius in the late seventeenth century and contains only one star brighter than fourth magnitude. ➤Table 4.

Lacertid   ➤*BL Lac object.*

lacus (pl. lacus)   Literally 'lake', a term used in the names of certain dark, isolated features on the Moon. Its use dates from a time when it was believed that the darker features on the Moon were of liquid water; this is now known to be untrue. In recognition of its use over a long period, the term has been retained in official place names on the Moon.

'Lacus' was also formerly used in the names of certain features on Mars. Though these names are still often used by amateur astronomers, they have been replaced officially by more precise descriptive terms. Thus, what was known as Solis Lacus is now properly called Solis Planum.

Lagoon Nebula (M8; NGC 6523)   A luminous nebula in the constellation of Sagittarius. It is a complex of ➤*ionized hydrogen*, gas and dust with hot, recently formed stars. A star cluster, NGC 6530, lies near the centre of the nebula. The light from two naked-eye stars in the cluster, 7 and 9 Sagittarii, is responsible for ionizing the gas. The nebula is estimated to lie at a distance of 4,500 light years.

Lagrangian points   Points in the orbital plane of two massive objects circling about their common centre of gravity where a particle of negligible mass can remain in equilibrium. There are five such points for two bodies in circular orbits around each other, but three are unstable to small perturbations. The other two, at points 60° either side of the less massive body and in the same orbit, are stable. The ➤*Trojan asteroids*, which share the orbit of Jupiter, are examples of masses trapped at the two stable Lagrangian points in an orbit. ➤*Roche lobe.*

Langrenus   A large lunar crater, 132 kilometres (82 miles) in diameter, on the eastern border of the Mare Fecunditatis. It is a bright crater with a central peak, terraced walls and a modest ray system.

La Palma   The island in the Canary Islands group where the ►*Observatorio del Roque de los Muchachos* is located.

Large Binocular Telescope (LBT)   A telescope consisting of two 8.4-metre (27.5-foot) mirrors on a single mount, to be constructed at the ►*Mount Graham International Observatory* in Arizona. The project is a collaboration between the University of Arizona and the Arcetri Astrophysical Observatory in Florence, Italy. The binocular arrangement will give the telescope a light-gathering power equivalent to a single 11.8-metre (39-foot) mirror, and a resolution corresponding to a 23-metre (75.5-foot) telescope.

Large Magellanic Cloud (LMC)   ►*Magellanic Clouds.*

Larissa   A satellite of Neptune (1989 N2) discovered during the flyby of ►*Voyager 2* in August 1989. ►Table 6.

Las Campanas Observatory   An observatory in the Sierra del Condor, Chile, at a height of 2,300 metres (7,500 feet), operated by the Carnegie Institution of Washington. The main instruments are 2.5-metre (100-inch) and 1-metre (40-inch) reflectors.

Laser Interferometer Gravitational-Wave Observatory (LIGO)   A project by California Institute of Technology and the Massachusetts Institute of Technology to construct a detector of ►*gravitational waves* in the USA. It will consist of detectors in Washington State and Louisiana. Each L-shaped installation will beam laser light down 4-kilometre-long (2.5-mile), evacuated tubes. Changes in the length of the light path caused by gravitational waves should register in the detectors.

laser ranging   ►*satellite laser ranging.*

La Silla Observatory   The observatory of the ►*European Southern Observatory*, located in the southern part of the Atacama desert, about 600 kilometres (370 miles) north of Santiago de Chile, at an altitude of 2,400 metres (7,900 feet). The instruments include a 3.6-metre (142-inch) telescope, the 3.5-metre (138-inch) ►*New Technology Telescope* and the 15-metre (50-foot) Swedish/ESO Submillimetre Telescope.

last quarter   The ►*phase* of the Moon when it is waning towards new Moon and appears as an illuminated semicircle. Last quarter is formally defined as the time when the Moon's celestial ►*longitude* is 270° greater than the Sun's. It occurs about seven days after full Moon.

late-type star   A relatively cool star, such as one of ►*spectral type* K or M. The nomenclature is misleading, dating from a time when the main sequence of the ►*Hertzsprung–Russell diagram* was thought to be an evolutionary track. It is, nevertheless, still in widespread use. Hot stars are described as 'early' in this nomenclature.

latitude Angular distance north or south of the equator in a spherical coordinate system. In celestial ►*equatorial coordinates*, the counterpart of latitude is ►*declination*.

LBT Abbreviation for ►*Large Binocular Telescope*.

LDEF Abbreviation for ►*Long Duration Exposure Facility*.

leap second ►*Universal Time*.

leap year A year containing 366 days instead of the normal 365. Leap years are a feature of the ►*Julian calendar* and the ►*Gregorian calendar* and were introduced to ensure that the calendar year keeps step with the seasons.

Leda A small satellite of Jupiter (number XIII), discovered by Charles Kowal in 1974. It is only about 15 kilometres across and belongs to a group of four satellites whose closely spaced orbits all lie between 11.1 and 11.7 million kilometres from Jupiter. (The others are Himalia, Lysithea and Elara.) ►Table 6.

Lemaître universe A model of the universe that begins with a ►*Big Bang*, has a static phase and then expands indefinitely. It is named after Georges Lemaître (1894–1966), who in 1927 published a major work on the expansion of the universe. He was the first to advocate expansion from a 'primeval atom', while Einstein was still propounding the merits of a static universe.

lens A transparent optical device that changes the directional properties of a beam of light passing through it. Different types of lens are designed for different purposes. *Convex* and biconvex lenses cause a parallel beam of light to converge at the focal point. *Concave* and biconcave lenses diverge a parallel beam. Lenses are frequently used in combination to achieve results unobtainable with a single lens, for example in a telescope eyepiece. Elements may be cemented together: *doublets* and *triplets* are lenses composed of two and three elements, respectively. Such combinations, made from lenses of different types of glass, are used to minimize ►*chromatic aberration*. The objective lens of a refracting telescope is its light-gathering element.

lenticular galaxy A galaxy of type So in the ►*Hubble classification,* intermediate between elliptical and spiral types, so named because the apparent shape of such galaxies resembles a convex lens.

Leo (The Lion) One of the twelve zodiacal constellations, included by Ptolemy in his list of 48 (*c.* AD 140). The pattern made by the brightest stars of this large and conspicuous constellation bear some resemblance to the outline of a lion in profile. The asterism outlining the head is known as the Sickle. There are ten stars brighter than fourth magnitude, the brightest being ►*Regulus* and ►*Denebola*. Leo also contains numerous galaxies, including five from the ►*Messier Catalogue* (M65, M66, M95, M96 and M105). ►Table 4.

Leo Minor (The Little Lion)   A small and very inconspicuous constellation between Leo and Ursa Major. It was introduced by Johannes Hevelius in the late seventeenth century and contains only one star brighter than fourth magnitude. ►Table 4.

Leonids   An annual ►*meteor shower*, the radiant of which lies within the 'sickle' in the constellation Leo. The peak occurs on 17 November and the normal limits are two days either side. Though a small number of meteors are detected each year, spectacular displays are occasionally seen. The Leonids in 1966 gave observers in the USA the richest shower ever recorded, with rates as high as 40 meteors a second.

The shower is associated with Comet 55P/Tempel–Tuttle, first recorded in 1865, which has a period of 33 years. The meteoric material is concentrated near the comet and is not evenly spread around the orbit. Good displays are possible only every 33 years, though they are not necessarily seen even then, if the comet passes too far from the Earth's orbit.

lepton era   The period from one-millionth of a second to one second after the primordial ►*Big Bang*. During this interval, the behaviour of the universe was dominated by the lighter elementary particles, such as electrons and neutrinos, together with a tiny proportion (one-billionth) of protons and neutrons.

Lepus (The Hare)   One of the original 48 constellations listed by Ptolemy (*c*. AD 140). It lies immediately south of Orion, possibly representing a hare pursued by the Hunter. It is small but distinctive and contains seven stars brighter than fourth magnitude. ►Table 4.

Lesser Dog   English name for the constellation ►*Canis Minor*.

Lesser Water Snake   English name for the constellation ►*Hydrus*.

Lexell's Comet   A comet discovered by C. Messier in 1770, but named after A. Lexell (1740–84) who investigated its orbit. He showed that a close approach to Jupiter in 1767 had caused a large change in its orbit that brought it close enough to the Earth to be visible. The comet passed within 1.2 million kilometres of the Earth, still the closest recorded approach of a comet. However, another close approach to Jupiter in 1779 further perturbed the orbit, so drastically that it was never seen again.

LHA   Abbreviation for ►*local hour angle*.

Libra (The Scales)   One of the twelve zodiacal constellations listed by Ptolemy (*c.* AD 140), though its stars were previously regarded as part of the Scorpion, which is next to it. Libra is one of the least conspicuous constellations in the zodiac, with just five stars brighter than fourth magnitude. ➤Table 4.

libration Any of several effects that alter precisely which hemisphere of the Moon's surface is visible from the Earth. Despite the fact that the Moon's rotation and orbital periods are equal, so that the Moon very nearly keeps the same face towards the Earth all the time, a total of 59 per cent of the Moon's surface can be viewed from the Earth at some time or other as a result of libration. *Physical libration* is a real irregularity in the Moon's rotation; the larger effect is *geometrical libration*, in both latitude and longitude. Libration in latitude results from the Moon's orbit being inclined to the ecliptic by an angle of 5° 9'. The elliptical shape of the Moon's orbit means that its orbital velocity is not constant, and this produces libration in longitude of 7° 45'. Additionally, *diurnal libration* is a small effect that results from observing the Moon at different times of day.

Lick Observatory An observatory belonging to the University of California. The observatory site is on Mount Hamilton in the Californian Diablo Range at a height of 1,300 metres (4,200 feet). This is now only an observing station, and it is administered from the University of California's Santa Cruz campus.

The funds for the observatory were provided to the University by a millionaire businessman, James Lick (1796–1876). The building and a 92-centimetre (36-inch) refracting telescope were completed in 1888, twelve years after Lick's death. He is buried at the base of the telescope.

The main research telescope is now the Shane 3-metre (120-inch) reflector, in operation since 1959. The 92-centimetre (36-inch) Crossley reflector, built by Andrew Common, was presented to the observatory in 1895 by its English owner, Edward Crossley. The most recent instrument is the Nickel 1-metre (40-inch) reflector, a modern automated instrument completed in 1980. There is also a 50-centimetre (20-inch) ➤*astrograph*, with twin tubes designed to take photographs simultaneously in the blue and yellow regions of the spectrum.

light bucket A colloquial expression for a ➤*flux collector*.

light curve A graph on which the light output from a ➤*variable star* (or other varying astronomical object) is plotted against time.

light echo A reflection of the burst of light from a ➤*supernova* or a ➤*nova* by neighbouring interstellar clouds, resulting in a ring of light surrounding the supernova or nova. The ring of light is seen to expand over time.

light pollution The scattering of light from man-made sources into the night sky, which increases the background brightness of the sky above its natural level and interferes with astronomical observations. Light pollution is worst close to major centres of civilization. Legislation has been enacted in part of the USA to protect important observatory sites from the damaging effects of unnecessary artificial lighting in nearby cities. However, the problem is a growing one and a matter of worldwide concern for both amateur and professional astronomers.

light-time   The time it takes for light, or any other form of electromagnetic radiation, to travel a particular distance.

light year (l.y.)   The distance travelled through a vacuum by light (or any other form of electromagnetic radiation) in one year. A light year is equivalent to 9.4607×10^{12} kilometres, 63,240 astronomical units or 0.306 60 parsecs.

LIGO   Abbreviation for ►*Laser Interferometer Gravitational-Wave Observatory*.

limb   The extreme edge of the visible disc of a body such as the Sun, Moon or a planet.

limb brightening   An increase in the intensity of emitted radiation between the centre and the limb of the visible disc of the Sun, or other astronomical body.

limb darkening   A decrease in the observed brightness of the disc of the Sun, or other astronomical body, between the centre and the limb.

limiting magnitude   The ►*magnitude* of the faintest object that can be detected with a given imaging system.

linea (pl. lineae)   An elongated surface marking on a planet.

line profile   The variation in light intensity in a spectrum across a narrow wavelength range occupied by a single spectral line. When traced as a graph of intensity against wavelength, a typical stellar absorption line shows a bell-shaped profile, for example. If a line profile can be determined in some detail, information can be deduced about the physical conditions in the gas where the line originated.

liner   An acronym for low-ionization narrow emission line region. Such regions are found in a high proportion of spiral galaxies. Apparently they are a weak form of ►*active galactic nucleus*, which suggests that, at some level, all galaxies exhibit activity in their nuclei.

line spectrum   A ►*spectrum* that features ►*emission lines*, ►*absorption lines*, or both.

Linné   A small lunar crater, 2.4 kilometres (1.5 miles) in diameter, situated in the Mare Serenitatis. A mid-nineteenth-century claim that a close companion, a crater called Linné B, had disappeared seems likely to have been spurious since the observation was not confirmed. Though small, Linné is relatively conspicuous because it is surrounded by a bright area, presumably a small blanket of ejecta.

Lion   English name for the constellation ►*Leo*.

Lion Cub   English name for the constellation ►*Leo Minor*.

lithosiderite   Alternative name for a stony-iron ►*meteorite*.

lithosphere   The rigid outer layer of a planetary body, including the crust and part of the upper mantle, which lies above the weaker asthenosphere.

Little Bear   English name for the constellation ►*Ursa Minor*.

Little Dipper   A popular North American name for the constellation ►*Ursa Minor*, describing the figure formed by its main stars: Beta (β), Gamma (γ), Eta (η), Zeta (ζ), Epsilon (ε), Delta (δ) and Alpha (α).

Little Dog   English name for the constellation ►*Canis Minor*.

Little Dumbbell   Popular name for M76 (NGC 650), a ►*planetary nebula* in Perseus. It is the faintest object in the ►*Messier Catalogue*.

little green men   A name given by Jocelyn Bell Burnell to the first four ►*pulsars* discovered before their nature as rotating neutron stars was known.

Little Horse   English name for the constellation ►*Equuleus*.

Little Lion   English name for the constellation ►*Leo Minor*.

Lizard   English name for the constellation ►*Lacerta*.

LM   Abbreviation for Lunar Module, a term used to describe part of the craft used in the ►*Apollo programme* of Moon landings.

LMC   Abbreviation for Large ►*Magellanic Clouds*.

LMST   Abbreviation for ►*local mean solar time*.

Local Bubble   A low-density region of the ►*interstellar medium* within which the Sun is located. At the boundary of the 'bubble', the gas density rises sharply by a factor of at least 10. The bubble boundary is nearest to the solar system in a direction roughly towards the galactic centre, where an extended dust cloud (known as Tinbergen's Cloud) lies about 30 light years away. The nature of this cloud is not well understood but it appears that its relative motion has led to a considerable deformation of the Local Bubble. The furthest extent of the bubble is about 500 light years away in the direction of the constellation Canis Major, where the bubble wall is permeated by tunnel-like structures. The Local Bubble was probably blown out by the shock wave from one or more ►*supernova* explosions that took place in our part of the Galaxy in the remote past.

Local Cloud A small diffuse cloud of interstellar material in which the Sun is embedded. It is about 20–30 light years across and the Sun is located towards one edge. The Local Cloud lies within the ➤*Local Bubble*.

Local Group An assembly of galaxies to which our own Milky Way ➤*Galaxy* belongs. The dominant members are the ➤*Andromeda Galaxy* (M31), which is the largest and most massive, and our own Galaxy. Next in size are the spiral galaxy in Triangulum, M33, which is a near companion of M31, and the Large ➤*Magellanic Cloud*, near our Galaxy. The other members of the Local Group are small elliptical and irregular galaxies plus a number of dwarf spheroidal galaxies, resembling isolated globular clusters. These dwarf galaxies are so faint, it is very difficult to detect them at distances greater than the Andromeda Galaxy, so the total number is unknown. The four small elliptical galaxies (NGC 221, 205, 185 and 147) are satellites of M31; the Magellanic Clouds and various dwarf galaxies are satellites of our own. Thus the Local Group does not have a central condensation, but two subgroups centred around the two most massive members.

The Local Group occupies a volume of space with a radius of about 3 million light years (1 megaparsec). The next nearest galaxies are two or three times this distance away. ➤*Maffei galaxies*.

local hour angle (LHA) The ➤*hour angle* of a celestial object as measured by an observer at a particular locality. The local hour angle of an object at any instant varies according to the longitude of the observer.

local mean solar time (LMST) ➤*Mean solar time* at a particular locality. This will, in general, differ from both the ➤*civil time* at that place and ➤*apparent solar time* measured there, according to the ➤*time zone* in use and the ➤*equation of time*, respectively.

local sidereal time (LST) The ➤*sidereal time* at a particular locality. Local sidereal time differs from ➤*Greenwich Sidereal Time* by four minutes for each degree of longitude east or west of Greenwich. Times are later for locations to the east and earlier for those to the west.

local standard of rest (LSR) A frame of reference centred on the Sun in which the average velocity of the stars in the vicinity of the Sun is zero. Each star, including the Sun, is moving relative to the LSR. The Sun's motion, at about 20 km/s, is directed towards the solar ➤*apex*. The LSR's velocity of 250 km/s with respect to the centre of the Galaxy is a measure of the Galaxy's general rotation in the Sun's neighbourhood.

Local Supercluster A ➤*supercluster* of galaxies, centred on the ➤*Virgo Cluster*, which includes the ➤*Local Group* on its periphery. It is more than a hundred million light years in diameter and its existence was first suggested by G. de Vaucouleurs in 1956.

local time   Either the ►*apparent solar time* or the ►*mean solar time* at a particular locality. These will in general differ from both the ►*civil time* at that place and, for example, ►*Universal Time*, depending on the ►*time zone* in use and the longitude of the place, respectively.

Loki   A volcanic centre on ►*Io*, observed to be active at the encounters of both ►*Voyagers 1* and *2* in 1979.

long-baseline interferometry   A technique in radio astronomy in which two or more radio telescopes separated by up to 1,000 kilometres or so are linked in real time by signals transmitted via microwaves or cable in order to form a ►*radio interferometer.* ►*very-long-baseline interferometry.*

Long Duration Exposure Facility (LDEF)   A package of fifty experiments launched into Earth orbit from the Space Shuttle in 1984 for the study of the space environment over a period initially intended to be up to one year. Delays in the Shuttle programme meant that the experiments were not recovered until January 1990, just a few weeks before they would have re-entered the atmosphere and been lost.

longitude   In a spherical coordinate system, angular distance round the equator or a circle parallel to it from an arbitrary zero point. In the celestial ►*equatorial coordinate* system, the counterpart of longitude is ►*right ascension.*

longitude of perihelion   In orbital motion, the sum of the ►*longitude of the ascending node* and the ►*argument* of perihelion. ►*orbital elements.*

longitude of the ascending node (symbol Ω)   One of the principal ►*orbital elements* used to describe mathematically the shape and orientation in space of an orbit. It gives the point where the orbit intersects the reference plane as the object travels from south to north. For bodies orbiting the Sun, the reference plane is the ecliptic and the zero point is the ►*First Point of Aries* (the vernal equinox).

long-period comet   A ►*comet* with an almost parabolic orbit and a period of revolution round the Sun exceeding 200 years. Some have orbital periods of millions of years. ►*short-period comet.*

long-period variable   Variable stars with periods between about 100 and 1,000 days. Both the period and the amplitude, which is typically several magnitudes, are subject to considerable variation between cycles. These variables are ►*red giant* stars of which ►*Mira* is one of the best-known examples.

look-back time   The interval between the present and the time in the past when the radiation currently reaching us from a given galaxy was actually emitted. The look-back time increases with increasing ►*redshift*. Long look-back times take the observer to earlier eras in the history of the universe.

Lost City Meteorite A *►chondrite* that fell in Oklahoma in 1970. The *►fireball* observed as the meteorite passed through the atmosphere was photographed and the results used to locate the fall, which was recovered a few days later.

Lovell Telescope *►Nuffield Radio Astronomy Laboratories.*

Lowell Observatory A private observatory in Flagstaff, Arizona. It was founded in 1894 by Percival Lowell (1855–1916), who was particularly interested in the possibility of intelligent life on Mars. Among notable discoveries made at the Lowell Observatory was that of the planet Pluto by Clyde Tombaugh in 1930. A small staff of astronomers undertakes research in planetary science and other aspects of astronomy.

There are five telescopes at the Flagstaff site: the original 60-centimetre (24-inch) refractor, the 60-centimetre Morgan reflector, a 53-centimetre (21-inch) reflector, a 45-centimetre (18-inch) *►astrograph*, together with the 33-centimetre (13-inch) telescope used by Tombaugh to search for Pluto, restored and returned to its original dome in 1996. In addition, the observatory operates three other telescopes at Anderson Mesa, 24 kilometres (15 miles) south-east of Flagstaff, including the 1.8-metre (72-inch) Perkins Telescope of Ohio State and Ohio Wesleyan Universities.

LRV Abbreviation for *►Lunar Roving Vehicle.*

LSR Abbreviation for *►local standard of rest.*

LST Abbreviation for *►local sidereal time.*

luminosity (symbol L) The energy radiated per unit time by a luminous body. *►magnitude.*

luminosity class *►spectral type.*

luminosity function A mathematical or empirical expression describing the number of stars or galaxies per unit volume of space at each possible luminosity value, for a particular sample such as a galaxy or star cluster.

Luna A series of Soviet space probes to the Moon launched between 1963 and 1976. The first three were named *►Lunik. Luna 9* achieved the first soft landing, in the Oceanus Procellarum, in January 1966. *Luna 10* in March 1966 became the first lunar orbiting satellite. *Luna 16* in September 1970, *Luna 20* in 1972 and the last of the series, *Luna 24* in August 1976, returned soil samples. *Lunas 17* and *21* landed the *►Lunokhod* roving vehicles on the Moon. Successes were also achieved with *Lunas 11, 12* and *13* (1966), *14* (1968), *19* (1971) and *22* (1974).

lunar Pertaining to the Moon.

lunar eclipse *►eclipse.*

Lunar Orbiter   A series of American lunar probes launched in 1966 and 1967 with the primary objective of mapping the Moon and locating suitable landing sites for the manned ➤*Apollo programme*. It was the first systematic exploration of the Moon's surface and all five craft in the series were very successful. Photography was carried out using conventional photographic film that was developed automatically on board, then scanned so that the information could be transmitted back to Earth for reconstruction.

lunar parallax   The average value of the Moon's equatorial horizontal parallax (➤*diurnal parallax*), which is 3422.45 arc seconds.

Lunar Prospector   A NASA mission launched in January 1998. It was designed to spend one year in orbit around the Moon, mapping the chemical composition of the lunar surface and the Moon's magnetic and gravity fields.

Lunar Roving Vehicle (LRV)   Battery-powered vehicles for travelling on the Moon's surface taken with the last three ➤*Apollo programme* missions (15, 16 and 17). They were introduced because the objectives of the previous missions, especially as they became more ambitious, had been hampered by the astronauts' lack of mobility. The journeys made by the astronauts of *Apollos 15, 16* and *17* were 28, 27 and 35 kilometres (17.5, 16.5 and 22 miles), respectively.

lunation   A complete cycle of the phases of the Moon. The time taken is the synodic month, which lasts 29.530 59 days.

Lund Observatory   A Swedish observatory, dating originally from 1672. The present buildings in the centre of Lund were constructed in 1867. The observing station is now located 18 kilometres (11 miles) outside Lund, where the main instrument is a 61-centimetre (24-inch) reflector.

Lunik   The name of the first three Moon probes launched by the Soviet Union in January, September and October 1959. *Lunik 1* missed the Moon by 5,000 kilometres (3,000 miles). *Lunik 2* crashed near the crater Archimedes, but *Lunik 3* returned the first pictures of the lunar farside. Subsequent probes in the series were named ➤*Luna*, starting with *Luna 4*.

lunisolar precession   ➤*precession*.

Lunokhod   An automated roving vehicle landed on the Moon during two unmanned Soviet missions, *Luna 17* and *Luna 21*. Lunokhod 1 was landed at a site in the western part of Mare Imbrium by *Luna 17* on 17 November 1970 and operated for 10 months. Lunokhod 2 was delivered on 16 January 1973 by *Luna 21* to the eastern part of Mare Serenitatis, where it worked for four months. The total distances travelled were 10.5 kilometres and 37 kilometres (6.5 and 23 miles), respectively. Each eight-wheeled vehicle carried cameras, a communication system, a laser reflector, a magnetometer, solar panels and a cosmic ray detector.

Lupus (The Wolf)   One of the southern constellations included in Ptolemy's list of 48 (*c.* AD 140). It lies between Scorpius and Centaurus and contains eight stars brighter than fourth magnitude. ➤Table 4.

l.y.   Abbreviation for ➤*light year.*

Lyman–alpha forest   In the spectrum of a high-redshift ➤*quasar,* large numbers of closely spaced absorption lines originating in hydrogen gas clouds along the line of sight. The principal absorption lines are the Lyman alpha line of hydrogen at a range of redshifts corresponding to the velocities of the intervening clouds. ➤*Lyman series.*

Lyman series   A series of ➤ *spectral lines* in the ultraviolet spectrum of atomic hydrogen, resulting from transitions between the ground state and the various possible excited states. The lines are termed alpha, beta, gamma, and so on, corresponding to transitions to and from the first, second, third, and successive excited states.

The Lyman alpha line lies at 121.566 nanometres, the wavelengths of the other lines in the series all being shorter. It is a strong line in the emission spectra of ➤*active galactic nuclei* and ➤*quasars* and, if they are at high redshifts, appears shifted into the visible part of the spectrum.

Lyndon B. Johnson Space Center (JSC)   A large space technology complex at Houston, Texas, which is the headquarters for the astronauts of the US National Aeronautics and Space Administration (NASA). It is the site of 'mission control', responsible for the operations of the ➤*Space Shuttle* after launch and prior to landing. It has major offices responsible for the development and operation of the Space Shuttle Orbiter (the manned vehicle) as well as for the National Space Transportation System (the Space Shuttle as a recoverable rocket launching system).

JSC conducts major programmes related to manned space flight, space medicine and aviation. Exploratory studies of future ventures in manned and robotic space travel, such as lunar bases and the proposed Mars Rover Sample Return Mission, are undertaken there. It was named after the late US President, who took a strong interest in the space programme.

Lynx   An obscure northern constellation introduced in the late seventeenth century by Johannes Hevelius to fill a gap between Auriga and Ursa Major. It contains only two stars brighter than fourth magnitude. ➤Table 4.

Lyot filter   An interference filter with a very narrow wavelength passband of particular use for observing the Sun. It was devised by the French astronomer B. F. Lyot (1897–1952) in 1938. The filter consists of a sequence of plane polarizers separated by plates of quartz. By choosing the precise thickness and number of elements in the filter particular wavelengths of light can be isolated.

Lyra (The Lyre)   A small but prominent constellation in the northern hemisphere, listed by Ptolemy (*c.* AD 140). Its brightest star, ➤*Vega*, is zero magnitude and the fifth-brightest star in the sky. There are three other stars brighter than fourth magnitude. Epsilon Lyrae is a noted 'double double', consisting of a widely spaced pair of close double stars. Lyra also includes one of the best-known of all ➤*planetary nebulae*, the ➤*Ring Nebula*. ➤Table 4.

Lyre   English name for the constellation ➤*Lyra*.

Lyrids   An annual ➤*meteor shower*, sometimes called the *April Lyrids*, the radiant of which lies on the border between the constellations Lyra and Hercules. The shower peaks around 22 April and the normal limits are 19–25 April. The meteor stream is associated with Comet Thatcher (C/1861 G1). Though normally sparse, the shower has occasionally been good. Records of the shower date back 2,500 years.

Lysithea   A small satellite of Jupiter (number X), discovered by S. B. Nicholson in 1938. It is only 36 kilometres across and belongs to a group of four satellites whose closely spaced orbits all lie between 11.1 and 11.7 million kilometres from Jupiter. (The others are Leda, Himalia and Elara.) ➤Table 6.

M

M   A prefix given to objects in the ➤*Messier Catalogue.*

MACHOs   Invisible stellar-sized objects, such as dim stars or ➤*brown dwarfs,* which may account for the difference in mass between the luminous components of the Galaxy and the apparent total mass of the Galaxy implied by its gravitational behaviour. The term is said by those who invented it to stand for Massive Astrophysical Compact Halo Objects, but in part was contrived to contrast with another form of invisible matter, so-called ➤*WIMPs.* ➤*microlensing.*

Mach's principle   A philosophical idea propounded by Ernst Mach (1838– 1916) to the effect that the inertial properties of an individual piece of matter are determined by the distribution of distant matter in the universe. Einstein was influenced by this idea, though he and others have failed to incorporate it into his ➤*General Relativity* theory.

macula (pl. maculae)   A dark spot on the surface of a planet.

Maffei galaxies   Two galaxies discovered by Paolo Maffei in 1968. They are visible only in the red and infrared.

Maffei I is a giant elliptical galaxy about four million light years away, and might be an outlying member of the ➤*Local Group.* It is a large galaxy with a mass 200 billion times that of the Sun. Its position in the sky means that it is viewed through the dust clouds of the Milky Way, which dims its light by a factor of one hundred.

Maffei II, an average spiral galaxy, lies beyond the Local Group, five times further away than Maffei I.

Magellan   A US orbiting probe of ➤*Venus,* launched from the Space Shuttle *Atlantis* on 4 May 1989. Its objective was to map by means of ➤*synthetic aperture radar* at least 70 per cent of the surface of Venus to a resolution of several hundred metres. The radar technique is essential since Venus is perpetually covered by opaque cloud. *Magellan* arrived at Venus on 10 August 1990 and its first phase of operation, over a period of 243 days, was successfully completed in May 1991, 84 per cent of the surface having been mapped. The next phase of observation involved filling in gaps and making more detailed observations.

Previous studies had shown that volcanic lava flows cover about 80 per cent of Venus. The *Magellan* images have made it possible to examine these features in much closer detail. A number of large ➤*shield volcanoes* have been identified, along with impact craters, one of the largest of which is 275 kilometres (170 miles) in diameter, together with a variety of features unique to Venus. The wealth of data returned by *Magellan* makes a very significant addition to our knowledge of the surface of Venus.

Magellanic Clouds   Two small, irregular galaxies, which are satellites of our own ►*Galaxy*. They are visible as hazy patches in the southern sky. The *Large Magellanic Cloud* (LMC) is in the constellation Dorado and is about 170,000 light years away. The *Small Magellanic Cloud* (SMC), in Tucana, is about 210,000 light years distant.

Magellanic Stream   A long streamer of neutral hydrogen gas apparently spanning the 200,000 light years between the ►*Magellanic Clouds* and our own ►*Galaxy*. It forms an arc 150° long in the southern sky. A possible explanation is that it is gas dragged from the Magellanic Clouds in a tidal interaction with the Galaxy.

Magellan Project   A project involving a consortium led by the Carnegie Institution of Washington to construct two 6.5-metre (21-foot) telescopes at the ►*Las Campanas Observatory*.

magnetic monopole   A postulated defect in the fabric of ►*spacetime* that behaves like an isolated north or south pole of a magnet and has a mass 10^{16} times that of the proton. None has ever been detected.

The potential existence of magnetic monopoles is a serious problem that arises in ►*Grand Unified Theories* of the fundamental physical forces. If they exist in significant numbers, they would drastically slow the expansion of the universe, which is not what is observed. The problem of defects is avoided in the ►*inflationary universe*.

magnetic star   A star with an exceptionally strong magnetic field. Magnetic fields more than a thousand times stronger than the Sun's general field have been measured for a group of A stars, which also have peculiar spectra and so are classified as ►*Ap stars*. In the presence of a magnetic field, the lines produced in the stellar spectrum are split into polarized components (the *Zeeman effect*). Although the lines are usually too broad for the split components to be resolved, the change in polarization across the broadened spectral lines can be measured and interpreted in terms of a magnetic field strength.

In almost all cases, the fields and spectral line strengths vary regularly. This observation can be accounted for if the rotation and magnetic axes of the stars do not coincide. ►*peculiar star*.

magnetic storm   ►*geomagnetic storm*.

magnetograph   An instrument used in solar astronomy for mapping the strength, direction and distribution of magnetic field across the surface of the Sun.

magnetometer   An instrument for measuring the strength and direction of magnetic field.

magnetopause   A region between 100 and 200 kilometres (60 and 120 miles) thick that is the boundary layer between the ➤*magnetosphere* and the ➤*solar wind*.

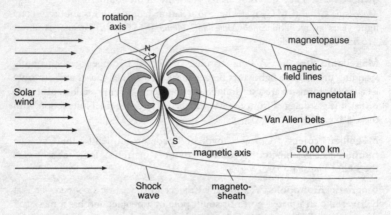

Magnetosphere. The structure of the Earth's magnetosphere.

magnetosphere   The region around the Earth, or any other planet, within which its natural magnetic field is constrained by the ➤*solar wind*.

magnetotail   The part of the ➤*magnetosphere* of the Earth, or of any other planet, which on the planet's nightside extends like a comet tail in the direction opposite the Sun. The Earth's magnetotail extends from around 8 to 10 Earth radii distant on the nightside, to 80, and perhaps as far as 1,000, Earth radii.

magnification (magnifying power)   The factor by which the angular size of an object is apparently increased when it is imaged by a particular combination of telescope and eyepiece. The magnification of an image made by a telescope is given by the focal length of the telescope divided by the focal length of the eyepiece.

A high magnification is not always an advantage. It results in a small field of view in which the effects of bad ➤*seeing* and any optical shortcomings are emphasized. The contrast between markings on planets and the Moon may also be reduced. Eyepieces should therefore be selected to be appropriate to the type of observations being made.

magnitude   A measurement of the brightness of a star or other celestial object. On the magnitude scale, the lowest numbers refer to objects of greatest brightness.

The magnitude system was initially a qualitative attempt to classify the apparent brightness of stars. The Greek astronomer Hipparchus (*c.* 120 BC) ranked stars on a magnitude scale from 'first' for the brightest stars to 'sixth' for those just detectable in a dark sky by the unaided eye. This qualitative description was standardized in the mid-nineteenth century. By this time it was understood that each arbitrary magnitude step corresponded roughly to a similar brightness ratio. (In other words, the magnitude scale is a logarithmic scale of brightness.)

In 1856, N. R. Pogson proposed that a difference in magnitude of 5 should correspond to a brightness ratio of 100:1, a system that is now universally accepted. If two stars differ by one magnitude, their brightnesses differ by a factor equal to the fifth root of 100, i.e. 2.512. This number is known as *Pogson's ratio.* The zero point of the scale was set by assigning standard magnitudes to a small group of stars near the north celestial pole, called the ►*North Polar Sequence.*

The brightness of stars as observed from the Earth, and hence their ►*apparent magnitude,* depends on both their intrinsic luminosity and their distance. ►*Absolute magnitude* is a measure of intrinsic luminosity on the magnitude scale, defined as the apparent magnitude an object would have at the arbitrary distance of ten ►*parsecs.*

The magnitude of an object varies with the wavelength range of the radiation observed. *Visual magnitude* corresponds to the normal sensitivity of the human eye. Photographic magnitude usually refers to the response of a standard photographic emulsion, which is chiefly in the blue and violet part of the spectrum. However, different photographic materials and detectors may have very different colour responses, and quoted magnitudes should include information about the method of measurement. *Bolometric magnitudes* take account of all radiation, both visible and outside the visible range.

Magnitudes measured over a defined wavelength range are often described as 'colours'. The accurate determination of such magnitudes is achieved by ►*photometry.*

magnitude of an eclipse ►*eclipse.*

main sequence A narrow band running from the upper left to lower right on the ►*Hertzsprung—Russell diagram,* on which stellar luminosity is plotted as a function of temperature, with temperature decreasing towards the right.

The temperature and luminosity of most stars places them on the main sequence. This reflects the fact that these two fundamental quantities are largely determined by the mass of the star, some variation being introduced by differences in chemical composition. The main sequence is thus a mass sequence. The points for the most massive stars lie at the upper left and those for the least massive at the lower right.

In stars on the main sequence, the fusion of hydrogen into helium in the stellar core is the source of energy. Stars in stages of ►*stellar evolution* before and after this phase are represented by points elsewhere on the HR diagram. For example, after hydrogen has been exhausted in the core, adjustments taking place internally cause a star to evolve in a way that carries it away from the main sequence towards the upper right of the HR diagram. Most stars spend about 90 per cent of their observable life on the main sequence.

main-sequence star   A star with a temperature–luminosity combination that places it on the ➤*main sequence* of the Hertzsprung–Russell diagram.

major planet   Any of the nine planets Mercury, Venus, Earth, Mars, Jupiter, Saturn, Uranus, Neptune or Pluto. The term 'minor planet' is used interchangeably with ➤*asteroid*.

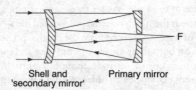

Shell and   Primary mirror
'secondary mirror'

Maksutov telescope. A schematic diagram of the optical arrangement.

Maksutov telescope   A reflecting telescope incorporating a deeply curved meniscus lens, which corrects the optical aberrations of the spherical primary mirror to give high-quality images over a wide field of view. It was invented by D. D. Maksutov (1896–1964).

The practical arrangement is typically a ➤*Cassegrain* system. A small secondary mirror is fixed to the back of the correcting lens and the image is formed just behind the primary mirror, which has a small central hole (see illustration).

The difficulty of making large correcting lenses limits the professional application of the design, but the compactness of the tube, the wide field of view and low focal ratio have made Maksutov telescopes popular with amateurs.

manganese star (mercury star; mercury–manganese star)   A star of ➤*spectral type* B with a peculiar spectrum in which the lines of manganese and a number of other exotic elements, notably mercury, gallium and yttrium, are exceptionally strong.

mantle   The layer in the structure of a planet or satellite lying below the crust and overlying the core. Earth's mantle contains 65 per cent of the planetary mass.

mare (pl. maria)   Literally 'sea', a term used for extensive dark areas on the Moon. Its use dates from a time when it was believed that the darker features on the Moon were of liquid water; this is now known to be untrue. In recognition of its use over a long period, the term has been retained in official names for these features on the Moon. The largest of the maria is called 'ocean' (*oceanus*) rather than 'sea' – Oceanus Procellarum.

The lunar maria are actually 'seas' of solidified lava, dating from the era shortly after its formation more than 4,000 million years ago when the Moon was volcanically active. The molten lava flowed into huge basins that had been excavated by the impacts of large meteorites. By this stage in the Moon's history, the frequency of meteoritic impacts had fallen. There is a noticeably lower density of craters on the lunar maria than on the brighter 'highland' areas ('terrae').

Maria Mitchell Observatory   An observatory in Nantucket, Massachusetts, founded in 1908 as a memorial to Maria Mitchell (1818–88), a pioneering scientist and teacher in an era when very few women took on such academic work. She won international fame after discovering a comet in 1847.

The observatory was established by the Maria Mitchell Association. It houses 18-centimetre (7-inch) and 20-centimetre (8-inch) telescopes used primarily for educational purposes. Variable-star work is undertaken.

Mariner   A series of spacecraft launched by the USA during the 1960s and 1970s in a programme to explore the planets Mercury, Venus and Mars.

Mariner 2 in 1962 achieved the first successful flyby of Venus, to be followed by *Mariner 5* in 1967. *Mariner 4*, launched in 1964, was the first successful probe to Mars and revealed the presence of craters on that planet. *Mariners 6* and *7* followed in 1969.

Mariner 9 was put into orbit around Mars in 1971 and returned over 7,000 images. *Mariner 10* in 1974 was the first two-planet mission. Its orbit allowed three separate encounters with Mercury, providing 10,000 images, as well as a flyby of Venus. There were a total of seven successful missions in the *Mariner* series. Numbers 11 and 12 were renamed ➤*Voyagers 1* and *2*.

Markarian galaxy   Any of the galaxies in the list drawn up by the Soviet astronomer B. E. Markarian in the 1970s, characterized by strong continuum emission of ultraviolet light.

Mars (1)   The fourth major planet from the Sun, often known as the Red Planet because of its distinctive colour, noticeable even to the naked eye.

Mars is one of the terrestrial planets with a diameter just over half that of the Earth. It had long been regarded as the planet (other than Earth) most likely to have life, a view encouraged by the presence of polar ice caps and observations of seasonal changes. Nineteenth-century observers, notably Percival Lowell, convinced themselves that they could make out systems of straight channels, ➤*canals*, that might be artificially constructed. Exploration of the planet by spacecraft has virtually eliminated the possibility that life exists currently on Mars. However, studies of meteorites believed to be of martian origin have fuelled speculation that microscopic life at least may have existed on Mars in the remote past when the climate was wetter and warmer.

Successful US probes to Mars include: ➤*Mariner 4* in 1965, *Mariners 6* and 7 in 1969, *Mariner 9* in 1971, and ➤*Vikings 1* and *2* in 1976. Following the failure of ➤*Mars Observer* in 1993, the USA launched ➤*Mars Global Surveyor* and ➤*Mars Pathfinder* for arrival in 1997. Mars is considered to be a realistic target for a manned landing in the early twenty-first century.

The relatively low density of Mars (3.95 times that of water) suggests that 25 per cent of its mass is contained in an iron core. There is a weak magnetic field, about 2 per cent the strength of the Earth's. The crust is rich in olivine and ferrous oxide, which gives the rusty colour.

The tenuous martian atmosphere is composed of 95.3 per cent carbon dioxide, 2.7 per cent molecular nitrogen and 1.6 per cent argon, with oxygen as a major trace constituent. The atmospheric pressure at the surface is only 0.7 per cent that at the surface of the Earth. However, strong winds in the atmosphere cause extensive dust storms, which occasionally engulf the entire planet.

A variety of clouds and mists occur. Early-morning fog forms in valleys and orographic clouds, formed when air masses cool as winds drive them from low to high ground, appear over the high mountains of the Tharsis region. In winter, the north polar cap is swathed in a veil of icy mist and dust, known as the *polar hood*. A similar phenomenon is seen to a lesser extent in the south.

The polar regions are covered with a thin layer of ice, thought to be a mixture of water ice and solid carbon dioxide. High-resolution images show a spiral formation and strata of wind-borne material. The north polar region is surrounded by stretches of dunes. The polar ice caps grow and recede with the seasons, which arise – as they do on Earth – because the planet's rotation axis is tilted (by 25°) to the orbital plane.

The martian year is about twice the length of the Earth year, so the seasons are also longer. However, the relatively high eccentricity of Mars's orbit makes them of unequal duration: southern summers, which occur when Mars is near perihelion, are shorter and hotter than those in the north. Seasonal changes in the appearance of features as observed from Earth are explained as physical and chemical changes.

There is a marked difference in the nature of the terrain between the two halves of Mars divided roughly by a great circle tilted at 35° to the equator. The more southerly part consists largely of ancient, heavily cratered terrain. The major impact basins – the Hellas, Argyre and Isidis planitiae – are located in this hemisphere. The north is dominated by younger, more sparsely cratered terrain, lying 2–3 kilometres lower. The highest areas are the large volcanic domes of the Tharsis and Elysium planitiae. Both areas are dominated by several huge extinct volcanoes, the largest of which is ➤*Olympus Mons*.

These volcanic areas are located at the east and west ends of an immense system of canyons, the Valles Marineris, which stretches for more than 5,000 kilometres (3,000 miles) around the equatorial region and has an average depth of 6 kilometres. It is believed to have been caused by faulting associated with the upthrust of the Tharsis dome.

There is evidence, in the form of flow channels, that liquid water once existed on the surface of Mars. Channels from the Valles Marineris appear to have been created in some kind of sudden flood. There are also sinuous, dried-up river beds with many tributaries, found only in the heavily cratered terrain.

Mars has two small natural satellites, ►*Phobos* and ►*Deimos*, which are in near circular orbits in the equatorial plane, close to the planet. They are very difficult to see from Earth. They are so different from Mars that it seems likely they are captured asteroids. ►Tables 5 and 6.

Mars 96   A Russian-led, international Mars mission which was lost when the launch failed in November 1996.

Mars Global Surveyor   A NASA mission to Mars, launched on 7 November 1996 for arrival in September 1997, effectively as a replacement for the failed ►*Mars Observer*. It was successfully put into a high elliptical orbit around Mars on 11 September 1997, then gradually manoeuvred into an almost circular near-polar orbit from which to carry out systematic mapping. The spacecraft's orbit was modified by 'aerobraking', in which the drag of the martian atmosphere was used to reduce the orbital speed of the craft. Mapping operations were due to start in March 1998 but were delayed by about a year owing to problems with aerobraking. Mars Global Surveyor was designed to serve as a communications satellite for future missions after completion of its mapping mission.

Mars (2)   A series of Soviet spacecraft intended to return data about the planet Mars. Useful information was returned by *Mars 2* and *3* in 1971 and *Mars 5* in 1974, but the others were not successful.

Mars Observer   An unmanned NASA mission to Mars, successfully launched in 1992, but lost (presumed destroyed) on its arrival at Mars in September 1993. The probe had been intended to orbit around Mars, undertaking detailed mapping and remote sensing for one martian year. Replacements for six of the eight instruments lost were subsequently carried on ►*Mars Global Surveyor*.

Mars Pathfinder   A NASA Mars mission which was launched on 4 December 1996 and arrived at the planet on 4 July 1997. The main objective was to test a low-cost means of sending a spacecraft and a surface 'rover' vehicle to land on the martian surface. It carried a 10-kg (22-lb) miniature 'rover', named *Sojourner*, equipped to measure the elemental composition of the surface rocks and soil and take images around the landing site in Ares Vallis.

In a new technique, the lander's impact was cushioned by airbags, which bounced several times before coming to rest. The performance of both the lander and the rover exceeded expectations and they operated for 12 weeks rather than the minimum target of 7 martian days (sols) originally envisaged. Panoramic views of the landscape were returned and *Sojourner* successfully travelled on expeditions covering about 80 metres. Instruments on the lander monitored atmospheric conditions at the surface. Measurements of the martian atmosphere also were made during the parachute descent.

After arrival at Mars, the lander base station was renamed the Sagan Memorial Station in honour of the American planetary scientist, Carl Sagan, who died in 1997.

Mars Surveyor 1998 A two-spacecraft NASA Mars mission scheduled for launch in December 1998 and early January 1999. It comprises both an orbiter, which will be launched first, and a lander, to be launched about a month later. The orbiter is intended to continue the global reconnaissance started by ➤*Mars Global Surveyor*, while the second craft is to land near the south polar region. The emphasis of the mission is on studying the martian climate and interactions between the atmosphere and the surface.

mascon An area of anomalously strong gravitational field on the Moon. The word is a contraction of 'mass concentration'. Mascons are presumed to indicate the presence of rocks denser than average, though there is no consensus about exactly how they formed. The areas are roughly circular and are associated with ➤*mare* areas.

maser In astronomy, an emission process in ➤*molecular clouds* whereby certain spectral lines in the microwave emission of particular molecules are strongly amplified by natural processes similar to those exploited in lasers. Maser action in an astronomical source was first discovered in 1965 in the emission from hydroxyl (OH) molecules in a source in the ➤*Orion Nebula*. Other molecules subsequently found to show a similar effect include water (H_2O), silicon monoxide (SiO), formaldehyde (H_2CO) and methyl alcohol (CH_3OH). The word 'maser' is an acronym for 'microwave amplification by the stimulated emission of radiation'. ➤*OH source*.

mass function For a group of related objects, such as a cluster of stars or galaxies, a graph or mathematical expression describing the relative numbers of objects found in different ranges of mass. ➤*initial mass function*, *Salpeter function*.

mass–luminosity relation The simple relation between the mass (M) and luminosity (L) of a star on the ➤*main sequence* that takes the form $L \propto M^n$, where the index n has the value 3.5 for stars of 7 solar masses or less. For stars with masses in the range 7 to 25 solar masses, n falls to 3.0 and is nearer 2.7 for masses in excess of 25 times the Sun's.

mass transfer   The flow of material from one star to another in a close binary system. The process can occur when a star has expanded during its evolutionary process to the extent that its outer layers are pulled towards the companion. The material transferred may stream directly on to the star's surface or form an ➤*accretion disc.* ➤*contact binary, dwarf nova, Roche lobe, semidetached system, X-ray binary.*

Mathilde   Asteroid 253, which was imaged by the ➤*Near-Earth Asteroid Rendezvous* (NEAR) mission in a close flyby on 27 June 1997. Mathilde is a uniformly dark ➤*C-type asteroid* with an albedo of only 3 per cent. NEAR found its mean diameter to be 52 km (33 miles). Five craters with diameters in excess of 20 km (12 miles) were identified on the side of the asteroid in sunlight at the time of the encounter. Its rotation period was also measured and found to be unexpectedly long at 17.4 days.

Mauna Kea Observatories   Observatories on a mountain-top site on the island of Hawaii at a height of 4,200 metres (13,800 feet). It is one of the best sites in the world for optical, infrared and sub-millimetre-wave astronomy and has been developed for this purpose since 1970. The first large telescope to be installed was a 2.24-metre (88-inch) reflector of the University of Hawaii. In 1979 three more major telescopes began operation at the observatory: the ➤*United Kingdom Infrared Telescope* (UKIRT), NASA's ➤*Infrared Telescope Facility* (IRTF) and the ➤*Canada–France–Hawaii Telescope.* The ➤*James Clerk Maxwell Telescope* working in the millimetre wave region, and the Caltech 10.4-metre (34-foot) Submillimeter Array were opened in 1987. The first of the two ➤*Keck Observatories* was completed in 1992, and the second in 1996. Two 8-metre (26-foot) instruments, the Japanese ➤*Subaru Telescope* and one of the international ➤*Gemini Telescopes,* are due for completion in 1999.

Maunder diagram   ➤*butterfly diagram.*

Maunder minimum   An interval of about 70 years, starting around 1645, during which ➤*solar activity* was consistently at a low level and sunspots were rare. For 37 years no aurora was recorded.

Max-Planck-Institut für Astronomie   A German astronomy research institute with headquarters at Heidelberg and an observatory at ➤*Calar Alto* in Spain.

Max-Planck-Institut für Radioastronomie   A German institute for research in radio astronomy which has its headquarters in Bonn and an observing station at Effelsberg, 40 kilometres (25 miles) to the south-west of Bonn. The observatory operates a 100-metre (328-foot), fully steerable radio dish on a computer-controlled altazimuth mounting, which was opened in 1971.

Maxwell Montes   The highest mountain peaks on Venus, located on Ishtar Terra. They rise to 11.5 kilometres (38,000 feet) above the venusian datum level.

Maxwell Telescope ➤*James Clerk Maxwell Telescope.*

Mayall Telescope A 4-metre (160-inch) optical reflecting telescope at ➤*Kitt Peak National Observatory*, belonging to the US ➤*National Optical Astronomy Observatories*. It has been in operation since 1973.

McDonald Observatory An observatory belonging to the University of Texas located on Mount Locke in the Davis Mountains near Fort Davis. It was established in 1932, financed by a bequest from a wealthy Texan banker and amateur astronomer, William J. McDonald. The original instrument, completed in 1938 and still in use, is a 2.08-metre (82-inch) reflector, known as the Otto Struve Telescope after the observatory's first director. In 1969, a 2.72-metre (107-inch) reflector came into use. The latest addition is the ➤*Hobby*Eberly Telescope*, completed in 1996. There are also 91-centimetre (36-inch) and 76-centimetre (30-inch) reflectors dating from 1956 and 1970 respectively. The 76-centimetre mirror is the central portion of the 2.08-metre, cut away to create the hole for the ➤*Cassegrain* system. Additionally, there is a 76-centimetre telescope dedicated to ➤*satellite laser ranging* work and a 5-metre (16-foot) millimetre-wave dish.

McMath–Pierce Solar Telescope Facility A large solar observatory located at ➤*Kitt Peak* and belonging to the US ➤*National Solar Observatory*. The main telescope, completed in 1962, consists of a 1.6-metre (60-inch) mirror, mounted on a tower, which directs the sunlight along a long light shaft inclined at 32° to the horizontal, much of which is below ground. A high-resolution image of the Sun 75 centimetres (30 inches) in diameter is produced by this system. The entire building is encased in copper, and coolants are piped through the outer skin to maintain a uniform temperature inside.

mean anomaly For a body travelling in an undisturbed elliptical orbit, the product of its mean motion and the interval of time since it passed ➤*pericentre*. The mean anomaly is thus the angular distance from pericentre of a hypothetical body travelling with a constant angular speed equal to the ➤*mean motion*. ➤*anomaly*.

mean equinox The ➤*equinox* determined by considering only ➤*precession* and ignoring small variations of short period. Star positions are normally referred to the mean equinox of a standard ➤*epoch*.

mean motion The constant angular speed that would be required for a body travelling in an undisturbed elliptical orbit of specified semimajor axis to complete one revolution in the actual orbital period. It is thus the angular speed a body would have if it were following a circular orbit with a radius equal to the semimajor axis of its actual elliptical orbit.

mean parallax The same as ➤*statistical parallax*.

mean solar day ➤*mean solar time.*

mean solar time A system of time measurement based on the rotation of the Earth, which is assumed to be constant. The Earth's rotation rate is not in fact precisely constant when checked against ➤*atomic clocks.* Mean solar time has therefore been superseded by ➤*International Atomic Time* (TAI). 'Leap seconds' are occasionally introduced to allow TAI to keep pace with the Earth's rotation.

Since the Earth's rotation axis is inclined to the plane of the ecliptic and its orbit around the Sun is elliptical rather than circular, the Sun's apparent motion through the sky is not uniform over the course of a year. Apparent solar time, as measured directly by a sundial, differs from mean solar time by an amount that varies through the year and is known as the ➤*equation of time.* To define mean solar time, the abstract concept of the *mean Sun* was introduced. This hypothetical object follows a circular orbit around the celestial equator at a constant speed, completing one circuit in a tropical year.

mean Sun A concept used to define ➤*mean solar time.*

mechanics The branch of applied mathematics that deals with the motion and equilibrium of bodies. It is subdivided into *dynamics* (motion under the influence of forces), *statics* (equilibrium conditions) and *kinematics* (motion without reference to mass or force). *Celestial mechanics* is the application of mechanics in astronomical situations.

megaparsec (symbol Mpc) A unit of distance equal to one million ➤*parsecs.*

Melpomene Asteroid 18, diameter 162 km, discovered by J. R. Hind in 1852.

Mensa (The Table or Table Mountain) A faint southern constellation introduced in the mid-eighteenth century by Nicolas L. de Lacaille with the longer name Mons Mensae, the Table Mountain. It contains no stars brighter than fifth magnitude, but part of the Large ➤*Magellanic Cloud* lies within its boundaries. ➤Table 4.

mensa (pl. mensae) Literally, 'table'. The term is used to describe a flat-topped, elevated feature on the surface of a planet.

Merak (Beta Ursae Majoris; β UMa) One of the two stars of the ➤*Plough* in Ursa Major (with Dubhe) called the ➤*Pointers.* The stars in the Plough have been designated by position rather than in brightness order. It is therefore actually the fifth-brightest star in the constellation, with a magnitude of 2.4. Merak is an ➤*A star* and its name, of Arabic derivation, means 'the loin'.

Mercury (1) The nearest major planet to the Sun and the smallest of the terrestrial planets.

Telescopic observation of Mercury from the Earth is very difficult, partly because of its small size and partly because it can never be more than 28° from the Sun on the celestial sphere since its orbit lies well inside the Earth's. For the same reason, Mercury (like Venus, the other inferior planet) exhibits a cycle of phases, similar to those of the Moon. Hardly any surface detail can be discerned and very little was known about the planet until the flybys of *Mariner 10* in 1974 and 1975. The spacecraft was put in an orbit around the Sun such that it encountered Mercury three times before it ran out of attitude-control gas. The images returned have allowed about 35 per cent of the surface of Mercury to be mapped.

Ancient, heavily cratered terrain accounts for 70 per cent of the area surveyed. The most significant single feature is the Caloris Basin, a huge impact crater with a diameter of 1,300 kilometres – a quarter the diameter of the planet. The basin has been filled by a relatively smooth plain, and terrain of the same type covers parts of the ejecta blanket. The impact took place 3,800 million years ago and produced a temporary revival of the volcanic activity that had mostly ceased 100 million years earlier, creating the smoother areas inside and around the basin. At the point on Mercury diametrically opposite the impact site, there is curious chaotic terrain that must have been created by the shock wave.

Characteristic features found on Mercury are lobate scarps (rupes), which take the form of cliffs between a few hundred and 3,000 metres high, believed to have formed when the planetary crust shrank as it cooled. In places they cut across craters.

The planet's rotation period is such that a 'day' on Mercury lasts two 'years'. This leads to immense temperature contrasts: at perihelion, the subsolar point reaches 430°C; the night-time temperature plunges to −170°C.

The high daytime temperatures and the small mass of the planet make it impossible for an atmosphere to be retained. The small amounts of helium detected may be the product of radioactive decay of surface rocks or have been captured from the ➤*solar wind*.

The average density of Mercury is only slightly less than that of the Earth. Taking account of its smaller size and lower interior pressure leads to the conclusion that Mercury has a substantial iron core accounting for 70 per cent of its mass and 75 per cent of its total diameter. There is also a magnetic field of about 1 per cent the strength of the Earth's field, providing further evidence for the metallic core. ➤Table 5.

Mercury (2)   A series of American spacecraft capable of carrying one astronaut. The first American experiments in manned spaceflight used Mercury capsules, and the first suborbital flight, to an altitude of 187 kilometres (116 miles), was in 1961. The first orbital flight was made by John Glenn in February 1962.

mercury star (mercury−manganese star)   ➤*manganese star.*

meridian (1)   The great circle on the celestial sphere passing through the poles and the zenith.

meridian (2)   A line of longitude on the Earth, or on another astronomical body. On the Earth, the meridian through Greenwich marks the zero of longitude and is sometimes referred to as the *prime meridian*.

meridian circle   ➤*transit circle*.

meridian passage   ➤*culmination*.

MERLIN   Acronym for Multi-Element Radio Linked Interferometer Network, a network of radio telescopes at various locations in the UK, operated by the University of Manchester from Jodrell Bank. MERLIN is used for ➤*long-baseline interferometry* and can be linked to other telescopes and networks worldwide to carry out ➤*very-long-baseline interferometry*.

Merope   One of the brighter stars in the ➤*Pleiades*.

mesosiderite   A class of stony-iron ➤*meteorite*.

mesosphere   A region of the Earth's atmosphere above the stratosphere, between heights of about 50 and 85 kilometres (30 and 50 miles), in which the temperature decreases with height to $-90°C$ at its upper boundary, the *mesopause*.

Messier Catalogue   A catalogue of about a hundred of the brightest galaxies, star clusters and nebulae, compiled by the French astronomer Charles Messier (1730–1817). His initial list, published in 1774, contained 45 objects but it was supplemented later with additional discoveries and contributions from Messier's colleague, Pierre Méchain. Objects in the catalogue, which is still widely used, are identified by the prefix 'M' and their catalogue number.

The list was not compiled systematically. Messier's prime interest was searching for comets, and he noted hazy objects spotted during comet searches. Some were first recorded by Messier, but others were already known.

There are some errors and discrepancies in the list as published. M40 is a double star and M73 a group of four stars, but not a true cluster. The identification of M91 is uncertain from the original source, and M102 was a duplication of M101. Messier's own list stopped at number 103 but a further seven have been added in the twentieth century.

metagalaxy   The entire observable universe.

metal abundance   The proportion of elements heavier than helium in the composition of a star. In astronomical terminology, the expression 'metal' is used to encompass all the elements heavier than helium. Though many are not metals in the usual chemical sense of the word, the term probably arose because lines of metals such as iron and nickel tend to dominate stellar spectra.

metallic-line star An ➤*A star* that has unusually strong lines of many of the heavier elements, including rare earths, in its spectrum. The lines of the elements calcium and scandium are generally weaker than usual.

metamorphic rock Igneous or sedimentary rock that has been deeply buried and changed by the action of heat or pressure, or both.

meteor The brief luminous trail observed as a particle of dust or piece of rock from space enters the Earth's upper atmosphere. The popular name for a meteor is *shooting star* or *falling star*.

The Earth is constantly bombarded with material from space. The individual objects range in size from rocks of several kilograms down to microscopic particles weighing less than one millionth of a gram. It is estimated that more than 200 million kilograms of meteoric material are swept up by the Earth in the course of a year. One-tenth of this mass reaches the ground, in the form of ➤*meteorites* and ➤*micrometeorites*. The remainder burns up in the atmosphere, becoming visible as *meteor trails*.

The meteoric material enters the atmosphere at typical speeds of around 15 km/s. Frictional heating causes medium-sized particles to vaporize, creating visible light and leaving a temporary trail of ionized gas. This trail is capable of reflecting radar signals, and radar has been used to detect meteors too faint to be seen by eye and meteors occurring during the daytime.

Much of the meteoric material in the solar system orbits the Sun in distinct *streams*. The orbital characteristics can be calculated from observations of the meteor trails. In this way it has been shown that many such meteor streams have the same orbits as known comets. The particles may be strung out all along the orbit or concentrated in a particular swarm. When the Earth's orbital motion causes it to cut through a stream, a ➤*meteor shower* is observed. The effect of perspective makes the meteors, which are travelling along parallel paths, appear to radiate from a single point in the sky, the ➤*radiant*.

In addition to the dozens of regular meteor showers, a background of *sporadic meteors* is observed throughout the year. They may come from any direction. ➤*fireball*.

Meteor Crater ➤*Arizona meteorite crater*.

meteorite The recovered fragment of a ➤*meteoroid* that has survived passage through the Earth's atmosphere. Individual meteorites are normally named after the place where they fell. Studies of the paths of a small number of meteorites observed as ➤*fireballs*, and subsequently recovered, show they were in orbits originating in the ➤*asteroid belt*. The chemical and mineralogical composition of meteorites is studied with considerable interest since they appear to be samples from remote parts of the solar system and thus provide clues to its origin and evolution.

There are three main classes of meteorite: *irons* (siderites), *stony-irons* (siderolites or lithosiderites) and *stones* (aerolites). Stony meteorites are further divided into two important categories: the *achondrites* and the *chondrites*. Chondrites are characterized by the presence of chondrules, small spherical inclusions, which may be of metal or of silicate or sulphide materials. Chondrules are not present in achondrites.

The chemical composition of the chondrites is very similar to that of the Sun, except that they contain no free hydrogen and helium and there is more lithium and boron. The interpretation is that the chondrites represent primitive solar system material that has not been altered significantly by heating, though there is evidence for some metamorphism and for alteration by water. *Carbonaceous chondrites* have the highest proportion of volatiles and the composition closest to the Sun's. 'Ordinary' chondrites have the lowest proportion of volatiles, while *enstatite chondrites* are intermediate between the two.

Among the achondrites, numerous subtypes are identified according to detailed chemical and mineralogical composition. In Antarctica, where large numbers of meteorites have been preserved and concentrated in certain areas of the ice sheet, examples have been found with compositions very close to those of lunar samples returned by the Apollo missions.

Stony-iron meteorites contain free metal and stony material in roughly equal proportions. *Pallasites* consist of olivine grains enclosed in metal; *mesosiderites* are agglomerates of metal and silicates.

Iron meteorites consist almost entirely of iron and nickel. Over forty different minerals have been identified in them, though the basic constituents are two forms of iron–nickel alloy, kamacite and taenite. Iron meteorites are categorized according to the proportion of nickel, which determines the crystalline structure. ►*Hexahedrites* contain up to 6 per cent nickel, ►*octahedrites* between 6 and 14 per cent nickel, and ►*ataxites* up to 66 per cent.

meteoroid   A piece of rock or dust in space with the potential to become a ►*meteor* or ►*meteorite*.

meteor shower   ►*Meteors* observed to radiate from a single point in the sky, seen over a limited period, usually of several hours or days. Meteor showers occur when the Earth's orbital motion causes it to cut through a stream of meteoric material. Dozens of annual showers are known, though only a handful provide significant regular displays. Very occasionally, if the Earth encounters a particularly dense swarm of particles, an exceptional shower may occur with tens or hundreds of meteors every minute. A more typical, average rate for a good regular shower would be around 50 meteors per hour.

Shower members are identified by the fact that their trails, if traced back on a sky map, all appear to intersect at a single point, called the *radiant*. The effect is one of perspective. The meteors are in fact caused by material entering the upper atmosphere along parallel tracks.

meteor stream An extended swarm of meteoric material in orbit around the Sun. Many meteor streams are known to be associated with particular comets, sharing the same orbits. The material may be spread out evenly along the orbit or it may be in a swarm concentrated at one place. In particular, a young meteor stream may still be concentrated near the parent comet. ➤ *meteor, meteor shower*.

Metis (1) Asteroid 9, diameter 190 km, discovered in 1848 by A. Graham.

Metis (2) A small satellite of Jupiter (number XVI) , discovered by S. P. Synnott in 1979. It is about 40 kilometres in diameter, irregularly shaped and reddish in colour. ➤Table 6.

Metonic cycle A period of 19 tropical ➤*years*, after which the phases of the Moon recur on the same days of the year. This happens because 19 tropical years equals 6,939.60 days, which is almost exactly 235 ➤*synodic months* (6,939.69 days). The discovery of the cycle is attributed to the Greek astronomer Meton, who worked in the fifth century BC.

metric A measure of distance in space or ➤*spacetime* that is the same for all observers regardless of their state of motion.

In ordinary Euclidean geometry, distances (s) are expressed in terms of the coordinates x, y and z, by $s^2 = x^2 + y^2 + z^2$. This is the metric in three-dimensional Euclidean space.

For astronomy and cosmology, the term metric has special importance when the geometrical properties of the universe at large are considered. By introducing time as the fourth dimension, Einstein set out the concept of an 'interval' in the spacetime continuum, given by the metric $s^2 = t^2 - (x^2 + y^2 + z^2)/c^2$. The inclusion of time in this metric ensures that measurements of intervals do not change (are *invariant*) in frames of reference that are in relative motion.

Matter in the universe causes spacetime to be curved. Various ways of describing different sorts of curvature mathematically lead to different metrics. The simple metric given above is the *Minkowski metric*, which would apply in an infinite universe containing no matter. The metrics used in more realistic models, such as the ➤*Robertson–Walker metric* or ➤*Kerr metric*, are more complex.

Meudon Observatory ➤*Observatoire de Paris*.

Mice A popular name for the pair of interacting galaxies, NGC 4676 A and B. Long tail-like streamers of material extend from the galaxies, giving them shapes reminiscent of a pair of mice.

Michelson interferometer An instrument in which interference fringes in visible light are produced by splitting and subsequently recombining a beam of light. A half-silvered mirror is used to split the beam between two 'arms' at right angles to each other; the beams are then reflected back from plane mirrors and recombined. The appearance and quality of the fringes produced provide information on the optical path differences in the two arms. This instrument is not to be confused with the ➤*Michelson stellar interferometer*.

Michelson stellar interferometer   Any one of a series of instruments constructed by A. A. Michelson (1852–1931) in an attempt to measure, by means of interferometry, the diameters of stars, none of which can be resolved directly by ground-based telescopes. In the simplest form, the telescope objective is covered by a screen containing two holes. If the star were a point source of light, the image would appear as a circular ring pattern crossed by dark fringes. For a source of finite angular diameter α and light of wavelength λ, the fringes disappear when the separation of the holes is $1.22\ \lambda/\alpha$, a result that can be used to determine α.

Though the theory is simple, the practical difficulties are very great. In the most successful experiment, Michelson used the 2.54-metre (100-inch) Hooker Telescope at Mount Wilson and mounted a pair of mirrors on racks to achieve a large, variable separation. (In fact, with this design, the size of the telescope objective is irrelevant. Michelson used the Hooker Telescope because it was rigid enough to bear the weight of the extra structures.)

The use of the instrument was limited by the fact that no more than a handful of stars were bright enough. The angular diameters of six stars were measured, notably that of ➤*Betelgeuse*.

microlensing   The action of a stellar-sized object as a ➤*gravitational lens*. Microlensing has been used as a technique to search for invisible dim stars or ➤*brown dwarfs* (objects sometimes described as ➤*MACHOs*), which may account for as much as 90 per cent of the total mass of the Galaxy. It produces a temporary brightening of a star that happens to pass behind one of the dark objects as viewed from Earth. A systematic programme to search for such events has observed several which may be attributable to microlensing.

micrometeorite   A particle of meteoritic material so small that its energy is dissipated before it burns up in the Earth's atmosphere. Micrometeorites fall to Earth as a rain of minute dust particles. It is estimated that four million kilograms fall to Earth each year in this form. The size of particle is typically less than 120 micrometres. Such particles have been collected by experiments in space, and ferrous particles can be detected at the Earth's surface through their magnetic properties. ➤*F corona, meteor, meteorite, zodiacal light*.

micrometeoroid   A very small ➤*meteoroid*.

micrometer   In general, any instrument for measuring small distances accurately. In astronomy, micrometers attached to telescopes are used by visual observers to measure the angular separations of pairs of objects, such as ➤*binary stars*.

In the *filar micrometer*, the distance between two fine wires or markers viewed in an eyepiece can be adjusted and measured by means of a screw mechanism. In the *double-image micrometer*, the adjustment required to a split lens or birefringent crystal to bring images into coincidence or arrange them in a particular pattern is calibrated in terms of the angular separation of the pair of objects.

micrometre (micron, symbol μm)   A unit of measurement equal to one millionth of a metre.

Microscope   English name for the constellation ➤*Microscopium*.

Microscopium (The Microscope)   A small and insignificant southern constellation introduced in the mid-eighteenth century by Nicolas L. de Lacaille. Its brightest star is magnitude 4.7. ➤Table 4.

microwave astronomy   The study of radio waves from astronomical sources across a wide band of the electromagnetic spectrum from the far infrared at 1 millimetre wavelength to short-wave radio at about 6 centimetres. At the shorter wavelength end, these waves are absorbed by the Earth's atmosphere. ➤*Cosmic Background Explorer, millimetre-wave astronomy, radio astronomy, submillimetre-wave astronomy.*

microwave background radiation   ➤*cosmic background radiation.*

Midcourse Space Experiment (MSX)   A spacecraft launched in April 1996 by the US Ballistic Missile Defense Organization, with an expected working lifetime of 4 years. It carried instruments which were used for a wide range of astronomical observations as well as for military purposes. They included a sensitive infrared telescope and instruments capable of imaging and spectroscopy in the ultraviolet and visible parts of the spectrum.

midsummer   The summer ➤*solstice.*

Milankovic cycles   Small variations in the tilt of the Earth's rotation axis and the eccentricity of its orbit around the Sun that have been linked with long-term oscillations in climate and the incidence of ice ages.

Milk Dipper   An ➤*asterism* formed by the stars Zeta (ζ), Tau (τ), Sigma (σ), Psi (ψ) and Lambda (λ) in the constellation Sagittarius, named presumably for its ladle shape and the fact that the most intense part of the Milky Way lies in Sagittarius.

Milky Way   A band of hazy light circling the sky. It results from the combined light of vast numbers of stars in our own ➤*Galaxy*. The term Milky Way is also used as a synonym for the Galaxy.

The band of light around the celestial sphere is the disc of the Galaxy viewed from within. The Sun is situated two-thirds of the way out towards the edge of the galactic disc, and the Milky Way appears brightest in the direction of the bulge around the galactic centre, which lies in the constellation Sagittarius. Clouds of obscuring dust, such as the ➤*Coalsack* near the Southern Cross, give the Milky Way a patchy appearance in places.

The main constellations through which the Milky Way passes are Perseus, Cassiopeia, Cygnus, Aquila, Sagittarius, Scorpius, Centaurus, Vela, Puppis, Monoceros, Orion, Taurus and Auriga.

Millimeter Array (MMA)   A projected millimetre-wave telescope, planned for completion in about 2005. The proposal is for an array of forty 8-metre (26-foot) antennas located in the Atacama desert in northern Chile.

millimetre-wave astronomy   Astronomical observations of radio waves in the region of the electromagnetic spectrum from about 1 to 10 millimetres. This part of the spectrum is rich in lines from complex molecules and is important in the study of ➤*molecular clouds*, regions of star formation, circumstellar discs and ➤*comets*. In order to improve on the limited resolution achievable with single-dish telescopes, arrays consisting of several dishes have been constructed, such as the ➤*BIMA Array*. ➤*Millimeter Array, submillimetre-wave astronomy.*

Mills Cross   A type of ➤*radio interferometer*, first built in Australia in 1957, in which two arrays of parabolic reflectors intersecting at right angles are used to obtain a reasonable resolving power. It is named after the radio astronomer B. Y. Mills. ➤*Molonglo Observatory.*

Milne universe   A model of the ➤*expanding universe*, published in 1948 by Edward Milne, that attempts an explanation without using ➤*General Relativity*. It is an expanding, isotropic and homogeneous universe containing no matter. It has negative curvature and is open.

Mimas   A satellite of Saturn, discovered by William Herschel in 1789, spherical in shape and 390 kilometres in diameter. Its heavily cratered surface shows no evidence of subsequent change. The largest crater, Herschel, is 130 kilometres in diameter, one-third the size of the satellite, and has a central peak. It has been calculated that the impact that created Herschel must very nearly have shattered Mimas. ➤Table 6.

Mimosa (Beta Crucis; β Cru)   The second-brightest star in the constellation Crux. It is a giant ➤*B star* of magnitude 1.3, though it is a slightly variable ➤*Beta Canis Majoris star*, with an amplitude of 0.1 magnitude and a period of about 6 hours. ➤Table 3.

mineral   A natural material, found in rocks, composed of a single chemical substance.

Minkowski's Footprint   A double-lobed nebula surrounding a star in Cygnus, interpreted as an early stage in the formation of a ➤*planetary nebula*. It was discovered by Rudolph Minkowski in 1946.

minor planet   An alternative name for an ➤*asteroid*. ➤*major planet*.

Mintaka (Delta Orionis; δ Ori)   One of the three stars forming the belt of Orion. At magnitude 2.2, it is the seventh-brightest star in the constellation. It is actually an ➤*eclipsing binary* and varies in brightness by 0.1 magnitude in a period of 5.7 days. There is also a seventh magnitude visual companion. The primary star is a ➤*supergiant* ➤*O star*. Mintaka, a name of Arabic origin, means 'the belt'.

minute   A unit of time equal to sixty ➤*seconds*.

minute of arc   ➤*arc minute*.

Mir   A Russian space station, launched into Earth orbit in 1986.

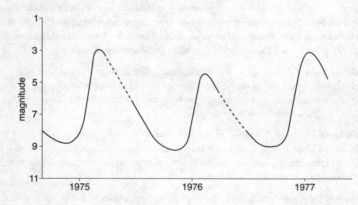

Mira. A schematic version of part of the light curve of Mira. The dashed parts correspond to periods when the star is not visible.

Mira   (Mira Ceti; Omicron Ceti; ο Cet) The prototype of a class of long-period variable stars. The name is Latin for 'wonderful'.

Mira was the first variable star to be discovered: the Dutch astronomer David Fabricius noted it at third magnitude in 1596, but found it invisible to the naked eye a few months later. He noted it again at third magnitude in 1609. Mira is a ➤*giant* ➤*M star* that varies between second and tenth magnitudes in a period of about 332 days, though there are irregularities in the period and in the maximum and minimum brightnesses reached. It is shedding large amounts of gas and dust, which form a strong ➤*stellar wind*.

Ultraviolet and visible images taken with the Hubble Space Telescope have resolved Mira's ➤*white dwarf* companion, and have shown that the primary star has an elongated asymmetrical shape. It appears that material is being transferred to the white dwarf.

Mirach (Beta Andromedae; β And)   The second-brightest star in the constellation Andromeda. It is a ➤*giant* ➤*M star* of magnitude 2.1. The name, of Arabic derivation, means 'girdle'.

Miranda   A satellite of Uranus, the smallest of those known prior to the ➤*Voyager 2* encounter in 1986. It was discovered by Gerard Kuiper in 1948. The spacecraft passed Miranda at a distance of only 3,000 kilometres (1,800 miles), returning very detailed images of its surface.

Though only 470 kilometres in diameter, Miranda has several contrasting types of terrain. Alongside cratered areas, typical of planets and satellites, there are large tracts of grooves and ridges. It seems unlikely that such variety could have been caused by geological activity in such a small satellite. One theory suggests that the satellite was once shattered by a massive impact into several parts that subsequently coalesced again. ➤Table 6.

Mira star   A member of a class of long-period variable stars, of which ➤*Mira* is the prototype.

Mirfak (Alpha Persei; α Per)   The brightest star in the constellation Perseus. It is a yellow ➤*supergiant* ➤*F star* of magnitude 1.8. The name, of Arabic origin, means 'the elbow'. ➤Table 3.

mirror   An element in an optical system, designed for the purpose of reflecting light or other electromagnetic radiation.

The high degree of accuracy and reflectivity needed for astronomical mirrors used in optical telescopes is achieved by grinding and polishing to the required surface shape a piece of glass which is then coated with a thin layer of aluminium. Formerly, silver was used rather than aluminium, but it is less durable and oxidizes more quickly than aluminium.

When light falls on any glass/air interface, part is reflected, part transmitted and part absorbed. Mirrors are designed to achieve the maximum reflection. Strong internal reflection may also take place at a surface in a glass prism. This phenomenon is used, for example, in prismatic binoculars.

Mirzam (Beta Canis Majoris; β CMa)   The second-brightest star in the constellation Canis Major. It is a giant ➤*B star* of magnitude 2.0 and the prototype of the class of the slightly variable ➤*Beta Canis Majoris stars*. It changes in brightness by a few hundredths of a magnitude every six hours. This low level of variability cannot be detected by the naked eye alone. ➤Table 3.

missing mass   Matter, as yet undetected, that would need to be present in the universe to stabilize it against infinite expansion. Missing mass, if it exists, must be in the form of ➤*dark matter*.

mixmaster universe   A chaotic model for the early universe in which giant convulsions and oscillations allow light to circumnavigate the universe and transform a non-uniform universe into a uniform one. This was found not to work.

Mizar (Zeta Ursae Majoris; ζ UMa)   The fourth-brightest star in the constellation Ursa Major, an ➤*A star* of magnitude 2.3. It forms an optical pair (i.e. not a true binary system but a mere coincidence in the line of sight) with the fourth magnitude star Alcor. Mizar also has a true fourth magnitude companion, and both stars are also ➤*spectroscopic binaries*. The Arabic name means 'girdle'.

MK classification (MKK classification)   A system for classifying the spectra of stars. ➤*spectral type*.

MMT   Abbreviation for ➤*Multiple Mirror Telescope*.

mock Sun (parhelion; sundog)   A circular patch of light in the sky, 22° away from the real Sun. Mock Suns usually appear in pairs, one either side the real Sun, on a circular halo of light, though one may be much brighter than the other according to circumstances. The effect is caused by refraction by ice crystals in the Earth's atmosphere.

model   In science, an attempt to build a description of a physical situation by means of numerical values for physical parameters (e.g. temperature or pressure), and mathematical expressions for physical laws linking them. A satisfactory model should be able to account for existing observations and also predict any changes or additional observations by which means the validity of the model can be tested. Often, more than one model may account for the limited observations available. In reality, none of those suggested may be correct, though any model that successfully predicts future events has some value, whether strictly correct or not.

molecular cloud   A cloud of interstellar matter in which the gas is predominantly in molecular form. There are two distinct types, both found close to the plane of the Galaxy, within the Milky Way. These are termed *small molecular clouds* and *giant molecular clouds* (GMCs).

The small clouds are typically a few light years in diameter, with 1,000 to 10,000 molecules per cubic centimetre and temperatures of around 10–20 K. They may contain even colder condensed 'cores', with densities ten or a hundred times greater. These small clouds contain mostly molecular hydrogen (H_2). They are very cold because there is no radiation from stars within to heat them.

Giant molecular clouds are made up primarily of molecular hydrogen and carbon monoxide (CO), but they also contain many other ➤*interstellar molecules*. They are the most massive entities within our Galaxy, containing up to ten million solar masses, and are typically 150 to 250 light years across.

The density is as high as ten million molecules per cubic centimetre. Infrared emission from such clouds is evidence that they are regions of star formation. GMCs are nearly always found to be associated with clusters of hot, massive, young stars. Luminous clouds of ►*ionized hydrogen* (H II regions) may be produced by such stars formed near the edges of a GMC. In the ►*Orion Nebula*, a notable example, a GMC lies behind the optically visible nebula. Another is associated with the ►*Omega Nebula* (M17). A GMC containing between three and five million solar masses of material is located near the galactic centre, in front of the radio source Sagittarius B2. It contains many of the known types of interstellar molecule. Up to 4,000 GMCs are thought to exist in the Galaxy.

molecular line radio astronomy   The study of spectral lines emitted by molecules in interstellar clouds. The data obtained may be used to derive the densities and temperatures of the clouds. Lines are generally studied as ►*emission lines* from dense clouds, but are sometimes seen as ►*absorption lines* as well. Emission lines of certain molecules are also observed in the spectra of ►*diffuse clouds*. ► *interstellar molecules, molecular cloud*.

Molonglo Observatory   An Australian radio astronomy observatory, located near Canberra, belonging to the University of Sydney. A large ►*Mills Cross* telescope was constructed there in 1966. The east–west arm of the cross has been converted into the Molonglo Observatory Synthesis Telescope (MOST).

momentum   The product of the velocity and mass of a body. Momentum is conserved within a system as long as no external forces act on it.

monocentric eyepiece   A solid type of telescopic ►*eyepiece* consisting of three lenses cemented together to make a triplet.

Monoceros (The Unicorn)   A faint constellation, but one rich in stars and nebulae by virtue of its location in the ►*Milky Way*, straddling the celestial equator next to Orion. It is not one of the ancient constellations but seems to have gained general acceptance in the mid-seventeenth century. Its brightest stars are two of the third magnitude; it contains the ►*Rosette Nebula*, the ►*Cone Nebula* and ►*Hubble's Variable Nebula*. ►Table 4.

monochromator   An instrument for producing a beam of light covering a very narrow wavelength range. This may be achieved, for example, by producing a spectrum with a diffraction grating then isolating the required part of the spectrum by means of an exit slit.

mons (pl. montes)   A mountain. A term used in the names of planetary surface features.

month   The period of time taken by the Moon to complete one orbit of the Earth. The length of the month varies according to the reference point.

Type of month	Reference point	Length in days
Anomalistic	Apse	27.554 55
Draconic	Node	27.212 22
Sidereal	Fixed stars	27.321 66
Synodic	Phase	29.530 59
Tropical	Equinox	27.321 58

moon A natural ➤*satellite*.

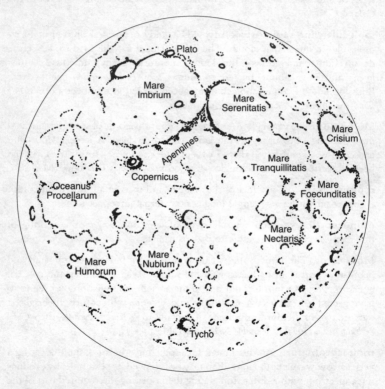

Moon. Identification of some of the principal maria and craters on the Earth-facing side of the moon.

Moon The Earth's only natural satellite. It is a barren, cratered world, lacking liquid water or an atmosphere. It was explored at first hand by the American astronauts who travelled there during the ➤*Apollo programme* landings of 1969–

72 and has been extensively mapped from orbiting craft and from the Earth.

Tidal forces have ensured that the same side of the Moon now always faces the Earth, apart from the minor effects of ➤*libration*. As the Moon travels round the Earth in the course of a month, it undergoes the familiar cycle of ➤*phases*. The Moon shines only by reflected sunlight; the proportion of the sunlit side visible from Earth depends on the relative alignment of the Sun, Earth and Moon, which changes continuously over the Moon's orbital period.

The terrain on the nearside falls into two basic types: the heavily cratered, light-coloured highlands (or 'terrae' meaning 'land'), and the darker, more sparsely cratered *maria* ('seas'). The maria have roughly circular outlines, a relic of their formation in the early history of the Moon by the impact of large meteorites. A further type of surface terrain is formed by ➤*ejecta*. Significant areas are marked by material ejected from the large Imbrium and Orientale basins.

The way the Moon formed is uncertain, though the balance of evidence suggests it resulted from an impact on the newly formed Earth by a body perhaps as large as Mars. It has existed as a separate body for around 4,500 million years. Early in its life it became hot and molten. As it cooled, the crust formed but it was heavily cratered by the impact of large numbers of meteorites, the largest of which created the mare basins. These subsequently filled with dark basaltic lavas. Significant volcanic activity then ceased, at least 2,000 million years ago.

The farside of the Moon differs from the nearside in that it lacks any large lava-flooded mare areas. ➤Table 6.

Morgan, Keenan and Kellman classification (MKK classification) A system for classifying the spectra of stars. ➤*spectral type*.

morning star A term applied to the planet Venus (or occasionally Mercury) when it appears in the eastern sky in the early morning before sunrise.

Mount Bigelow ➤*Steward Observatory*.

Mount Graham International Observatory An observatory site located on Mount Graham near Safford in south-eastern Arizona. The first two tele-scopes to be installed at the site were the 1.8-metre (6-foot) Vatican Advanced Technology Telescope (VATT) and the ➤*Heinrich Hertz Submillimeter Telescope*. It will also be the site of the ➤*Large Binocular Telescope*.

mounting (mount) The support for a telescope that allows it to be pointed at the desired position in the sky. Two main types are commonly used: the ➤*altazimuth mounting* and the ➤*equatorial mounting*.

Mount Lemmon ➤*Steward Observatory*.

Mount Stromlo and Siding Spring Observatories Optical observatories belonging to the Australian National University (ANU) and operated by the university's Institute of Advanced Studies. Observing facilities are located at Mount Stromlo, near Canberra, and at Siding Spring in New South Wales. It was founded in 1924 as the Commonwealth Solar Observatory. After World War II, it changed to being an astrophysical observatory and became part of the Institute of Advanced Studies of the ANU. The administration is located at Mount Stromlo, together with 1.9-metre (74-inch), 76-centimetre (30-inch) and 66-centimetre (26-inch) telescopes. At the Siding Spring site, which was established in 1962, the Observatories operate a 2.3-metre (90-inch) altazimuth reflector, and 1-metre (40-inch), 66-centimetre (26-inch) and 61-centimetre (24-inch) telescopes.

The development of the city of Canberra made it necessary to find a site more suitable for optical astronomy than Mount Stromlo. Siding Spring Mountain is at an altitude of 1,000 metres (3,200 feet) in the Warrumbungle range, most of which forms a national park. This observatory site now also houses a number of telescopes owned by organizations other than the ANU. These are the ➤*United Kingdom Schmidt Telescope*, the ➤*Anglo-Australian Telescope* and a Swedish ➤*Schmidt camera.*

Mount Wilson Observatory   An observatory near Pasadena, California, located on Mount Wilson at an altitude of 1,750 metres (5,700 feet). The first instrument built there was a horizontal solar telescope in 1904, a project initiated by George E. Hale. Two solar tower telescopes were added in the next few years, first a '60-foot' then a '150-foot' in 1910. A 1.5-metre (60-inch) reflecting telescope was begun in 1904 and brought into service in 1908. The mirror blank had been a birthday present to Hale from his father. It was the largest telescope in the world until the opening of the '100-inch' (2.5-metre) ➤*Hooker Telescope* in 1917.

Until 1985, the observatory was operated by the Carnegie Institution. From 1948 to 1970 it was administered jointly with ➤*Palomar Observatory* under the title 'Mount Wilson and Palomar Observatories'. From 1970 to 1980 the name was changed to 'Hale Observatories'. Between 1980 and 1985, Mount Wilson Observatory became part of the 'Mount Wilson and Las Campanas Observatories'. Carnegie withdrew from Mount Wilson in 1985, when the Hooker Telescope temporarily ceased operation. From 1985, the solar towers and the 60-inch telescope were operated by Harvard University and the astronomy departments of the University of Southern California and the University of California at Los Angeles. The Hooker Telescope was subsequently renovated and brought back into use in 1993.

The Mount Wilson site also has several optical and infrared ➤*interferometers.* The largest, scheduled to start operation in 1998, is Georgia State University's CHARA array. (CHARA stands for Center for High Angular Resolution Astronomy.) It consists of five 1-metre (40-inch) telescopes arranged in a 'Y' shape on a 400-metre-diameter circle. ➤*Las Campanas Observatory.*

moving cluster   An open star cluster, the distance to which can be derived from measurements of the ►*radial velocity* and ►*proper motion* of individual members. The prime example is the ►*Hyades*. The members of the cluster are presumed to share a common motion through space. Perspective makes their paths appear to converge towards or diverge from a single point. The direction towards this point is parallel to the direction of motion of the stars. This provides sufficient information for the distance to be determined.

MRAO   Abbreviation for ►*Mullard Radio Astronomy Observatory*.

Mrkos, Comet   ►*Comet Mrkos*.

msh   Abbreviation for millionths of a solar hemisphere, a unit of area used for ►*sunspots* and other solar features. One msh is approximately equivalent to 3.036 million square kilometres (1.173 million square miles).

MSSL   Abbreviation for ►*Mullard Space Science Laboratory*.

M star   A star of ►*spectral type* M. M stars have surface temperatures in the range 2,400–3,480 K and are red in colour. Molecular bands are prominent in their spectra, particularly those of titanium oxide (TiO). Examples of M-type stars include the nearest star, ►*Proxima Centauri*, which is a dwarf, and the supergiant ►*Antares*.

MSX   Abbreviation for ►*Midcourse Space Experiment*.

M-type asteroid   A fairly common type of asteroid with moderate albedo, presumed to have a metallic composition similar to that of iron meteorites.

Mu Cephei   ►*Garnet Star*.

Mullard Radio Astronomy Observatory (MRAO)   The radio astronomy observatory of the University of Cambridge. The main instrument is an ►*Earth rotation synthesis* interferometer, which consists of eight 13-metre (43-foot) dishes on an east–west baseline 5 kilometres (3 miles) long. This is called the Ryle Telescope after Sir Martin Ryle who founded the observatory in 1946 and was its first director. He was joint winner of the Nobel Prize for Physics in 1974 for the development of the principle of Earth rotation synthesis. In addition, there is a low-frequency deep-survey telescope of ►*Yagi antennas* operating at 151 MHz and using the same principle.

The observatory has specialized in cataloguing radio sources, producing the Third, Fourth, Fifth, Sixth and Seventh Cambridge catalogues (abbreviated to 3C, 4C etc.) at different frequencies. These have led to the discovery of many ►*quasars* and ►*radio galaxies*. The first ►*pulsars* were detected at the MRAO in 1967. A development of the 1990s has been the construction of an optical ►*interferometer* (COAST).

Mullard Space Science Laboratory (MSSL)   An institute in Surrey, south of London, which forms part of the Department of Physics and Astronomy of University College London, a constituent college of the University of London. Its work is particularly concerned with building instruments for astronomical satellites.

Multiple Mirror Telescope (MMT)   A telescope of unique design which operated at the Fred Lawrence Whipple Observatory on Mount Hopkins in Arizona between 1977 and 1997 as a joint venture of the Smithsonian Institution Astrophysical Observatory and the University of Arizona. It combined six individual 1.8-metre (72-inch) mirrors in a circular array on an altazimuth mounting that together had the light-gathering power of a single mirror 4.5 metres (176 inches) in diameter. The cost of construction was very much lower than for the single mirror equivalent at the time. Though such multi-mirror telescopes had been proposed for many years, the concept became a practical reality with the development of the computer control technology necessary to keep each mirror element pointing accurately at the same place in the sky.

In 1989, taking into consideration new developments in mirror technology, the decision was made to substitute a single light-weight mirror of 6.5 metres (21 feet) diameter in order to double the light-gathering power and substantially increase the field of view. The conversion was carried out in 1998.

multiple star   A group of three or more stars orbiting in a system in which they are bound together by mutual gravitational attraction. A well-known example is the four-star system, Epsilon (ε) Lyrae.

mural quadrant   ➤*quadrant.*

Murzim   Alternative spelling of the star name ➤*Mirzam.*

Musca (The Fly)   A small southern constellation containing one second magnitude star and three of third magnitude. Its origin is obscure, but it has been attributed to Johann Bayer. ➤Table 4.

MUSES-B   ➤*HALCA.*

MUSES-C   A proposed US/Japanese mission to obtain a sample from a small near-Earth asteroid, Nereus. A Japanese vehicle is to provide the launch in January 2002 and a miniature roving vehicle contributed by NASA will land to make measurements on the asteroid surface. The asteroid samples will be returned to Earth in January 2006 by MUSES-C via a parachute-borne capsule.

N

nadir   The point on the celestial sphere diametrically opposite the ▸*zenith*.

Nagler eyepiece   A specialist telescope ▸*eyepiece* with an exceptionally wide (80°) field of view.

Naiad   A satellite of Neptune (1989 N6) discovered during the flyby of ▸*Voyager 2* in August 1989. ▸Table 6.

naked-eye star   A star visible in principle without the aid of a telescope. In principle, stars down to about sixth magnitude are visible to the naked eye under ideal conditions, but this is rarely achieved in practice except on moonless nights under transparent skies remote from artificial illumination.

naked singularity   A ▸*singularity* without an ▸*event horizon*. According to the ▸*cosmic censorship* hypothesis, this should not be possible. However, if the universe began as a singularity, the ▸*Big Bang* was a naked singularity.

nakhlite   A rare type of achondritic meteorite made of calcic pyroxene and olivine. Together with shergottites and chassignites, nakhlites belong to the class of ▸*SNC meteorites*, believed to have originated on the surface of Mars.

Nançay   The location in France of the radio astronomy station of the ▸*Observatoire de Paris*.

nanometre (symbol nm)   One thousand-millionth (10^{-9}) of a metre.

nanosecond   One thousand-millionth (10^{-9}) of a second.

Narrabri   ▸*Culgoora*.

narrow-band photometry   The measurement of the magnitudes of stars in a set of narrow wavelength bands, typically less than 30 nanometres wide. The most commonly used narrow-band system is ▸*uvby photometry*. The wavebands are isolated by means of standard filters. The narrow-band magnitudes can be used to determine basic physical properties of stars more simply and quickly than through spectroscopy, and are also useful for statistical studies on groups of stars.

NASA   Abbreviation for ▸*National Aeronautics and Space Administration*.

Nasmyth focus   A focal point to one side of the tube of an altazimuth-mounted reflecting telescope, created by placing a third deflecting mirror in

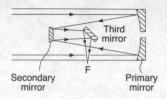

Nasmyth focus. A third mirror in the optical system of a telescope is used to bring the light to a focus at F, where in practice it is possible to mount heavy equipment on a platform forming part of the altazimuth mounting of a large telescope.

the optical system. This extra mirror is arranged to direct the beam along the altitude axis, and through a hole in the supporting trunnions. It was first used by the inventor James Nasmyth in the nineteenth century.

With the new generation of large altazimuth telescopes, made possible by computer control of the drive mechanism, the use of the Nasmyth focus has been revived. It has the advantage of remaining at a fixed position relative to the telescope wherever the instrument is pointed, and bulky or heavy instruments can be mounted there on a permanent platform, which rotates only in ►*azimuth*. In practice, there are two possible Nasmyth foci, one either side of the telescope tube. In the ►*William Herschel Telescope*, for example, the Nasmyth flat can be motor-driven into position to deflect the light beam to either of the Nasmyth foci. ►*altazimuth mounting*.

National Aeronautics and Space Administration (NASA) The US government agency responsible for civilian manned and robotic activities in space, including launch vehicle development and operations, scientific satellites (such as orbiting observatories) and space probes (such as missions to the Moon and planets), and advanced satellite technology development. Weather and communications satellites may be turned over to other agencies upon successful installation in space.

NASA was created on 29 July 1958, when US President Dwight D. Eisenhower signed the National Aeronautics and Space Act of 1958. This legislation was widely acknowledged to have been passed by the US Congress in response to the technological challenge perceived in the successful and unexpected launching of the first artificial Earth satellite (Sputnik 1) by the Soviet Union. The headquarters of NASA are in Washington, DC, and it operates field centres and other facilities at various locations in the USA as well as several tracking stations around the world. ►*Goddard Space Flight Center, Jet Propulsion Laboratory, Kennedy Space Center, Lyndon B. Johnson Space Center*.

National Optical Astronomy Observatories (NOAO) An organization formed in the USA in 1984 to bring under one administration the national facilities for optical astronomy at ►*Kitt Peak National Observatory*, ►*Cerro Tololo*

Inter-American Observatory and the ➤*National Solar Observatory*. NOAO is also responsible for the operation of the ➤*WIYN Telescope* at Kitt Peak, and the US contribution to the international ➤*Gemini 8-meter Telescopes* project. It is operated by the Association of Universities for Research in Astronomy (AURA), Inc., under contract to the National Science Foundation.

National Radio Astronomy Observatory (NRAO)   The combined radio astronomy facilities operated in the USA by a private consortium of universities, Associated Universities Inc. It obtains its funding under a cooperative agreement with the National Science Foundation. The telescopes operated by NRAO are at three different locations: the ➤*Very Large Array* (VLA) in New Mexico, a millimetre-wave telescope at ➤*Kitt Peak*, Arizona, and the Green Bank Telescope, a 42-metre (140-foot) dish and an interferometer, at ➤*Green Bank*, West Virginia. The NRAO administration is in Charlottesville, Virginia.

National Solar Observatory (NSO)   The solar observation facilities of the US ➤*National Optical Astronomy Observatories*. They consist of the ➤*McMath–Pierce Solar Telescope Facility* at ➤*Kitt Peak*, Arizona, and the ➤*Sacramento Peak Observatory* in New Mexico. NSO also administers the US involvement in the ➤GONG project.

Nautical Almanac   An abbreviated form of *Nautical Almanac and Astronomical Ephemeris*, the name prior to 1960 of the main annual compilation of astronomical data published by the UK's Nautical Almanac Office. This subsequently became the ➤*Astronomical Ephemeris* and more recently the ➤*Astronomical Almanac*. The current *Nautical Almanac* contains abridged data suitable for navigation purposes.

nautical twilight   Formally defined as the interval of time during which the Sun is between 96° and 102° below the ➤*zenith* point. ➤*twilight*.

n-body problem   A general term for calculations involving the gravitational interaction of an arbitrary number (*n*) of masses. The motion of two bodies is easily analysed and solutions exist for certain cases of three-body interaction. For values of *n* greater than 2, there are no general solutions and particular cases necessitate vast amounts of computation. Examples in astronomy requiring the solution of *n*-body problems are the motion of a space probe in the solar system, and the orbits of stars within a cluster.

NEA   Abbreviation for near-Earth asteroid. ➤*near-Earth object*.

NEAR   Abbreviation for the ➤*Near-Earth Asteroid Rendezvous mission*.

Near-Earth Asteroid Rendezvous mission (NEAR)   A NASA mission, launched on 17 February 1996, to rendezvous with the asteroid ➤*Eros* in January 1999. It is intended that the spacecraft will orbit around Eros, enabling it to study the asteroid for a period of about one year, from distances as close as 24 km (15 miles). On the way, it flew past asteroid 253 ➤*Mathilde* in June 1997.

Near-Earth Asteroid Tracking system (NEAT)   A NASA survey for
➤*asteroids* approaching relatively near the Earth. It started in 1995, using a CCD
camera installed on a 1-metre (39-inch) telescope operated on Mount Haleakala
in California by the US Air Force.

near-Earth object   An ➤*asteroid* or ➤*comet* in an orbit which makes it possible
for the object to approach exceptionally close to the Earth. Most such objects
are in orbits around the Sun with perihelia less than 1.3 AU and aphelia greater
than 0.983 AU. Observed objects range in size down to tens of metres. They
are believed to be asteroids from the main ➤*asteroid belt* which have been
perturbed by the gravitational attraction of the major planets, and the inactive
nuclei of comets. ➤*Amor, Apollo, Aten, Eros.*

NEAT   Abbreviation for ➤*Near-Earth Asteroid Tracking system.*

nebula (pl. nebulae or nebulas)   A cloud of interstellar gas and dust. The term
was formerly also used for objects now known to be galaxies (e.g. the 'great
nebula' in Andromeda, now usually called the Andromeda Galaxy).

 An ➤*emission nebula* glows in presence of ultraviolet radiation; a ➤*reflection
nebula* shines by reflecting starlight. An ➤*absorption nebula* is dark and is usually
evident only in silhouette against the background of a luminous nebula or
starfield.

 Other objects consisting of luminous gas are also known as nebulae, in
particular ➤*planetary nebulae* and ➤*supernova remnants.*

nebular hypothesis   The theory first put forward in 1755 by the philosopher
Immanuel Kant that the solar system formed from a primeval nebula around
the Sun.

nebulium   A hypothetical element, the existence of which was postulated in
the nineteenth century to account for unidentified emission lines in the spectra
of some luminous nebulae. It is now known that these lines are attributable to
known elements, but they are not usually observable under laboratory conditions. For that reason they are described as *forbidden lines* by physicists.

NEO   Abbreviation for ➤*near-Earth object.*

Neptune   A major planet of the solar system, normally the eighth in order
from the Sun. (Between 1979 and 1999 the eccentric orbit of Pluto brings it
temporarily closer than Neptune.) It is one of the four 'gas giant' planets, having
a small rocky core surrounded by an icy mantle of frozen water, methane and
ammonia. Its diameter is almost four times the Earth's. The outer atmosphere
is mainly molecular hydrogen with 15–20 per cent helium (by mass) and some
methane.

 Neptune was discovered by J. G. Galle of the Berlin Observatory on 23
September 1846 following predictions made independently by John Couch

Adams in England and Urbain J. J. Leverrier in France. Their calculations were based on discrepancies between the observed and predicted orbits of Uranus since its discovery in 1781, which were attributed to the gravitational perturbations of an unknown planet.

Viewed from Earth, Neptune is a seventh or eighth magnitude object and so not visible to the naked eye. With high magnification and larger telescopes, it is seen as a faintly bluish disc, the colour coming from methane in the upper atmosphere. Surface features are not detectable by ground-based optical observation, though bright spots are observed in the infrared.

Close-up images were obtained by ► *Voyager 2* during its flyby of Neptune in August 1989. Observations with the Hubble Space Telescope (HST), capable of resolving atmospheric detail, began in 1994. In many ways, such as size and structure, Neptune is similar to Uranus. But by contrast with Uranus, Neptune has distinctive and varying cloud features in a highly dynamic atmosphere. The most prominent feature found by *Voyager 2* was termed the *Great Dark Spot*, and appeared to be somewhat similar in nature to Jupiter's ►*Great Red Spot*. Located about 20° south of the equator, it rotated anticlockwise in a period of about 16 days. Bright cirrus-like clouds had formed over this and other small dark spots. However, it had completely disappeared when observations were made with the HST in 1994. Meanwhile, another dark spot, not seen by *Voyager*, had formed in the northern hemisphere. It too was accompanied by bright clouds. Subsequent observations with the HST revealed that the pattern of clouds was changing, though the underlying banded structure of the atmosphere remained stable.

There are two main cloud layers in Neptune's upper atmosphere. The highest consists of crystals of methane ice, and this lies over a lower opaque blanket of cloud that may contain frozen ammonia or hydrogen sulphide. There is also a high-altitude haze of hydrocarbons produced by the action of sunlight on methane.

Regular radio bursts detected by *Voyager 2* revealed that Neptune has a magnetic field and is surrounded by a magnetosphere. The bursts occurred at intervals of 16.11 hours, apparently the rotation period of the planetary core. The atmospheric features rotate at different rates, also moving in latitude. Wind speeds up to 2,200 km/hour were measured. The magnetic axis is tilted at 47° to the rotation axis and it is thought that the asymmetric field may originate in the mantle rather than the core.

Based on the total radiated energy, the average temperature is 59 K. It is not understood why Neptune radiates 2.7 times more energy than it receives from the Sun.

Observations made from the ground during occultations by Neptune had suggested the presence of incomplete ring 'arcs'. *Voyager 2* detected four tenuous rings, one of which is 'clumpy' in a way that can account for the occultation observations. The mission also discovered six new moons around Neptune, bringing the total number known, with ►*Triton* and ►*Nereid*, to eight. One of moons found by *Voyager*, Proteus, is more than twice the size of Nereid with a diameter of about 400 kilometres. ►Tables 5, 6 and 7.

Nereid   The small outermost moon of Neptune, discovered by Gerard Kuiper in 1949. The best *Voyager 2* image obtained was from a distance of 4.7 million kilometres (2.9 million miles), not sufficiently close to reveal surface detail but good enough to obtain a more precise radius of 170 kilometres (105 miles). ►Table 6.

Net   English name for the constellation ►*Reticulum*.

neutral hydrogen (H I or H°)   Un-ionized atomic hydrogen gas. It is an important component of the ►*interstellar medium*, accounting for perhaps half its mass, even though its density – typically 50 atoms per cubic centimetre – is very low.

The temperature of neutral hydrogen ranges between 25 and 250 K, which is too cold for it to emit visible radiation. However, its radio emission at a wavelength of 21 centimetres has made it possible to map the distribution of neutral hydrogen in the spiral arms of our own Galaxy and other nearby galaxies.

neutrino astronomy   The attempt to detect neutrinos from cosmic sources, especially the Sun. Neutrinos are elementary particles with no electric charge and almost no mass, and they interact only very weakly with other matter. They travel essentially at the velocity of light and are produced in vast quantities by the nuclear reactions that take place in the centres of stars and in ►*supernova* explosions.

Because they hardly interact with matter at all, neutrinos are very difficult to detect. The longest-running experiment to search for solar neutrinos, at Homestake Mine, South Dakota, was designed to use the fact that occasional neutrinos interact with a chlorine atom, converting it to a radioactive isotope of the gas argon. The detector consisted of a tank containing 400,000 litres of the cleaning fluid, carbon tetrachloride. Such experiments need to be underground to avoid confusion from events due to ►*cosmic rays*. Theoretical considerations suggested that one interaction should be detected per day by this set-up. In practice, only one-third that number have been seen. This discrepancy is known as the *neutrino problem*.

In another form of neutrino detector shown to work successfully, detectors in a large tank of water pick up ►*Cerenkov radiation* generated by the interaction of electrons with solar neutrinos. Detectors of this type, the Kamiokande experiment in Japan and a similar detector in Ohio, made the first observation of neutrinos from a supernova – those from S N 1987A. In 1996 Kamiokande was superseded by a larger version, ►*Super-Kamiokande*. A European collaboration (GALLEX) and a Russian experiment have been designed to make use of the interaction of neutrinos with gallium.

As results become available from new, more sensitive experiments, it is hoped to determine whether the current theories of solar physics are faulty or whether unknown physical processes are the cause of the 'neutrino problem'.

neutron star A star with a mass between about 1.5 and 3 solar masses that has collapsed under gravity to such an extent that it consists almost entirely of neutrons. Neutron stars are only about 10 kilometres across and have a density of 10^{17} kg/m^3. They are formed in ➤*supernova* explosions and observed as ➤*pulsars*.

Once nuclear fuel is exhausted in a star, the core starts to cool and the internal pressure falls, leading to contraction. This is a sudden and catastrophic event for stars of more than 1.8 solar masses, which implode until the pressure between neutrons balances the inward pull of gravity. In the resulting supernova, much of the original mass of the star is blown off into space.

A stellar remnant of three solar masses or more will collapse into a ➤*black hole* rather than a neutron star.

New General Catalogue of Nebulae and Star Clusters (NGC) A catalogue of non-stellar objects compiled by J. L. E. Dreyer of Armagh Observatory and published in 1888. It listed 7,840 objects. A further 1,529 were listed in a supplement that appeared seven years later, called the *Index Catalogue* (IC). The *Second Index Catalogue* of 1908 extended the supplementary list to 5,386 objects. The NGC and IC numbers are widely used to identify non-stellar astronomical objects.

The NGC got its title because the project was seen as a development of John Herschel's *General Catalogue of Nebulae*, which had been published in 1864.

new generation telescope A telescope incorporating design features that take advantage of the most recent technology. Such features include computer control, a lightweight primary mirror, an altazimuth mounting, automatic or remote operation, and special attention to the thermal environment.

new Moon The Moon's phase when it is at the same celestial ➤*longitude* as the Sun and thus totally unilluminated as seen from Earth. ➤*phase*.

New Style Date The system of date determination currently in use, introduced on 14 September 1752 in Britain and its American colonies when they adopted the ➤*Gregorian calendar*. The eleven days 3 – 13 September 1752 were eliminated and the day on which the count of years changes was moved from 25 March to 1 January. Dates according to the Julian calendar in use previously are known as 'Old Style Dates'.

New Technology Telescope (NTT) A 3.5-metre (138-inch) reflecting telescope of the ➤*European Southern Observatory*, located at the ➤*La Silla Observatory* in Chile. Regular observations with the telescope started in 1990.

The name reflects a number of innovative features incorporated in the design. The relatively thin mirror is kept to the required shape by means of an ►*active optics* system which analyses the shape of a stellar image and controls the mirror supports about once per second. The altazimuth mount and special enclosure are designed for maximum stability and pointing accuracy, and minimum disturbance of images from air turbulence. The telescope can be operated remotely from ESO's headquarters in Germany via satellite link.

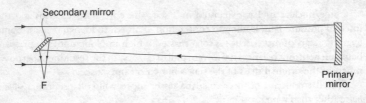

Newtonian telescope. Schematic optical arrangement.

Newtonian telescope   A simple type of reflecting telescope, designed by Isaac Newton (1642–1727), who demonstrated it to the Royal Society in London in 1671. The primary mirror is a paraboloid (or spherical for small apertures), and the secondary is a flat mirror positioned in the converging reflected beam at an angle of 45° to the optical axis to form an image just outside the main tube (see illustration). The design is widely used for small amateur instruments, but is not suitable for large telescopes.

Next Generation Space Telescope (NGST)   A project to launch a successor to the ►*Hubble Space Telescope*. It is being proposed that such a telescope should be primarily for infrared astronomy and placed in orbit around the Sun, rather than around the Earth. Such an orbit would have the advantage of minimizing stray light and temperature changes. The suggestion is that the telescope would be a 6- or 8-metre reflector, its mirror possibly constructed in segments.

N galaxy   An active galaxy with a compact and extremely bright nucleus, sometimes associated with a radio source.

NGC   Abbreviation for ►*New General Catalogue*.

NGST   Abbreviation for ►*Next Generation Space Telescope*.

nightglow   An alternative name for ►*airglow*.

Nix Olympica   The former name of the extinct martian volcano now known as ►*Olympus Mons*.

NLC   Abbreviation for ►*noctilucent clouds.*

NNTT   Abbreviation for ►*National New Technology Telescope.*

NOAO   Abbreviation for ►*National Optical Astronomy Observatories.*

noctilucent clouds (NLC)   A high-atmosphere phenomenon appearing as luminous bluish clouds in the summer twilight sky. They occur at heights of about 80 kilometres (50 miles) and show a range of structures.

Noctilucent clouds are very thin and scatter only a very small proportion of the sunlight incident on them, so they cannot be seen from the ground during daytime or in bright twilight. Since they are a summer phenomenon, they cannot be observed from the highest latitudes, where the sky never gets dark enough. However, they are also a high-latitude phenomenon, so the range of latitudes from which they can in practice be seen, between 50° and 65°, is quite narrow.

The clouds form in the presence of nuclei on to which water can condense and freeze into ice. It is not known just what these nuclei are, but they may be ions created by solar ultraviolet radiation or micrometeoritic particles. The main prerequisite is a low enough temperature, which is calculated to be 120 K (−150°C) at a height of 80−90 kilometres (50−56 miles). This is created by the pole-to-pole air flow at these heights and is not affected by the level of solar radiation.

There is observational evidence to suggest an overall increase in the occurrence of NLCs over recent decades, and this has been linked to an increase in water vapour in the upper atmosphere as a consequence of increasing amounts of methane. The number of sightings also varies inversely with the ►*solar activity cycle.*

Noctis Labyrinthus   A system of short, narrow canyons on the surface of Mars, east of the Tharsis region. The canyons are apparently ►*grabens* that formed as a result of the uplift of the Tharsis 'bulge'.

nocturnal   A simple instrument for telling the time at night by observation of the position of the two stars in the constellation Ursa Major known as the ►*Pointers.* The continuation of the line joining these stars passes very close to the north celestial pole, and can effectively be used as a giant clock-hand in the sky as the Earth's rotation causes them to sweep out a circle daily.

The nocturnal consists of two concentric discs and a sighting arm, fixed at the centre by an eyelet through which the Pole Star can be sighted. The lower disc is graduated with the days of the year and the upper one with the 24 hours of the day. In use, the sighting arm is aligned with the Pointers. The dials are needed to convert the direct measurement, which is effectively the ►*sidereal time,* to ►*mean solar time.*

node   Either of the points on the celestial sphere where the plane of an orbit intersects a reference plane. The position of a node is one of the ➤*orbital elements* used to specify the orientation of an orbit.

noise   In a radio receiver, the characteristic hiss produced by the random motion of electrons in the components of the receiver. Cosmic radio waves are also produced by random motions of electrons, so the signal detected by a radio telescope is indistinguishable in nature from the noise generated locally. For this reason, very-low-noise receivers are required for radio astronomy.

non-thermal radiation   In astronomy, the ➤*electromagnetic radiation* produced by an electron travelling at a speed close to that of light when it is forced to change its velocity. The commonest form is ➤*synchrotron radiation*, caused when electrons spiral round a magnetic field. The detection of non-thermal radiation from an object is a clear indication that high-energy processes are at work. More generally, the term is used for any electromagnetic radiation not produced by thermal processes.

Nordic Optical Telescope (NOT)   A 2.56-metre (101-inch) reflecting telescope at the ➤*Observatorio del Roque de los Muchachos* in the Canary Islands, operated jointly by Denmark, Finland, Norway and Sweden since 1989.

Norma (The Rule)   A small and insignificant southern constellation introduced by Nicolas L. de Lacaille in the mid-eighteenth century. It contains no star brighter than fourth magnitude. ➤Table 4.

normal galaxy   Any spiral or elliptical ➤*galaxy* that does not have unusual structure, a disturbed or active nucleus or non-thermal radio emission.

normal reflectivity   For a surface, the fraction of light reflected straight back from a beam of light falling on that surface in a direction perpendicular to it.

North America Nebula (NGC 7000)   A complex region of nebulosity in the constellation Cygnus. It includes ➤*emission nebulae*, ➤*reflection nebulae* and ➤*absorption nebulae*. It was discovered by William Herschel in 1786 and first photographed in 1890 by Max Wolf. The photographs revealed the shape, reminiscent of the North American continent. It is about 1° across and is just detectable with the naked eye under ideal conditions. Its estimated distance is 2,300 light years.

Northern Cross   A name sometimes given to the constellation ➤*Cygnus* and, in particular, the cross formation made by the stars Alpha (α), Beta (β), Gamma (γ), Delta (δ), Epsilon (ε) and Eta (η).

Northern Crown   English name for the constellation ➤*Corona Borealis*.

northern lights   A popular name for an ➤*aurora* when observed from northern latitudes.

North Galactic Spur   A region of radio and X-ray emission extending northwards from the plane of the Galaxy. It appears to be the most prominent segment of a huge ring of gas, probably a very old ➤*supernova remnant*.

north polar distance (NPD)   The angular distance on the celestial sphere between an object and the north celestial pole, measured along a great circle.

North Polar Sequence   A list of about a hundred stars, all lying within two degrees of the north celestial pole and covering a wide range of brightness, which are used as standards for the astronomical ➤*magnitude* scale. They were selected so they can be observed for comparison purposes at any time of the year from northern observatories.

North Star   A popular name for the star ➤*Polaris*, which lies within one degree of the north celestial pole.

nova (pl. novae)   A star that suddenly increases in brightness by about ten magnitudes, then declines gradually over a period of months. The word nova is a shortening of the Latin 'nova stella' – a new star.

Observations have demonstrated that novae are close binary stars of which one component is a ➤*white dwarf*. When the companion star evolves and expands to fill its ➤*Roche lobe*, material streams towards the white dwarf, forming an ➤*accretion disc* around it. The accepted theory of nova outbursts is that material accumulates in a layer on the surface of the white dwarf until the temperature and pressure at the base of the layer become high enough for the ➤*carbon cycle* nuclear reactions to be initiated. The energy produced is unable to escape as more material is deposited in the overlying layers. The temperature may rise to 100 million degrees and, at some point, explosive nuclear reactions are triggered, producing the observed nova outburst.

Some novae have been seen surrounded by an expanding envelope of gas. At speeds of up to 1,500 km/s, the envelope soon disperses into space. It is estimated that the mass of material lost is about one ten-thousandth (10^{-4}) the mass of the Sun, and the energy released is only a millionth of that released in a ➤*supernova*. The ejected material is rich in the elements carbon, nitrogen and oxygen, and the observed ratios of the isotopes $^{13}C/^{12}C$ and $^{15}N/^{14}N$ are consistent with the theory.

Classical novae are observed to erupt only once, though it is believed that outbursts may recur every 10,000 to 100,000 years. Recurrent novae, such as ➤*P Cygni*, have been observed to repeat their outbursts on timescales of ten to a hundred years. In any one galaxy, typically a few tens of novae occur in a year. ➤*dwarf nova*.

NPD   Abbreviation for ➤*north polar distance*.

NRAO   Abbreviation for ➤*National Radio Astronomy Observatory*.

NSO   Abbreviation for ➤*National Solar Observatory*.

N star   An obsolete stellar classification now incorporated into the group of ➤*carbon stars*. The former N-type stars fall into classes C6 to C9.

NTT   Abbreviation for ➤*New Technology Telescope*.

Nubecula Major   The Latin name for the Large ➤*Magellanic Cloud*.

Nubecula Minor   The Latin name for the Small ➤*Magellanic Cloud*.

nucleosynthesis   The making of chemical elements by means of naturally occurring nuclear reactions. Nucleosynthesis takes place in the interiors of stars, in ➤*supernova* explosions and other astrophysical situations where high-energy collisions can take place between atomic nuclei and elementary particles. ➤*carbon cycle, r-process, s-process*.

Nuffield Radio Astronomy Laboratories   The radio astronomy department of the University of Manchester, England. It is located at Jodrell Bank in Cheshire. The main instrument is the 76-metre (250-foot) fully steerable dish, known since 1987 as the Lovell Telescope. It was completed in 1957, its construction having been conceived and directed by Bernard Lovell, who had begun radar experiments on the site in 1945 with the hope of detecting cosmic ray showers.

An elliptical dish (38 × 25 metres) was built in 1964. Since 1980, it has been possible to link the Jodrell Bank telescopes with others at distant sites to form a ➤*radio interferometer* known as ➤*MERLIN*.

Nunki (Sigma Sagittarii; σ Sgr)   The second-brightest star in the constellation Sagittarius. It is a ➤*B star* of apparent magnitude 2.0. ➤Table 3.

nutation   A relatively short-period oscillation superimposed on the ➤*precession* of the rotation axis of a rotating body under turning forces (torque) from external gravitational influences. The nutation of the Earth's axis, a maximum of 15 arc seconds with a period of about 18.6 years, is caused by changes in the Moon's orbit over this time.

Nysa   Asteroid 44, diameter 68 km, discovered in 1857 by H. Goldschmidt. It is notable for its high albedo of nearly 40 per cent. It is one of the two large members of the Nysa ➤*Hirayama family*, the other being 135 Hertha.

O

OAO   Abbreviation for ➤*Orbiting Astronomical Observatory*.

O association (OB association)   ➤*association*.

Oberon   One of the larger satellites of the planet Uranus, discovered by William Herschel in 1787. Its surface is covered by numerous impact craters, many surrounded by bright rays and blankets of ejecta. Several have very dark material within them. ➤Table 6.

object glass (OG)   An old term for the ➤*objective* lens of a refracting telescope. Its abbreviation, OG, is still often used when observations are recorded, to signify that the telescope used was a refractor.

objective   The main light-collecting lens in a refracting telescope.

objective grating   A ➤*diffraction grating*, of the transmission type, placed over the aperture of a telescope in order to produce spectra of all the stars in the field of view.

objective prism   A thin prism placed over the aperture of a telescope in order to produce spectra of all the stars in the field of view.

oblateness   ➤*ellipticity*.

obliquity of the ecliptic (symbol ε)   The angle between the planes of the Earth's equator and the ➤*ecliptic*. Its present value is approximately 23° 26′. The effects of ➤*precession* and ➤*nutation* cause it to change between extreme values of 21° 55′ and 24° 18′.

Observatoire de Paris   The French national astronomical research institute, based at the original site in Paris where it was founded in 1667. This is the oldest astronomical observatory still in use for research. There is an astrophysics section, located at the Observatoire de Meudon, just outside Paris, and a radio astronomy station at Nançay. Research is carried out in many branches of astronomy.

At the Paris site, there are three nineteenth-century instruments, including the telescope built for the ➤*Carte du Ciel* project and a 38-centimetre (15-inch) refractor, which is occasionally used for positional work. Systematic astrometric measurements are made with a ➤*prismatic astrolabe*.

The Observatoire de Meudon was founded in 1876. It became the Astrophysics Section of the Observatoire de Paris in 1926 when the two institutions were merged. The instruments there include an 83-centimetre (33-inch) refractor dating from 1893, a 1-metre (40-inch) reflector, also from 1893 but modernized in 1969, a solar tower telescope used for spectroscopic studies of the Sun, a ►*spectroheliograph* and a large ►*siderostat* used in conjunction with a solar ►*magnetograph* and instruments to monitor the solar ►*chromosphere*.

The Nançay radio astronomy station, established in 1953, is a large site with many instruments. Observations of solar radio emission are made with a ►*radioheliograph*, a multi-channel instrument for spectral observations of the Sun in the radio band and telescopes for monitoring solar activity. A special array is used for observations of the Sun and the planet Jupiter at wavelengths between 3 and 300 metres. The largest radio telescope at the site is of a unique design, consisting of two immense reflecting surfaces, one flat and one concave, which face each other. The flat reflector consists of ten panels, each 20 × 40 metres (65 × 130 feet), which can be rotated about a horizontal axis. The concave reflector is 300 metres (980 feet) long and 35 metres (115 feet) high. Radio signals are reflected from the plane reflector on to the concave reflector, from which they are brought to a focus and collected by receivers mounted in a movable cabin. This telescope is used for studies of the 21-centimetre emission from ►*neutral hydrogen*, emission from the hydroxyl (OH) molecule, and other work.

Observatorio del Roque de los Muchachos   An observatory on the island of La Palma in the Canary Islands Group. The observatory site, regarded as one of the best in the world, is operated by the Instituto de Astrofisica de Canarias, which is host to a number of different countries that have telescopes there. The UK's Isaac Newton Group consists of the ►*William Herschel Telescope*, the ►*Jacobus Kapteyn Telescope* and the ►*Isaac Newton Telescope*. Other instruments at the observatory include a Swedish solar telescope, a 2.56-metre (101-inch) telescope belonging jointly to Sweden, Norway, Finland and Denmark (the Nordic Telescope, or NOT) and the Carlsberg ►*meridian circle* operated by Denmark, the UK and Spain. The observatory occupies an area of nearly two square kilometres at an altitude of 2,400 metres (7,900 feet).

observatory   A place or building at which astronomical observations are or were formerly carried out, or the administrative centre for such work.

occultation   The passage of one astronomical object directly in front of another so as to obscure it from view as seen by a particular observer. ►*grazing occultation, eclipse*.

oceanus   ►*mare*.

octahedrite   A class of metallic ►*meteorites* containing between 6 and 17 per cent nickel by weight. Octahedrites contain two different forms of iron–nickel alloy, called kamacite and taenite. The intergrowth of crystals of the two types produces an octahedral form and a characteristic pattern, called the ➤*Widmanstätten figures*, that shows up clearly on the surface of a slice of meteorite that has been polished and etched with dilute acid.

Octans (The Octant)   A faint and obscure constellation containing the south celestial pole. It was introduced in the mid-eighteenth century by Nicolas L. de Lacaille and contains only one star brighter than fourth magnitude. ➤Table 4.

Octant   English name for the constellation ➤*Octans*.

ocular   An alternative term for ➤*eyepiece*.

Of star   An ➤*O star* with emission lines, mainly of helium and nitrogen, in its spectrum. ➤*spectral class*.

O G   Abbreviation for ➤*object glass*.

OH source   An astronomical source emitting microwave radiation characteristic of the hydroxyl (OH) molecule, especially one showing a ➤*maser* effect. OH sources are found in ➤*molecular clouds* in interstellar space and in the cool envelopes of evolved stars. The first detection of OH was in 1963. Four microwave spectral lines, near a wavelength of 18 centimetres, were seen in absorption from the direction of the galactic centre against the radio source ➤*Sagittarius A*. In 1965 emission was detected from a source in the ➤*Orion Nebula* in which the relative strengths of the four lines were not as predicted by theory and as seen in normal sources. A line at a frequency of 1,665 MHz was found to be fifty times too strong; this was explained in terms of maser action. The OH molecules absorb infrared radiation then re-radiate the energy at the particular frequency of the 1,665-MHz spectral line.

Olbers' Paradox   The question 'Why is the sky dark at night?' In 1826, Heinrich W. M. Olbers (1758–1840) drew attention to the fact that, in a large universe of infinite age, filled more or less uniformly with stars, the sky should be a continuous blaze of light, because every line of sight from an observer would ultimately encounter a star. The paradox is that, despite this, the sky is dark at night.

The resolution of the paradox lies in identifying an underlying assumption that is wrong. Since the paradox makes assumptions about the properties of the universe, it is clear that the darkness of the night sky is of cosmological importance. The paradox is resolved in the ➤*Big Bang* cosmology: the finite age of the galaxies means that there has been insufficient time to fill the universe with light. Furthermore, the expansion of the universe leads to yet more dilution of the observed sky brightness from remote objects.

Old Style Date   ➤*New Style Date*.

olivine   Magnesium iron silicate, the most abundant mineral in chondritic ➤*meteorites*.

Oljato   Asteroid 2201, diameter 2.8 km, discovered in 1947 by H. Giclas then lost until recovered in 1979. It is in a highly elliptical Earth-crossing orbit and has a unique spectrum that does not resemble that of any other known asteroid, meteorite or comet. Its nature is not known, but it could be the 'dead' nucleus of a comet that has ceased to be active.

Olympus Mons   The highest peak on Mars, and the largest volcano in the solar system. It rises to a height of 27 kilometres (17 miles) above the datum level (selected on the basis of atmospheric pressure). This gigantic ➤*shield volcano*, 700 kilometres (435 miles) across, is similar in nature to volcanoes on Earth but its volume is at least fifty times greater than its nearest terrestrial equivalent. The caldera at the summit is 90 kilometres (60 miles) across and a cliff at least 4 kilometres high rings the mountain. Older volcanic rocks, fractured and eroded by the wind, surround the main peak, forming an area called the *aureole*. Olympus Mons is located to the north-west of the Tharsis region and was formerly known as *Nix Olympica* (Olympic Snow) because the clouds over the area appeared as a light-coloured spot to Earth-based observers.

Omega Centauri (ω Cen; NGC 5139)   A particularly bright ➤*globular cluster* of stars in the southern constellation Centaurus. Its diameter, 620 light years, is the largest of any globular cluster known.

Omega Centauri lies at a distance of 16,500 light years. Through a large telescope, it is seen to spread over as much as 1° of the sky. Its total magnitude is 3.6 and it is easily visible to the naked eye. A curious feature is its shape, which is distinctly elliptical with axes in the ratio 5:4.

The name Omega Centauri is of a type normally given to single stars. The cluster was mistaken for a single star by early observers at Mediterranean latitudes, for whom it never rises more than about 10° above the horizon.

Omega Nebula (M17; NGC 6618)   A luminous nebula in the constellation Sagittarius, also known as the *Horseshoe Nebula* and the *Swan Nebula*. It is 4,800 light years away and 27 light years in diameter. It is a region of ➤*ionized hydrogen*, excited by a group of at least five hot stars. A dark dust cloud lies on the western edge of the luminous region. ➤*molecular cloud*.

Omicron Ceti (o Ceti)   The designation of the variable star more generally known as ➤*Mira*.

Oort cloud (Oort–Öpik cloud)   A spherical shell, surrounding the solar system between about 2,000 and 20,000 AU from the Sun, which contains billions of ➤*comets* with a total mass about that of the Earth.

The cloud is invoked as a source for long-period comets, observed in the inner solar system, which could be deflected there by the gravitational influence of a passing star. The idea was first put forward by E. Öpik in 1932 and developed by J. Oort in the 1950s. (The term Öpik–Oort cloud is sometimes used.) There is no observational evidence for the existence of the cloud, but the orbits of long-period comets and dynamical studies of the formation of the solar system provide strong circumstantial evidence. It is argued that incipient comets were formed near the present location of the outer planets and ejected to a much greater distance later.

Oort limit The mass density in the plane of the ►*Galaxy* near the Sun's locality, as calculated from the velocities and distribution of stars in relation to the gravitational field of the galactic disc.

At one time it was believed that the mass density deduced in this way exceeded the amount that could be seen directly in the form of stars and interstellar material. However, the discovery of fainter stars and more interstellar material has brought the observed and calculated values of local mass density closer together, implying little, if any, 'hidden' matter in the solar neighbourhood.

The name is that of the Dutch astronomer, Jan Oort, distinguished for his work in the field of galactic dynamics.

Oort's constants Two empirical quantities contained in mathematical expressions derived by the Dutch astronomer, Jan Oort, for the ►*radial velocity* and ►*proper motion* of stars resulting from their orbital motion around the centre of the Galaxy, which is assumed to be circular. The expressions take the form

$$v = A \; r \sin 2l \text{ and } \mu = 0.211(B + A \cos 2l)$$

where v is radial velocity, μ proper motion, r distance from the Sun and l the star's galactic longitude. A and B are Oort's constants; generally accepted values are 15 kilometres per second per kiloparsec for A and -10 kilometres per second per kiloparsec for B.

opacity A measure of the extent to which a material absorbs and scatters incident electromagnetic radiation. It is dependent on wavelength, though a mean value may be used to simplify calculations.

open cluster A type of star cluster containing several hundred to several thousand stars distributed in a region a few light years across. The member stars are much more spaced out than in ►*globular clusters*.

Open clusters are relatively young, typically containing many hot, highly luminous stars. They are located within the disc of the ►*Galaxy* and so appear to lie within the ►*Milky Way*. Well-known open clusters include the ►*Pleiades*, the ►*Hyades* and the ►*Jewel Box*. ►*Trumpler classification*.

open universe A model of the universe in which expansion continues for ever and reaches infinity with a non-zero speed. ►*oscillating universe, closed universe*.

Ophelia   A small satellite of Uranus discovered by the ➤*Voyager 2* spacecraft in 1986. Ophelia is one of two satellites that act as 'shepherds' of the planet's Epsilon ring (the other being Cordelia). ➤Table 6.

Ophiuchus (The Serpent Bearer)   A large constellation straddling the celestial equator. It was one of the 48 constellations listed by Ptolemy (*c.* AD 140). The mythological figure of the serpent holder is sometimes identified with the healer Aesculapius. Though Ophiuchus is not traditionally a zodiacal constellation, the ➤*ecliptic* passes through its southern part. It contains five stars of second magnitude and seven of third magnitude. ➤*Barnard's Star* also lies in Ophiuchus. ➤Table 4.

Öpik–Oort cloud   ➤*Oort cloud.*

opposition   The position of one of the superior planets when it is opposite the Sun in the sky, i.e. when its ➤*elongation* is 180°. At opposition planets are at full phase and reach their highest point in the sky at midnight. At the same time, they achieve their closest approaches to the Earth. As the orbits of the planets are elliptical rather than perfectly circular, some oppositions bring the planets closer to Earth than others. This effect is particularly marked with Mars.

opposition effect   The additional brightening of an asteroid observed at full phase in excess of the brightness that would be expected from measurements made at partial phases.

optical depth (symbol τ)   A measure of the degree of absorption that takes place when electromagnetic radiation travels along a specified path through a gaseous or dusty medium, such as the atmosphere of a star or planet.

optical double star   A pair of stars that lie close to each other in the sky by chance, but are not physically associated with each other as with a true ➤*binary star.*

optical interferometer   ➤*interferometer.*

orbit   The path followed by a body moving in a gravitational field. For bodies moving under the influence of a centrally directed force, without significant perturbation, the shape of the orbit must be one of the 'conic section' family of curves – it must be a circle, an ellipse, a parabola or a hyperbola.

orbital elements   A set of parameters that together completely define the shape, orientation and timing of orbital motion. Those most commonly used for the orbits of planets and comets around the Sun are ➤*semimajor axis*, ➤*perihelion distance*, ➤*eccentricity*, ➤*inclination*, ➤*argument of perihelion*, ➤*longitude of the ascending node* and ➤*period*. In order to determine the position of an object at a specific time, it is also necessary to know the time when a particular point in the orbit is reached, such as perihelion passage. Analogous elements are used to describe the orbits of stars in binary systems or of satellites (natural or artificial) orbiting planets.

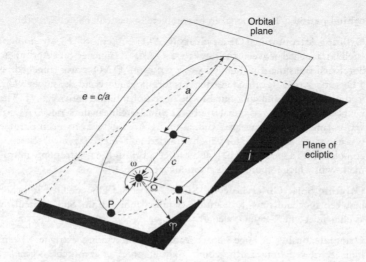

Orbital elements. The elements of the elliptical orbit of a planet or comet around the Sun: (N) Ascending node (♈), First Point of Aries.

The standard orbital elements are detailed in the table and the illustration, which depicts an elliptical orbit in the solar system. Here P is the perihelion point, N the ascending node and ♈ the direction of the ▸*First Point of Aries*.

Name	Symbol	Function
Semimajor axis	a	Defines the size of an elliptical orbit
Minimum distance from pericentre	q	Defines the size of a parabolic or hyperbolic orbit
Eccentricity	e	Defines the distance of the foci from the centre for an elliptical orbit
Inclination	i	Angle between the orbital plane and the reference plane
Longitude of the ascending node	Ω	Direction in space of the line where the orbit intersects the reference plane
Argument of pericentre	ω	Defines the orientation of the orbit with respect to the reference plane
Period	P	Time to complete one revolution of a closed orbit
Time of pericentre passage	T	Sets the time frame of the orbital motion

orbital period   The time taken to complete one circuit of a closed orbit.

Orbiting Astronomical Observatory (OAO)   A series of US astronomical satellites launched between 1966 and 1972. OAO-1, launched on 8 April 1966, developed faults the day after and was a write-off. OAO-2 was launched on 7 December 1968. It carried instruments for photometry and sky survey work in the ultraviolet, and continued to be operational until February 1973. The next launch, in 1970, was a total failure. The final satellite in the series, OAO-3, was renamed Copernicus after launch on 21 August 1972. Its main purpose was ultraviolet astronomy, for which it carried a 0.8-metre (31-inch) telescope complete with a spectrograph. It also carried a British X-ray astronomy instrument with which important observations were made.

Orbiting Solar Observatory (OSO)   A series of US satellites, launched between 1962 and 1975, for various scientific studies of the Sun, particularly at ultraviolet and X-ray wavelengths.

Orientale Basin   A huge impact feature on the Moon's extreme western limb as viewed from Earth, visible only at times of favourable ➤*libration*. Photographs taken from lunar orbit by spacecraft show a structure of at least three concentric rings. Unlike many other impact basins on the Moon, it is not extensively filled by dark ➤*mare* material.

Orion (The Hunter)   A brilliant constellation straddling the celestial equator, widely considered to be the most magnificent and interesting in the sky. Its pattern is interpreted as the hunter brandishing a raised club and a shield. Three bright stars mark his belt, and several fainter ones a sword hanging from it. Orion contains five stars of the first magnitude or brighter and a further ten brighter than fourth magnitude. The most spectacular diffuse nebula in the sky, the ➤*Orion Nebula*, is faintly visible to the unaided eye in the 'sword'. ➤Table 4.

Orion Arm   The spiral arm of the ➤*Galaxy* in which the Sun is located.

Orionids   An annual ➤*meteor shower*, the multiple radiant of which lies on the border of Orion and Gemini, near the star Gamma Geminorum (γ Gem). The peak of the shower occurs around 22 October, and the normal limits are 16–27 October. The shower is produced by meteoroids that have come from ➤*Halley's Comet*.

Orion Molecular Cloud   ➤*Orion Nebula*.

Orion Nebula (M42 and M43; NGC 1976 and NGC 1982)   A bright emission nebula surrounding the multiple star Theta^1 Orionis (θ^1 Ori) in the 'sword' of Orion.

   The luminous nebula is just part of a complex region of interstellar matter at a distance of 1,300 light years occupying much of the constellation of

Orion. The Orion cloud is the largest such dark cloud known in the Galaxy. Millimetre-wave observations of the emission from the molecules C O (carbon monoxide), H C H O (formaldehyde), and many others, reveal the presence behind the visible part of a large ➤*molecular cloud*, known as the Orion Molecular Cloud (O M C - 1). This is an important region of star formation, and the group of four young hot stars that make up θ^1 Orionis, also known as the *Trapezium*, are believed to be less than 100,000 years old. The ➤*Becklin–Neugebauer object* and the ➤*Kleinmann–Low Nebula*, both detected through their infrared emission, are sites of current star formation.

The Trapezium stars are creating an expanding spherical cavity near the edge of the dark cloud. Their ultraviolet radiation is ionizing the gas and blowing away the dust. Relatively recently, in astronomical terms, the bubble broke through on our side of the dark cloud, revealing the stars and ionized hydrogen within. The sharper edges of the nebula are produced by remnants of dust. M43 (N G C 1982) is a northern section of the nebula separated from the larger part (M42; N G C 1976) by a lane of dust.

In photographs, the dominant colour of the luminosity is red from the ➤*hydrogen alpha* light. Observed visually, the nebula appears greenish because of the eye's low sensitivity to red light. The green emission is due to oxygen. The nebula occupies an area of sky about one degree across and is faintly visible to the naked eye. It has the highest surface brightness of all nebulae.

Orion's Belt   ➤*Belt of Orion*.

orrery   A working model of the solar system showing the planets, possibly with some of their moons, in their orbits around the Sun. The term 'orrery' was first applied to such a model in 1713 when one was made for the Fourth Earl of Cork and Orrery.

The gear systems in such models are usually made so that the orbital periods of the planets are in the correct ratios. However, it is impossible to construct a demonstration model of this kind in which the planet sizes and the distances between them are both to scale because of the great range of distance scales.

orthoscopic eyepiece   A good general-purpose telescopic ➤*eyepiece*, giving a high-quality image and good ➤*eye relief*. There are various forms, but the eyepiece typically consists of a convex or planoconvex eyelens with an achromatic triplet as the field lens (see illustration).

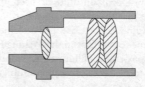

Orthoscopic eyepiece.

Oschin Telescope   The 1.2-metre (48-inch) ➤*Schmidt camera* at ➤*Palomar Observatory*. It has been in operation since 1948.

oscillating universe A model of the universe in which it goes through cycles of expansion, collapse to a so-called 'Big Crunch', ➤*Big Bang*, expansion, and so on, in an infinite sequence.

oscillator strength (*f*-value) A measure of the intrinsic strength of a ➤*spectral line*. It depends on the probability that the transition between atomic energy levels responsible for the line will take place in a particular time. It is independent of the physical conditions under which the transition occurs.

osculating elements The ➤*orbital elements* describing the instantaneous, purely elliptical orbit of a body subject to perturbations. If the perturbing influences were to disappear at some particular time, the body would continue to follow an orbit described by the osculating elements for that moment in time.

OSO Abbreviation for ➤*Orbiting Solar Observatory*.

O star A star of ➤*spectral type* O. O stars have surface temperatures in the range 28,000 – 50,000 K and are bluish white in colour. Their spectra are characterized by lines of both neutral and ionized helium; emission lines are also commonly present. The four brightest O stars in the sky are Delta (δ) and Zeta (ζ) Orionis, the easternmost stars in Orion's belt, and the southern stars Zeta (ζ) Puppis and Gamma^2 (γ^2) Velorum.

outer planets The planets beyond the ➤*asteroid belt*, namely Jupiter, Saturn, Uranus, Neptune and Pluto.

outflow channel A channel on ➤*Mars*, typically several kilometres wide and hundreds of kilometres long, that once drained large amounts of liquid water from the southern uplands to the northern lowlands.

outgassing The release of gas, possibly creating an entire atmosphere, from the surface material of a planet.

Owens Valley Radio Observatory The radio astronomy observatory of the California Institute of Technology (Caltech), located 400 kilometres (250 miles) north of Los Angeles at an altitude of 1,200 metres (4,000 feet). The instruments there are a five-element ➤*radio interferometer* (which includes two 27-metre (88-foot) dishes, dating from 1960), a 40-metre (130-foot) dish built in 1965 and a millimetre-wave interferometer consisting of three 10.4-metre (34-foot) dishes. The interferometer is used particularly for solar observation and the 40-metre dish is used for ➤*very-long-baseline interferometry*.

Owl Nebula (M97; NGC 3587) A ➤*planetary nebula* in the constellation Ursa Major. It is one of the largest planetary nebulae known, with a diameter of 1.5 light years, and lies at a distance of 1,600 light years.

Ozma, Project ➤*Project Ozma*.

ozone   A form of molecular oxygen containing three atoms instead of the normal two. It is created by the action of ultraviolet light on air or oxygen. A layer of ozone in the Earth's upper atmosphere at a height of about 25–40 kilometres (15–25 miles) serves as a protective barrier against ultraviolet light from the Sun, which is harmful to life.

P

PA (p.a.)   Abbreviation for ➤*position angle*.

Painter's Easel (or Painter)   English name for the constellation ➤*Pictor*.

palimpsest   A circular spot on an icy moon believed to be the 'ghost' of a former impact crater.

Pallas   Asteroid 2, mean diameter 533 km, discovered by Heinrich W. M. Olbers in 1802. It is the second-largest asteroid and of the carbonaceous type, which is similar to the largest asteroid, ➤*Ceres*. Its orbit is at the unusually steep inclination of 35° to the plane of the solar system.

pallasite   A class of stony-iron ➤*meteorite*.

Palomar Observatory   The observatory on Palomar Mountain in California where the 5-metre (200-inch) ➤*Hale Telescope* is sited. It is owned and operated by the California Institute of Technology (Caltech). The other instruments at the observatory are the 1.2-metre (48-inch) Oschin Telescope (a ➤*Schmidt camera*), a 46-centimetre (18-inch) Schmidt camera and the 1.5-metre (60-inch) reflector owned jointly by Caltech and the Carnegie Institution of Washington.

Palomar Sky Survey   A photographic atlas of the whole sky north of declination −30° consisting of wide-field plates taken with the 1.2-metre (48-inch) ➤*Schmidt camera* (the Oschin Telescope) at ➤*Palomar Observatory*. There are two plates, one taken in red light and one in blue, for each area of sky.

palus   Literally 'swamp' or 'marsh', a term used in the names of certain dark features on the Moon. Its use dates from a time when it was believed that the darker features on the Moon were of liquid water; this is now known to be untrue. In recognition of its use over a long period, the term has been retained in certain official names for features on the Moon. ➤*mare*.

Pan   A small satellite of Saturn (1981 S13) orbiting in the ➤*Encke Division* in the planet's ring system. It was found by Mark R. Showalter in 1990 from studies of images taken by the spacecraft ➤*Voyager 1* and ➤*Voyager 2*. Its existence had been predicted as an explanation for observed structure in the rings around the Encke Division. Pan is about 20 kilometres in diameter.

Pandora   A small satellite of Saturn, measuring about 110 × 70 kilometres, discovered by ➤*Voyager 2* in 1980. ➤Table 6.

parabola A particular form of open curve belonging to the family of curves known as ►*conic sections*. It is encountered in astronomy in two main contexts. First, it is one of the shapes that can be taken by the orbit of a body moving under the influence of a central gravitational force, such as a comet when it approaches the Sun. Second, the surface obtained by rotating a parabola, a *paraboloid*, is commonly used as the shape for the primary mirror in a reflecting telescope. Images formed by such parabolic mirrors, in contrast to spherical mirrors, do not suffer from ►*spherical aberration*.

parabolic mirror A mirror whose surface is figured to the shape of a paraboloid. ►*parabola*.

parallax (symbol π) The change in the relative positions of objects when they are viewed from different places. The actual angular shift measured when a viewpoint is changed is also described as trigonometric parallax. In the case of astronomical objects, such changes are measurable only for relatively nearby objects in relation to the more distant stars. However, the measurement of parallaxes, where possible, is important since it is one of the most direct methods of determining astronomical distances. In astronomy, the word 'parallax' is often used synonymously with 'distance'.

The rotation of the Earth produces a ►*diurnal parallax* effect, and the Earth's orbital motion around the Sun causes ►*annual parallax*. ►*statistical parallax*.

Paranal Observatory The site in Chile of the ►*European Southern Observatory*'s ►*Very Large Telescope* (VLT). It is situated in the Atacama desert, 120 km (75 miles) south of Antofagasta, at an altitude of 2,632 metres (8,500 feet).

parent body An ►*asteroid*, ►*comet* or other body of which a ►*meteorite* is a fragment.

parhelion (pl. parhelia) An alternative name for a ►*mock Sun* or sundog.

Paris Observatory ►*Observatoire de Paris*.

Parkes Observatory An Australian radio astronomy observatory located 20 kilometres (12 miles) north of Parkes, New South Wales. The instrument is an altazimuth mounted, prime-focus, 64-metre (210-foot) single dish. It was commissioned in 1961 and operated by the Commonwealth Scientific and Industrial Research Organization (CSIRO) Division of Radiophysics as the Australian National Radio Astronomy Observatory. In 1988 it became a unit of the ►*Australia Telescope*. It can serve as a stand-alone telescope or as a member of a long-baseline array.

The Parkes dish was the first to be built in the southern hemisphere. It was used for the identification of the first quasar in 1963, and to discover many ►*interstellar molecules* and over half the known ►*pulsars*. It has also been used as an additional element in the Deep Space Network for tracking spacecraft, for example during the ➤*Voyager 2* encounters with Uranus and Neptune and the *Giotto* mission to Halley's Comet.

parsec (symbol pc)   A unit of distance used in professional astronomy. It is defined as the distance at which an object would have an ➤*annual parallax* of one arc second. It is equivalent to 3.0857×10^{13} kilometres, 3.2616 light years or 206,265 astronomical units.

Multiples in common use are the kiloparsec (kpc, 1,000 parsecs) and the megaparsec (Mpc, 1,000,000 parsecs).

particle horizon   The limit of the visible universe, within which light has had time to reach us since the expansion of the universe began.

Pasiphae   One of the small outer satellites of Jupiter (number VIII), discovered in 1908 by P. J. Melotte. ➤Table 6.

patera (pl. paterae)   A shallow crater with a scalloped or complex edge.

Patientia   Asteroid 451, diameter 230 km, discovered by A. Charlois in 1899.

Paul–Baker system   An optical design for a reflecting telescope that has an exceptionally wide field of view with good definition. It uses a paraboloid primary of ➤*focal ratio f/4* or less, a convex spherical secondary mirror, and a concave spherical third mirror with curvature equal but opposite to that of the secondary. It was devised by the French optician Maurice Paul in 1935 and discovered independently by James Baker in about 1945. ➤*Willstrop telescope.*

Paul Wild Observatory   ➤*Australia Telescope.*

Pavo (The Peacock)   A southern constellation introduced in the 1603 star atlas of Johann Bayer. It contains one first magnitude star, sometimes itself called 'Peacock'. ➤Table 4.

Pavonis Mons   One of the three giant ➤*shield volcanoes* of the ➤*Tharsis Ridge* on Mars. It is about 400 kilometres (250 miles) in diameter and rises to a height of 27 kilometres (17 miles), 17 kilometres above the level of the surrounding ridge.

pc   Abbreviation for ➤*parsec.*

P Cygni   An unusual variable star, the spectral lines of which have a particularly characteristic profile interpreted in terms of an expanding envelope around the star.

P Cygni is a ➤*recurrent nova.* It was recorded as third magnitude in August 1600 and stayed at this brightness for six years before fading slowly. A second outburst occurred in about 1655, which was again followed by slow fading. It subsequently fluctuated in brightness around sixth magnitude and has been about fifth magnitude, with only small variations, since 1715.

The lines in the spectrum of P Cygni are all double, consisting of a broad ➤*emission line* with a narrower ➤*absorption line* adjacent on the blue side. The absorption comes from starlight passing through surrounding shells of material, while the emission comes from the portions of the shells either side of the central star as viewed from the Earth. The emission and absorption components are displaced from each other by the Doppler effect because the shells are expanding. Detailed analysis has shown that there are three distinct shells, the outermost of which is pulsating with a 114-day period.

Similar line profiles are observed in the spectra of other objects surrounded by expanding envelopes and are described as *P-Cygni profiles*.

Peacock (1)   English name for the constellation ➤*Pavo*.

Peacock (2)   The brightest star in the constellation ➤*Pavo*.

peculiar galaxy   A term loosely applied to any ➤*galaxy* that does not readily fit into the ➤*Hubble classification*, shows signs of unusual energetic activity or is interacting tidally with other neighbouring galaxies.

peculiar motion (1)   The motion of an individual star relative to the ➤*local standard of rest* or to some specified group of stars.

peculiar motion (2)   The individual motion of a galaxy or cluster of galaxies over and above that associated with the general expansion of the universe. ➤*Hubble flow*.

peculiar star   A star whose spectrum shows unusual features compared with the majority of stars of its ➤*spectral type*. A 'p' after the spectral type is used to indicate that a star is considered 'peculiar'. The term is most frequently applied to a group of stars in spectral classes A and B. ➤*Ap star*.

Pegasus (The Winged Horse)   A large northern constellation included by Ptolemy in his list of 48 (*c.* AD 140). It is noted for the prominent square – the *Square of Pegasus* – formed by its three brightest stars and Alpha Andromedae (Alpheratz), all of which are second magnitude. Alpha Andromedae used to be considered to belong to Pegasus and was known as Delta Pegasi. ➤Table 4.

Pele   An ➤*eruptive centre* on ➤*Io*. A large plume was seen by ➤*Voyager 1* in March 1979, but activity had ceased by the time of the ➤*Voyager 2* encounter, four months later.

Pelican Nebula   Popular name for the diffuse nebulae IC 5067 and 5070 in the constellation Cygnus, forming part of the ➤*North America Nebula* (NGC 7000) complex.

penumbra (1)   A region of partial shadow. During a solar ➤*eclipse*, when the Moon's shadow sweeps across the surface of the Earth, observers in the penumbral zone see only a partial eclipse.

penumbra (2)   The lighter periphery of a ►*sunspot*, surrounding the darker ►*umbra*. In the penumbra the magnetic field is horizontal and spreads radially.

PEP   Abbreviation for ►*photoelectric photometry*.

perfect cosmological principle   An extension of the ►*cosmological principle*, proposed in 1948 by Herman Bondi and Thomas Gold. It asserts that the universe appears essentially the same at all times for all observers everywhere. This principle is crucial to the ►*steady-state theory* but is violated by the ►*Big Bang* theory.

periastron   In the orbital motion of a ►*binary star* system, the point of closest approach of the two stars.

pericentre   The point in an orbit that is nearest the centre of force.

perigee   In orbital motion, the point of closest approach to the Earth of the Moon or an artificial satellite.

perihelion (pl. perihelia)   In orbital motion in the solar system, the point of closest approach to the Sun.

perihelion distance (symbol q)   The distance between the Sun and an object in orbit around it when they are at their closest approach. ►*orbital elements*.

period   The time after which a cyclical phenomenon repeats itself. ►*orbital elements*.

periodic comet   A ►*comet* in a closed, elliptical orbit within the solar system. Periodic comets are observed at their regular returns to the Earth's vicinity, as long as their orbit carries them near enough to be recoverable. The term is usually applied to comets with periods of less than 200 years, more strictly ►called *short-period comets*.

period—luminosity relation   The relationship between the absolute luminosity and the period of variability of ►*Cepheid variable* or ►*W Virginis* stars.

Perseids   A major annual ►*meteor shower*, the radiant of which lies near the star Eta Persei (η Per). The peak of the shower occurs on 12 August and the normal limits are 23 July–20 August. The meteor stream is associated with Comet 109P/Swift–Tuttle. This is one of the best, most reliable annual showers, with peak rates typically between 50 and 100 per hour. Records of it date back around 2,000 years.

Perseus   A large and interesting constellation of the northern hemisphere lying in a rich part of the ►*Milky Way*. It is one of the 48 listed by Ptolemy (*c.* AD 140) and contains ten stars brighter than fourth magnitude, including the noted variable star ►*Algol*. Perseus also includes a magnificent pair of ►*open* *clusters* visible to the naked eye, known as the ➤*Double Cluster in Perseus*. ➤Table 4.

Perseus Arm   One of the spiral arms of the Milky Way ➤*Galaxy*. It winds around from the far side of the galactic centre to the region of the Galaxy beyond the Sun.

personal equation   A systematic error in observations made by a particular individual.

perturbation   A temporary or localized gravitational disturbance in the otherwise uniform motion of a body under a stable gravitational force. For example, the motion of a comet, which is largely determined by the gravitational force of the Sun, suffers perturbations if the comet passes close enough to a planet for its pull to be significant in comparison with that of the Sun. An encounter with the most massive planet, Jupiter, can alter very significantly the orbit of a comet, to the extent that a comet moving in a parabolic orbit may be captured into a short-period elliptical orbit.

Petavius   A large lunar crater, 176 km (110 miles) in diameter, near the south-east limb of the Moon. A prominent rille runs across the crater floor between the multiple central peak and the terraced walls.

Phaethon   Asteroid 3200, diameter 6 km, discovered in 1983 by the ➤*Infrared Astronomical Satellite* (IRAS). It is in a highly eccentric, Earth-crossing orbit and appears to be the parent body of the ➤*Geminid* meteor shower. It may be the 'dead' nucleus of a former comet.

phase   The ratio of the illuminated area of the apparent disc of a celestial body to the area of the entire apparent disc, taken as a circle.
   The phases of the Moon are the recurring cycle of apparent forms of the Moon (see illustration on page 284). New Moon, first quarter, full Moon and last quarter are formally defined as the times at which the excess of the apparent celestial ➤*longitude* of the Moon over that of the Sun is 0°, 90°, 180° and 270° respectively.
   The Moon and planets show phases because bodies of this kind emit no light of their own, shining only by reflected sunlight. The hemisphere of a moon or planet facing the Sun is bright, its other hemisphere dark. The phase as seen from the Earth depends on the relative positions of the Earth and Sun with respect to the body, since this determines what proportion of its illuminated half is visible.

phase angle   The angle between the two lines formed by joining the centre of a planetary body to the Sun and to the Earth.

phase defect   The extent, measured as an angle, to which the illuminated area of the ➤*full Moon* differs from a complete circular disc because of the inclination of the Moon's orbit to the ➤*ecliptic*.

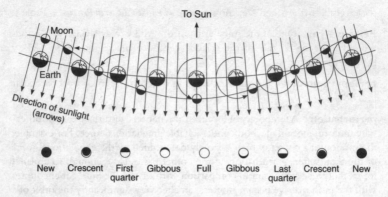

To Sun

Moon

Earth

Direction of sunlight (arrows)

New   Crescent   First   Gibbous   Full   Gibbous   Last   Crescent   New
quarter   quarter

Phase. The changing phases of the Moon during the course of a synodic month.

phase integral   The ratio between the ►*Bond albedo* and the ►*geometric albedo* of an asteroid or other planetary body.

Phekda (Phecda; Gamma Ursae Majoris; γ UMa)   The third-brightest star in the constellation Ursa Major. It is an ►*A star* of magnitude 2.4.

Phobos (1)   The inner of the two small satellites of Mars, discovered by Asaph Hall in 1877. Images from the ►*Viking* spacecraft of 1977 show Phobos to be ellipsoidal in shape (28 × 20 km) and covered with craters. The largest, Stickney, is 10 kilometres in diameter, one-third of the satellite's largest dimension. A series of striae emanating from Stickney appear to be fractures caused by the impact that created the crater. ►Table 6.

Phobos (2)   Two Soviet space missions to Mars launched in 1988. Contact with one was lost en route and the other operated only briefly, returning limited data.

Phocaea group   A group of asteroids at a distance of 2.36 AU from the Sun, with orbits inclined at 24° to the plane of the solar system. The group is separated from the main belt by one of the ►*Kirkwood gaps* and is not a true family with a common origin. The group is named after 25 Phocaea, diameter about 70 km (45 miles).

Phoebe   The outermost satellite of Saturn, discovered by W. Pickering in 1898. Phoebe appears to be spherical, 220 kilometres in diameter, and of very dark material that reflects less than 5 per cent of incident light. ►Table 6.

Phoenix   A southern constellation introduced in the 1603 star atlas of Johann Bayer. Though not particularly conspicuous, it does contain seven stars brighter than fourth magnitude. ►Table 4.

Pholus   Asteroid 5145, diameter 190 km, discovered in 1991. It follows a highly unusual, remote orbit, on which it ranges between 8.7 and 32 AU from the Sun. With ➤*Chiron*, and five other asteroids in orbits with similar characteristics, it forms the group termed ➤*Centaurs*. Pholus has a low ➤*albedo* of 4.4 per cent, and is very much redder in colour than typical asteroids.

photoelectric   Making use of the *photoelectric effect*, whereby electrons are liberated from the surface of a solid material when it is struck by photons of electromagnetic radiation. Since the number of electrons emitted is proportional to the intensity of radiation, photoelectric detectors provide an effective method of measuring light intensity.

photoelectric photometry (PEP)   The accurate measurement of the ➤*magnitudes* of stars by means of a ➤*photomultiplier* or ➤*CCD*. ➤*photometry*.

photographic magnitude   ➤*magnitude*.

photographic zenith tube (PZT)   A special telescope, mounted vertically, used for the very accurate determination of the positions of stars and for monitoring irregular changes in latitude and time, arising from ➤*polar motion* and the Earth's rotation.

photo–ionization   The ionization of an atom by the absorption of a photon of electromagnetic radiation. Ionization can take place only if the photon carries at least the energy corresponding to the *ionization potential* of the atom, i.e. the minimum energy required to overcome the force binding the electron within the atom.

photometry   The accurate determination of the ➤*magnitudes* of stars, or other astronomical objects, within specified wavelength bands. Photometric measurements can be used to deduce broad physical characteristics of stars, without the need for detailed study of their spectra. Several photometric systems are used for this purpose, most commonly ➤*UBV photometry* and ➤*uvby photometry*. Photometric measurements are also important for determining the ➤*light curves* of variable stars. Photometry is carried out using ➤*photoelectric* measurements; it is very difficult to determine magnitudes accurately from photographs or visually. Photoelectric measurements also allow monitoring of light intensity changes over very short timescales.

photomultiplier   A device for measuring the intensity of light. Electrons are liberated at a photocathode when light falls upon it, through the operation of the photoelectric effect. The current is amplified by a cascade process in which the electrons strike a series of secondary emitters. Photomultipliers are used particularly in ➤*photoelectric photometry*.

photon   A quantum of ➤*electromagnetic radiation*. Electromagnetic radiation shows both wave-like and particle-like properties, and a photon may be considered to be a discrete 'wave packet'. The energy, E, carried by a photon is related to the wavelength of its radiation, λ, by the formula $E = hc/\lambda = h\nu$, where h is ➤*Planck's constant*, c is the velocity of light and ν is frequency.

photosphere The visible surface of the Sun or a star. About 500 kilometres thick, the photosphere is a zone where the character of the gaseous layers changes from being completely opaque to radiation to being transparent. It is the layer from which the light we actually see is emitted. The temperature of the Sun's photosphere is about 6,000 K, falling to about 4,000 K at the base of the ➤*chromosphere*. The absorption lines in the spectrum are formed by net absorption and scattering of radiation in this layer. Phenomena such as ➤*sunspots*, ➤*granulation* and ➤*faculae* are located in the photosphere.

photovisual magnitude A ➤*magnitude* determined photographically using film sensitive to the same wavelength range as the human eye.

Pic du Midi Observatory An observatory located in the French Pyrenees, near the northern end of the mountain range, at an altitude of 2,877 metres (9,350 feet). A meteorological observatory was founded at the site in 1878. In 1903 it was affiliated to the Toulouse University Observatory and two telescopes – a 50-centimetre (20-inch) reflector and a 23-centimetre (9-inch) refractor – were installed. A 60-centimetre (24-inch) refractor was brought into operation in 1943 and a 1.06-metre (42-inch) lunar and planetary telescope in 1964, with the sponsorship of NASA. The most recent instrument is a 2-metre (79-inch) reflector, which started operation in 1980. There are also solar instruments: the first experiments with a ➤*coronagraph* were carried out here by Bernard Lyot in the 1930s.

Pico An isolated mountain peak in the Mare Imbrium on the Moon.

Pictor (The Painter's Easel or Painter) An inconspicuous southern constellation introduced in the mid-eighteenth century by Nicolas L. de Lacaille. Its original name was Equuleus Pictoris, the painter's easel, but this was subsequently shortened to Pictor. Its two brightest stars are third magnitude. ➤Table 4.

pincushion distortion ➤*distortion*.

Pinwheel Galaxy (M101; NGC 5457) The popular name for a large spiral galaxy in Ursa Major, which is seen face-on. It is estimated to be at a distance of 15 million light years.

Pioneer A series of 11 American spacecraft, launched between 1958 and 1973. *Pioneers 1* to *4* were directed towards the Moon and all failed. *Pioneers 5* to *9* were put in orbits around the Sun and were used to study the Sun and conditions in interplanetary space. *Pioneers 10* and *11* were highly successful flyby missions, *Pioneer 10* to Jupiter and *Pioneer 11* to both Jupiter and Saturn.

Pioneer 10 was launched on 3 March 1972 and passed Jupiter on 4 December 1973 at a distance of 132,000 kilometres (82,000 miles), returning the best pictures of the planet up to that time. It continued to transmit information continuously while heading out of the solar system. On 30 March 1996, when it was more than 66 AU from the Sun, it was permanently shut down. *Pioneer 11* was launched on 6 April 1973 to encounter Jupiter on 3 December 1974 at a distance of 42,800 kilometres (26,600 miles) and Saturn on 1 September 1979 at 20,800 kilometres (13,000 miles).

Pioneer Venus   Two American spacecraft sent to Venus in 1978. *Pioneer Venus 1* was an orbiter, launched on 20 May 1978, which obtained radar maps of the surface and returned visual images and other data. *Pioneer Venus 2*, launched on 8 August 1978, was an atmospheric probe, with five small landers not intended to transmit data after impact.

Pisces (The Fishes)   A large but faint zodiacal constellation, among those listed by Ptolemy (*c.* AD 140). Its three brightest stars are only fourth magnitude. ➤Table 4.

Piscis Austrinus (or Piscis Australis; The Southern Fish)   A small southern constellation, listed by Ptolemy (*c.* AD 140). It contains the first magnitude star ➤*Fomalhaut*, but no others brighter than fourth magnitude. ➤Table 4.

Piton   An isolated mountain peak in the Mare Imbrium on the Moon.

pixel   A contraction of 'picture element'. The term is used particularly in connection with electronic imaging techniques, such as the use of ➤*charge-coupled devices* (CCDs), where an extended image consists of a grid of many small individual elements. The pixel is thus the smallest area over which intensity variations can be resolved.

plage   A bright emission region in the solar ➤*chromosphere*. Plages coincide with ➤*faculae* in the photosphere beneath them and an enhancement of the ➤*corona* above, from which there is increased X-ray, extreme ultraviolet and radio emission. These are all regions where the vertical component of the magnetic field is strong. They are characteristic of ➤*active regions* on the Sun.

Planck era   The period earlier than 10^{-43} seconds after the ➤*Big Bang*. During this initial interval, the behaviour of the universe was dominated by gravitational interactions. ➤*quantum gravity*.

Planck length   A length of 10^{-35} metres obtained by combining the gravitational constant, the speed of light and ➤*Planck's constant* to yield a fundamental unit. This is the length-scale at which ➤*quantum gravity* operates.

Planck's constant (symbol *h*)   A fundamental constant of nature: the ratio between the energy of a photon and its frequency.

Planck Surveyor A mission selected in 1996 as part of the European Space Agency's programme. Its purpose is to image anisotropies in the ►*cosmic background radiation* over the whole sky with unprecedented sensitivity and resolution. The mission was formerly known as COBRAS/SAMBA, a merger of two proposals. The acronyms stood for Cosmic Background Radiation Anisotropy Satellite and Satellite for Measurement of Background Anisotropies.

Planck time A time of 10^{-43} seconds, obtained by combining the gravitational constant, the speed of light and ►*Planck's constant* to yield a fundamental unit of time. This is the time during which ►*quantum gravity* operated in the very early universe following the ►*Big Bang*.

planet An astronomical body orbiting the Sun, or another star, the mass of which is too small for it to become a star itself (less than about one-twentieth the mass of the Sun). Planets may be basically rocky objects, such as the inner planets (Mercury, Venus, Earth and Mars), or primarily gaseous (liquid in the high-pressure conditions of the interior) with a small solid core, like the outer planets (Jupiter, Saturn, Uranus and Neptune). These eight, together with Pluto, are the *major planets* of the solar system. Pluto, though more like the rocky planets, retains a significant amount of ices and is the sole example of an ice-dwarf major planet in the solar system. There are also within the solar system large numbers of *minor planets*, or ►*asteroids*, and a population of small ice dwarfs in the region beyond Neptune, known as the ►*Kuiper Belt*.

So far it has not been possible to image directly any planet orbiting another star. However, the presence of planets around a number of stars other than the Sun has been inferred indirectly from the measurement of small cyclical changes in ►*radial velocity*, as revealed through the ►*Doppler effect*. Observations of discs around newly forming stars, which could provide the material to form planetary systems, strengthen the argument that planetary systems probably accompany at least some stars comparable with the Sun. In addition, there is strong evidence (from small variations in pulse frequency) that at least one ►*pulsar* has planetary-sized companions. ►*extrasolar planet*.

planetarium A dome-shaped building housing a special projector that is used to simulate the appearance of the night sky. Planetaria are widely used for both educational purposes and entertainment.

The term is sometimes used for a mechanical model of the solar system, more normally called an ►*orrery*.

planetary nebula An expanding shell of gas surrounding a star in a late stage of ►*stellar evolution*. The name derives from the description given by William Herschel, who thought their circular shapes to be reminiscent of the discs of the planets as seen through a small telescope. There is no connection between planets and planetary nebulae.

Planetary nebulae are formed in the process of mass loss during which ►*red giant* stars ultimately become ►*white dwarfs*. Typically, the gaseous shell contains a few tenths of a solar mass of material and is moving outwards with a velocity of 20 km/s. The lifetime is perhaps 35,000 years before the shell becomes too tenuous to be visible. The spectra show emission lines from the glowing gas combined with the spectrum of the central star, which may contain absorption or emission lines, or both. These central stars are essentially burnt-out cores in the process of becoming white dwarfs, with temperatures of up to 125,000 K, and their diverse characteristics reflect those observed in white dwarfs.

The nebulae take a variety of forms – ring-shaped, circular, dumbbell-like or irregular. Notable examples are the ►*Ring Nebula*, the ►*Helix Nebula* and the ►*Dumbbell Nebula*.

planetary rings Ring structures, composed of numerous individual small bodies and dust, surrounding the four largest outer planets – Jupiter, Saturn, Uranus and Neptune.

The rings of Saturn were discovered as soon as Galileo became the first person to turn a telescope on the sky in 1610. In 1857, James Clerk Maxwell demonstrated theoretically that the rings must be made up of many unconnected particles, and this was later confirmed by spectroscopic observations showing that the inner particles orbit more quickly than the outer ones. In 1977, nine narrow rings around Uranus were detected when the planet occulted a star. In 1979, ►*Voyager 1* discovered a faint band around Jupiter and, in the early 1980s, what seemed to be incomplete ring arcs were detected around Neptune, again during an occultation. In 1989, ►*Voyager 2* showed that there are complete rings around Neptune and that some 'clumpiness' gave rise to the impression of incomplete arcs.

Virtually all the planetary rings lie within their ►*Roche limits*. In a disc of debris around a newly formed planet, material beyond the Roche limit could coalesce into satellites while, nearer the planet, the tidal forces would prevent the satellites from forming.

The rings around Jupiter are faint and tenuous, and their reflection qualities show that many of the particles can be no bigger than 1 or 2 micrometres. Dust of this size must be constantly replenished, perhaps by impacts on boulder-sized objects in the ring.

Saturn's rings are far more complex and extensive than any others. The rings readily visible from Earth were labelled A, B and C, C being the faint inner Crepe Ring. The A and B rings are separated by the *Cassini Division* and there is also a narrow but conspicuous gap towards the outer edge of the A ring, known as the *Encke Division* or *Encke Gap*. *Voyager 1* detected material

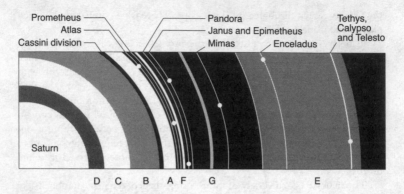

Prometheus — Pandora — Tethys, Calypso and Telesto
Atlas — Janus and Epimetheus
Cassini division — Mimas — Enceladus
Saturn

D C B A F G E

Planetary rings. Schematic representation of Saturn's rings, with the orbits of the inner satellites. Rings A, B and C contain most material; the others are very tenuous.

inside the C ring, which was called the D ring. Beyond the A ring lie further narrow, tenuous rings, known as the E, F and G rings. The ring particles are thought to consist of a mixture of water ice and dust, and range in size from a few micrometres up to a hundred metres. However, the composition is not uniform, as demonstrated in *Voyager* images showing marked colour variations. These images also show that the rings consist of thousands of narrow, closely spaced 'ringlets'. Many of the observed structures are attributable to the gravitational action of satellites. For example, Pandora and Prometheus act as 'shepherds', confining the F ring, and the Cassini Division lies where a satellite with an orbital period half that of Mimas would lie. (This is an example of a ➤*resonance* phenomenon.)

The nine rings of Uranus found in 1977 are labelled, in order of increasing distance from the planet: 6, 5, 4, α, β, η, γ, δ and ε. Two further rings were found by *Voyager 2* in 1986 and also a pair of satellites, Ophelia and Cordelia, shepherding the ε ring. The nine main rings appear to be composed of metre-sized boulders. But in backlighting, *Voyager 2* also saw many slender ringlets composed of dust.

Two main rings orbit Neptune (Leverrier and Adams rings), with a diffuse sheet of material extending outwards from the inner of the two (Lassell). There is also a tenuous third ring nearer the planet (Galle). The Lassell ring is bounded at its outer edge by the Arago ring. The outer ring, Adams, contains three bright arcs, about 8° in extent, which seem to be dominated by dust-sized particles. These have been called Liberty, Equality and Fraternity. It is thought that the gravitational influence of the moon Galatea, orbiting just inside the ring, acts to confine the arcs. There is evidence for a further ring between Arago and Adams. ➤Table 7.

Planet-B   A Japanese spacecraft to be launched to Mars in August 1998. It will enter orbit around Mars in October 1999 with the objective of studying Mars's atmosphere and ionosphere.

planetesimal   A body of rock and/or ice, under 10 kilometres in diameter, formed in the primordial ➤*solar nebula*. Larger planetary bodies are presumed to have formed from planetesimals coalescing.

planetographic coordinates   A system of longitude and latitude for defining the positions of features on the surface of a planet based on the ➤*right ascension* and ➤*declination* of the north pole of axial rotation of the planet and a selected ➤*prime meridian*.

planetoid   An ➤*asteroid* or minor planet.

planetology   The comparative study of planets and their natural satellites.

Planet X   A hypothetical major planet in the solar system lying beyond the orbit of Pluto. The 'X' can be construed as representing either the unknown or a Roman ten. The discovery of a population of small planetary bodies in the region beyond Neptune, constituting the ➤*Kuiper Belt*, suggests that it is very unlikely, on both theoretical and observational grounds, that such a major planet exists.

planisphere   A two-dimensional star map which is a projection of part of the celestial sphere. It may be equipped with a rotating overlay, an aperture in which shows what portion of the sky is visible at any given date and time. Such a planisphere is useful only over the limited latitude range for which it is made.

planitia (pl. planitiae)   A low plain on the surface of a planet.

planum (pl. plana)   A plateau or high plain on the surface of a planet.

Plaskett's Star   A sixth magnitude ➤*spectroscopic binary* star in the constellation Monoceros. The Canadian astronomer J. S. Plaskett discovered in 1922 that each component is of exceptionally high mass, estimated to be 55 times the mass of the Sun.

plasma   An ionized gas, consisting of a mixture of electrons and atomic nuclei. All of the matter in the interior of stars is in the plasma state, as is ➤*ionized hydrogen*. Astrophysical plasmas are important sources of radio emission.

plasmasphere   A region of high-density cold plasma surrounding the Earth above the ➤*ionosphere*, i.e. at altitudes greater than about 1,000 kilometres (600 miles). It extends out to between three and seven Earth radii. The upper boundary is marked by a sudden drop in plasma density, known as the *plasmapause*. The particles in the plasmasphere are almost all protons and electrons.

plasma tail ➤*ion tail*.

Platonic year The period of 25,800 years it takes the Earth's rotation axis to sweep out a complete cone in space as a result of ➤*precession*.

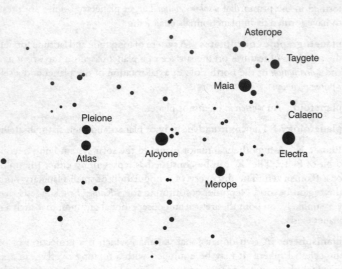

Pleiades. The brighter stars in the Pleiades, showing the appearance of the cluster in typical binoculars.

Pleiades (M45; NGC 1432) An ➤*open cluster* of stars in the constellation Taurus, clearly visible to the naked eye. It is estimated to contain between 300 and 500 members within a sphere 30 light years across, and is 400 light years away. The stars are embedded in a ➤*reflection nebula* of cold gas and dust that appears blue in colour photographs. The cluster is young by astronomical standards, about 50 million years old, and contains some very massive bright stars.

The popular name for the Pleiades is the *Seven Sisters*, but most people are able to distinguish only six stars with the naked eye.

Pleione One of the brighter stars in the ➤*Pleiades*. It is slightly variable in brightness and was observed in 1938 and 1970 to throw off shells of gas.

plerion A ➤*supernova remnant* with no clear shell structure and therefore similar to the ➤*Crab Nebula*. It appears that about 10 per cent of remnants are like this, about a dozen being known in our Galaxy.

Plössl eyepiece A telescopic ➤*eyepiece* composed of two achromatic doublets. It produces a high-quality image over a wide field of view, with good ➤*eye relief*, and is particularly effective in conjunction with telescopes of short focal length.

Plough (North America: Dipper or Big Dipper) The ➤*asterism* formed by the stars Alpha (α), Beta (β), Gamma (γ), Delta (δ), Epsilon (ε), Zeta (ζ) and Eta (η) in the constellation ➤*Ursa Major*.

plume eruption A type of volcanic activity observed on ➤*Io*. Plumes arise from rifts or vents on the surface. The eruptive centre is surrounded by a deposit of white or dark red material. The eruption may be violent and short-lived or a longer-lasting eruption of white material, rather like a geyser, consisting of liquid sulphur or sulphur dioxide. Hot liquid from underground changes to a gas as it rushes up through the eruptive centre then condenses again as it falls.

Evidence for similar eruptions on ➤*Triton* can be seen in images returned by ➤*Voyager 2*.

Plutino A ➤*transneptunian object* which, like Pluto, follows an orbit in a 2:3 ➤*resonance* with Neptune so that it completes two orbits of the Sun in the time it takes Neptune to make three orbits (248.5 years).

Pluto The ninth planet of the solar system, discovered as a fifteenth magnitude object on 18 February 1930 from the ➤*Lowell Observatory* by Clyde Tombaugh. Searches for a planet beyond Neptune had started in 1905, stimulated by apparent discrepancies between the calculated and observed orbits of Uranus and Neptune. However, it is now known that the mass of Pluto is less than one-fifth that of the Moon, insufficient to have any gravitational effect on Uranus and Neptune.

Pluto's orbit is more highly inclined to the ecliptic and more eccentric than that of any other planet. Its distance from the Sun ranges between 30 and 50 AU. Perihelion occurred in 1989 and, between 1979 and 1999, Pluto's orbit brings it nearer the Sun than Neptune.

The discovery of Pluto's satellite, ➤*Charon*, in 1978, made it possible to obtain improved values for the planet's diameter and mass. The diameter is 2,300±40 kilometres. Pluto's overall density is approximately twice that of water and it is thought likely to consist of a thick layer of water ice overlying a core of partially hydrated rock. Charon and Pluto are locked in synchronous rotation with a period of 6.39 days. Pluto's rotation axis is inclined at 122° to the plane of the ecliptic so that, like Uranus, it rotates in a retrograde sense, 'lying on its side'.

A rare series of mutual occultations and transits took place between 1985 and 1990. Such events, as viewed from Earth, take place only twice in the planet's 248-year orbital period. They made it possible to distinguish the spectral signatures of Pluto and Charon and to construct the first approximate albedo maps of Pluto's surface. These confirmed previous suspicions of a highly non-uniform and variable surface based on the change of brightness over the rotational period, and in the longer term. In contrast with Charon, which is grey, Pluto's surface is reddish in colour. Methane ice was detected on Pluto in 1976 by infrared spectroscopy. The occultation of a star by Pluto in 1988 revealed the presence of an extended tenuous atmosphere. Nitrogen and carbon monoxide ices were discovered on the surface in 1992. The surface temperature is about 40 K. In 1996 observations with the Hubble Space Telescope resolved broad light and dark features on Pluto's surface for the first time. ➤Tables 5 and 6.

Pogson's ratio   ➤*magnitude*.

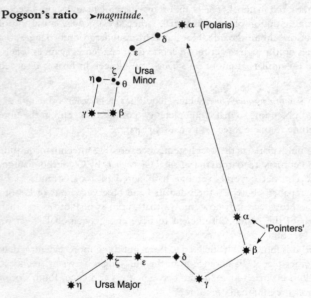

Pointers. A contiunation of the line from β Ursae Majoris to α Ursae Majoris points almost directly to the Pole Star (Polaris).

Pointers   The stars Alpha (α) and Beta (β) in the constellation Ursa Major, so called because the line joining them points almost directly to Polaris.

polar   A small class of short-period, variable ➤*binary stars* characterized by X-ray emission. Their light is strongly polarized and the polarization varies over the orbital period, which is between one and four hours. These close systems appear to consist of a normal star and a strongly magnetic ➤*white dwarf* with its spin locked in synchrony with the orbital period. Matter is transferred from the normal star to the white dwarf but, because of the strong magnetic field, an ►*accretion disc* cannot form. Instead, the material is channelled along the magnetic field lines and is deposited at the poles. Polars are also known as AM Herculis stars after the star considered to be the prototype for the class.

Intermediate polars are similar, but they have longer orbital periods of several hours. They emit pulsed radiation at the spin rate of the white dwarf, which is not locked to the orbital period in these systems and is typically less than one hour. Their white dwarfs are thought to have weaker magnetic fields, making it possible for an outer accretion disc to form, though material close to the white dwarf is channelled on to the magnetic poles. The pulsed emission is a searchlight effect seen as the accreting pole of the white dwarf sweeps across the line of sight. Intermediate polars are also known as DQ Herculis stars after their prototype.

polar axis   One of the two rotation axes about which a telescope on an ►*equatorial mounting* can turn. The polar axis must be accurately oriented parallel to the Earth's rotation axis, i.e. at an angle to the horizontal equal to the latitude of the place where it is located, and in the north–south plane. Rotation about the polar axis results in a change in the right ascension of the direction in which the telescope is pointing, but not in declination.

polar cap   A roughly circular area of limited extent around a pole of rotation of a planet. In the case of Earth and Mars, the term is applied to the areas covered by ice or frost in the two polar regions.

polar distance   The angular distance along a great circle on the celestial sphere between an object and either the north or south celestial pole.

polarimetry   The measurement of the state of ►*polarization* of a beam of electromagnetic radiation. The instrument used is called a *polarimeter*, and it must include optical elements capable of altering the state of polarization of the beam being tested. Such elements are described as *optically active*.

Polaris (Alpha Ursae Minoris; α UMi)   The brightest star in the constellation Ursa Minor, lying within one degree of the north celestial pole. It is a ►*Cepheid variable* and its magnitude changes between about 1.95 and 2.05 over a period of four days. ►Table 3.

polarization (of light)   The non-random distribution of electric field direction among the ►*photons* in a beam of ►*electromagnetic radiation*. In *linear polarization*, the vectors representing electric field are parallel. In *circular polarization*, the direction of polarization changes continuously in such a way that the electric field vector rotates with the frequency of the radiation. Elliptical polarization is similar to circular polarization except that the magnitude of the electric field vector also changes continuously, but at twice the frequency of the radiation. The properties of a beam of polarized light can be described by a set of four numbers known as the *Stokes parameters*.

polarizer   An optical component that transmits only light that is linearly polarized in a particular direction. ➤*polarization*.

polar motion   A slow and very slight movement of the Earth's geographic poles relative to the surface of the Earth (not relative to the stars). It does not affect the celestial coordinates of a star, but does affect the reduction of positional measurements, for example those made with a ➤*transit circle*. The origin is geophysical: principally that lack of exact coincidence between the Earth's axes of symmetry and rotation. The magnitude of the effect is typically 0.3 arc seconds, and there are periodicities of 433 days and one year. Much smaller variations also take place over short timescales, ranging between two weeks and three months, due to surface air pressure changes. ➤*Chandler wobble*.

Pole Star   Popular name for the star ➤*Polaris*.

Pollux (Beta Geminorum; β Gem)   The brightest star in the constellation Gemini. It was nevertheless given the ➤*Bayer letter* Beta rather than Alpha. It seems unlikely that Pollux has brightened since Bayer's time (1572–1625). Pollux is an orange-coloured, giant ➤ *K star*. ➤*Castor* and Pollux were the twin sons of Leda in classical mythology. ➤Table 3.

Poop   English name for the constellation ➤*Puppis*.

Population I   A collective name for stars and star clusters within the ➤*Galaxy* that exhibit characteristics suggesting they are relatively young and are confined to the ➤*galactic plane*, especially the spiral arms. The distinction between Population I and ➤*Population II* was first made by W. Baade in 1944, who introduced the terms. Typical Population I objects are hotter ➤*main-sequence stars*, ➤*open clusters* and ➤*associations*. Interstellar material is also associated with Population I. The stars are relatively abundant in heavier elements because the material from which they formed had been enriched by the products of ➤*nucleosynthesis* in earlier generations of stars. ➤*astration*.

Population II   A collective name for stars and star clusters within the ➤*Galaxy* that exhibit characteristics suggesting they are relatively old and occupy a spherical halo around the galactic centre, rather than being confined to the galactic plane. Population II stars typically contain significantly less of the elements heavier than helium than do other stars, have large space velocities relative to the Sun and other stars in the galactic disc, and have orbits in the Galaxy that are very elliptical and highly inclined to the galactic plane. ➤*Globular clusters* in particular belong to Population II. The characteristics are explained if the Population II stars formed before the Galaxy collapsed to its present flat structure and before the interstellar medium was enriched with heavier elements through mass loss from evolved stars. ➤*Population I*.

pore   A small ➤*sunspot* without a ➤*penumbra* that lasts for about a day.

Porrima (Gamma Virginis; γ Vir)   The second-brightest star in the constellation Virgo, and a noted visual binary. The combined magnitude is 2.8, but the system actually consists of two stars, each of magnitude 3.6. The orbital period is about 171 years. The elliptical orbit causes the observed separation of the two stars to vary between 0.4 and 6 arc seconds. Periastron occurs in late 2007. At a distance of 36 light years, it is relatively close to the solar system. The name, of Latin origin, is that of a Roman goddess of prophecy.

Porro prism   A right-angled glass prism with rounded corners, usually employed as one of a pair in an optical system. They are found in ►*binoculars* and may be used as a telescope accessory to provide an image that is both erect (the correct way up) and correctly oriented left–right.

Portia   A small satellite of Uranus, discovered in 1986 by the ► *Voyager 2* spacecraft. Its diameter is 110 kilometres. ►Table 6.

position angle (PA, p.a.)   An angle specifying the relative orientation of a pair of astronomical objects, such as a visual ►*binary star*, in relation to north. It is defined as the angle from north, measured in the sense north–east–south–west, on a scale from 0° to 360°. For double stars, the position of the fainter component with respect to the brighter is given.

Poynting–Robertson effect   The effect of solar radiation on small particles orbiting the Sun, which causes them to spiral slowly in. The particles absorb solar energy that is streaming out radially, but re-radiate energy equally in all directions. As a consequence there is a reduction in their kinetic energy, and thus in their orbital velocity, which has the effect of reducing the size of their orbit.

Praesepe (M44; NGC 2632)   An ►*open cluster* of stars in the constellation Cancer lying at a distance of 500 light years. Its brightest stars are about sixth magnitude, and more than 200 members are known. In many ways it is similar to the ►*Hyades*, even sharing the same direction and speed of motion through space; this suggests that the two clusters originated in the same interstellar cloud.

preceding (*p*)   A term used to distinguish a star that is at a lower ►*right ascension* than another, or a planetary or solar feature that leads another in its motion across the sky as the Earth rotates. Something coming into view later is described as ►*following*.

precession   The uniform motion of the rotation axis of a freely rotating body when it is subject to turning forces (torque) due to external gravitational influences.

Precession causes the Earth's rotation axis to sweep out, over a period of 25,800 years, a cone of angular radius about 23° 27′ around the perpendicular to the plane of the Earth's orbit (the ➤*ecliptic*). The main source of the torque is the combined effect of the gravitational pulls of the Sun and Moon on the Earth's equatorial bulge. (Precession would not occur if the Earth were a perfect sphere. Rotation, however, causes the equatorial radius to exceed the polar radius by about 0.3 per cent.) The combined effect of the Sun and Moon is called *lunisolar* precession. The Moon's contribution is about twice as large as the Sun's, because of its smaller distance. The gravitational force exerted on the Earth by the other planets causes small changes in the Earth's ➤*orbital elements*, giving rise to *planetary precession*. The sum of planetary and lunisolar precession is called *general precession*.

A consequence of the precession of the Earth's rotation axis is that the celestial poles trace out circles in the sky over 25,800 years. So, for example, about 13,000 years from the present, the nearest bright star to the north celestial pole will be Vega rather than Polaris.

The zero point of ➤*right ascension*, one of the ➤*equatorial coordinates*, normally used to define the positions of celestial objects, is based on one of the points, known as 'the First Point of Aries', where the celestial equator intersects the ecliptic. Because of precession, the equator is 'sliding around' the ecliptic so that the intersection points are constantly changing. In fact, the First Point of Aries is no longer in the constellation Aries, but has moved into Pisces and will soon be in Aquarius. This phenomenon is known as the *precession of the equinoxes*. Its effect on the right ascension and declination of an object is noticeable from year to year with the positional accuracy attainable with many telescopes. Tabulated values of right ascension and declination are therefore referred to a particular ➤*epoch*, at which they were precisely correct. ➤*equinox*, *nutation*.

pre-main-sequence star   A star in the process of formation, which has developed beyond the ➤*protostar* stage but has not reached the point where the steady fusion of hydrogen into helium in its core places it on the ➤*main sequence* of the ➤*Hertzsprung–Russell diagram*.

A protostar becomes a pre-main-sequence star when its core becomes hot and dense enough to produce conditions of ➤*hydrostatic equilibrium*. At this stage, the main mechanism by which energy is transported in the stellar interior changes from radiation to convection. The chief source of energy is gravitational collapse, until the temperature and density are high enough for hydrogen fusion to dominate, at which point contraction stops.

Stars in the process of forming are surrounded by thick shells of obscuring dust and can be detected only by their infrared or millimetre-wave emission. Eventually, the dust is blown away by ➤*stellar winds* or ➤*radiation pressure*, to be destroyed, reincorporated in the interstellar medium or form planetary systems.

primary mirror   The main light-collecting mirror in a reflecting telescope.

prime focus The point at which the primary mirror in a reflecting telescope brings incident light to a focus in the absence of a secondary mirror. The prime focus lies within the incident beam but, in very large telescopes, instruments may be positioned at the prime focus without significant loss of performance. Working at the

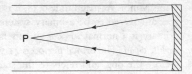

Prime Focus. Light striking the primary mirror of a reflecting telescope converges at the prime focus, P.

prime focus has the advantage that light loss and image imperfections introduced by secondary mirrors are eliminated, and a relatively low focal ratio and wide field of view can be obtained.

prime meridian The great circle on the surface of a planetary body taken as the zero of longitude measurement. On the Earth, the prime meridian is the circle of longitude passing through Greenwich, London – the *Greenwich Meridian*.

primeval atom The same as ➤*primeval fireball*.

primeval fireball The universe at a very early stage, when it was hot and dense and dominated by high-energy radiation. The ➤*Big Bang* theory of the universe requires that the universe was once like this. The existence of ➤*cosmic background radiation* provides strong support, since this can only be interpreted satisfactorily as the radiation from that era, now greatly cooled and expanded.

Principia The commonly used short version of the title *Philosophiae naturalis principia mathematica* ('The Mathematical Principles of Natural Philosophy'), by Isaac Newton (1642–1727), published by the Royal Society in 1687. In this comprehensive work of enormous importance, Newton set out his now familiar 'laws of motion' and, for the first time, analysed orbital motion in terms of the force of gravitation.

prismatic astrolabe An instrument for the accurate determination of the positions of stars. The telescope can be rotated in azimuth, and the time when a star reaches a particular predetermined ➤*zenith distance* automatically recorded.

An equilateral prism is placed in front of a telescope with one of its faces vertical. Two images of a star are formed from light passing through the prism, one from rays reflected directly off the lower internal face of the prism and one from rays first reflected from a horizontal mercury surface and then from the upper internal face of the prism. The two images coincide when the star's zenith distance is the predetermined value. The design was refined by the French astronomer André Danjon (1890–1967). A prototype was constructed in 1951 at the ➤*Observatoire de Paris*, where such an instrument remains in operation.

Procyon (Alpha Canis Minoris; α C M i) The brightest star in the constellation

Canis Minor. At magnitude 0.38, it is the eighth-brightest in the sky. Procyon was discovered to be a binary system by John M. Schaeberle in 1896. The primary is a normal ➤*F star* and its faint companion an eleventh magnitude ➤*white dwarf*. The orbital period is 41 years. The name, of Greek origin, means 'before the dog' and refers to the star's rising before the 'Dog Star', ➤*Sirius*. ➤Table 3.

prograde ➤*direct*.

Project Ozma The first serious scientific attempt to contact ➤*extraterrestrial intelligence* by radio waves. The experiment was conducted in 1960 at ➤*Green Bank* and unsuccessfully looked for radio signals from the nearby stars Tau Ceti (τ Cet) and Epsilon Eridani (ε Eri). ➤*SETI*.

Prometheus (1) A small satellite of Saturn, discovered in 1980 by ➤*Voyager 2*. ➤Table 6.

Prometheus (2) An eruptive centre on Jupiter's satellite ➤*Io*.

prominence A term used for a variety of cloud- and flame-like structures in the ➤*chromosphere* and ➤*corona* of the Sun, all of which have a higher density and lower temperature than their surroundings. When seen on the solar limb they appear as bright features in the corona, and when seen in projection against the disc they appear as dark ➤*filaments*.

Quiescent prominences occur away from active regions and are stable with lifetimes of many months. They may extend upwards for tens of thousands of kilometres. *Active prominences* are associated with ➤*sunspots* and ➤*flares*. They appear as surges, sprays and loops, have violent motions, fast changes and lifetimes of up to a few hours. Cooler material flowing back from prominences in the corona to the photosphere may be observed as coronal 'rain'.

promontorium Literally 'promontory' or 'cape', a term used in the names of a small number of features on the Moon where a brighter area protrudes into a darker area. Its use dates from a time when it was believed that the darker features on the Moon were of liquid water; this is now known to be untrue. In recognition of its use over a long period, the term has been retained in certain official names for features on the Moon.

proper motion The apparent motion of a star across the celestial sphere, measured as the angular shift in position per year, caused by the combination of the star's true motion through space and the relative motion of the solar system.

proplyd A recently formed star, surrounded by a cloud of gas and dust, which may be the progenitor of a planetary system. The term is a contraction of 'protoplanetary disk'.

proportional counter A detector for gamma rays and X-rays, similar to a Geiger counter except that it uses a lower voltage. The entry of ionizing radiation triggers an electrical discharge, resulting in a pulse of electric current, the strength of which is proportional to the energy of the interaction.

Proteus A satellite of Neptune (1989 N1) discovered during the flyby of ►*Voyager 2* in August 1989. ►Table 6.

protogalaxy A newly formed galaxy or galaxy fragment in which a first generation of stars is being created from a condensation of gas. The way in which galaxies began to form in the early universe is not fully understood, but there is some observational evidence that small protogalaxies merged to form larger galaxies.

proton–proton chain A series of nuclear reactions, believed to occur in stars, which convert hydrogen to helium and provide a major stellar energy source. The main process consists of the following stages:

$^1H + {}^1H \rightarrow {}^2H + positron + neutrino$
$^2H + {}^1H \rightarrow {}^3He + gamma\text{-}ray\ photon$
$^3He + {}^3He \rightarrow {}^4He + {}^1H + {}^1H.$

protoplanet The precursor of a planet during the ►*accretion* process that will ultimately lead to the formation of a planet.

protostar A star in the earliest observable stage of formation, when condensation is taking place in an interstellar cloud but before the onset of nuclear reactions in the interior.

Proxima Centauri The nearest star to the solar system, lying at a distance of 4.26 light years. It is an eleventh magnitude dwarf red ►*M star* in the constellation Centaurus. It appears to be associated physically with the bright binary star Alpha Centauri, which is two degrees away in the sky, and about 0.11 light years more distant from the solar system. It is estimated that Proxima may take a million years to orbit its companions.

Przybylski's Star The extremely peculiar ►*A star* HD 101065, usually linked with the name of the Polish astronomer who first drew attention to its unusual spectrum.

Psyche Asteroid 16, diameter 248 km, discovered by A. de Gasparis in 1852. It is of the metallic type and its surface appears to be an almost pure alloy of iron and nickel.

Ptolemaeus A large shallow crater, flooded with dark lava to form a walled plain in the Moon's southern upland area. The diameter is 153 km and there are many smaller craters within its walls.

Ptolemaic system   A ➤*geocentric model* of the solar system, described by the Greek astronomer Ptolemy (*c.* AD 100–170) in his book ➤*Almagest*. It was the model generally accepted in the Arab and Western worlds for more then 1,300 years, until superseded by the ➤*heliocentric model*. ➤ *epicycle*.

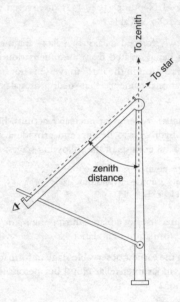

Ptolemy's rules. The angle between the vertical rod and the upper hinged rod is the star's zenith distance, which is 90 degrees minus altitude.

Ptolemy's rules   An early instrument for determining the altitude of a celestial body, also known as a *triquetrum*. Despite the name, it seems likely that its invention predates Ptolemy (*c.* AD 100–170). It consists of a vertical bar with hinged rods attached at the top and bottom (see illustration). The hinged rods are linked, so that the whole arrangement forms a triangle, but the upper arm can slide along the lower one, which is graduated. The upper arm is fitted with sights. When the object under observation is in view, the position of the upper arm against the graduated lower arm gives the measure of altitude.

P-type asteroid   A type of asteroid with low albedo, common in the outer main belt.

Puck   A satellite of Uranus, discovered by the ➤*Voyager 2* spacecraft in 1985. Its diameter is 150 kilometres, and a *Voyager* image shows several relatively large craters. ➤Table 6.

Pulkovo Observatory An observatory near St Petersburg in Russia, originally established in 1718. The present site dates from 1835; the buildings were destroyed during World War II and subsequently rebuilt in their old style. The observatory is associated particularly with the Struve family, six members of which became well-known astronomers. F. G. W. Struve was director from 1839 to 1862 and his son, Otto, director from 1862 to 1889. Today the observatory is concerned mainly with astrometry and instrument development. Its official title is the Main Astronomical Observatory of the Russian Academy of Sciences at Pulkovo.

pulsar A stellar source of radio waves, characterized by the rapid frequency and regularity of the bursts of radio waves emitted. The time between successive pulses is milliseconds for pulsars in binary systems and up to 4 seconds for the slowest. Some pulsars emit pulsed radiation in other bands of the electromagnetic spectrum, including visible light, as well as radio waves.

A pulsar is a rotating ►*neutron star*, with a mass similar to the Sun's but a diameter of only about 10 kilometres. The pulses occur because the neutron star is rotating very rapidly: a beam of radio emission sweeps past an observer once per rotation. The pulses are very regular, apart from the occasional ►*glitch*, and all single pulsars are slowing down as they lose rotational energy.

Some X-ray pulsars are in binary systems where complex dynamical effects cause the spin rate to speed up, and these *millisecond pulsars* are the fastest known. Millisecond pulsars not currently in binary systems are thought to have once belonged to pairs that have been split apart. Most have been discovered in ►*globular clusters*, where stars are densely packed and gravitational interactions can easily occur. At least one pulsar appears to have another neutron star as a companion, and another has two or three planet-sized companions. Their presence is deduced from variations in the arrival time of pulses.

Pulsars are formed in ►*supernova* explosions, though only two – the ►*Crab Pulsar* and the ►*Vela Pulsar* – are within currently observable supernova remnants.

pulsating star A ►*variable star* with an unstable internal structure that causes it to pulsate in a regular way. ►*Cepheid variable, RR Lyrae star*.

pulse nulling A drop to a low level in the intensity of radio emission from a ►*pulsar*. The phenomenon, which is common, lasts for a few pulses, after which the intensity of emission returns to normal.

Pump English name for the constellation ►*Antlia*.

Puppis (The Poop or Stern) A large southern constellation lying in rich starfields of the ►*Milky Way*. It is the largest of the three sections into which the ancient constellation Argo Navis was divided by Nicolas L. de Lacaille in the mid-eighteenth century. It contains ten stars of the second and third magnitudes. ►Table 4.

Purkinje effect The change in the colour sensitivity of the human eye that occurs when ►*dark adaption* takes place.

In daylight, a normal eye detects light in the wavelength range 400–750 nanometres, the peak sensitivity lying in the yellow to green region. Under low levels of illumination, the eye adapts to increase overall sensitivity to light, but is less able to detect the redder end of the spectrum; the peak sensitivity moves further towards the green.

Purple Mountain Observatory An observatory of the Chinese Academy of Sciences, located on the Purple Mountain near the city of Nanking. It was constructed between 1929 and 1934. The instruments located there include a 43-centimetre (17-inch) ►*Schmidt camera*, a 1.5-metre (5-foot) radio dish for solar observations, a 20-centimetre (8-inch) refractor, a 60-centimetre (24-inch) reflector and a 40-centimetre (16-inch) twin refractor used for planetary work.

pyrheliometer An instrument for measuring the total heat energy received from the Sun per unit area per unit time.

Pythagoras A large lunar crater near the north-west limb of the Moon. It is 129 km in diameter and has high walls and a central peak.

Pyxis (The Compass) A small and insignificant southern constellation introduced by Nicolas L. de Lacaille in the mid-eighteenth century. It contains only one star that is brighter than fourth magnitude. ►Table 4.

PZT Abbreviation for ►*photographic zenith tube.*

Q

QSO Abbreviation for quasi-stellar object. ➤*quasar*.

Q-type asteroid A rare type of asteroid, interpreted as being similar to the class of meteorites known as ➤*chondrites*. Only ➤*Apollo* and a few other Earth-approaching asteroids are known to belong to this class.

quadrant An instrument for measuring the altitudes of stars and the angular separation between celestial objects, consisting of a graduated quarter-circle and a movable sighting arm. Prior to the invention of the telescope, such instruments were the only ones available to astronomers for positional measurements. Tycho Brahe (1546–1601), for example, used a large *mural quadrant*, built on to a specially constructed wall lying along the north–south direction. On instruments made after the invention of the telescope, telescopic sights were often employed. The modern equivalent is the ➤*transit circle*.

Quadrantids An annual ➤*meteor shower*, the radiant of which lies in the constellation Boötes, near the border with Hercules and Draco. The name dates from the time when this area of sky was identified as the constellation Quadrans Muralis, now no longer used. The peak of the shower occurs on about 3 January and the normal limits are 1–6 January. The narrow stream of meteors is not associated with any known comet, and the shower is somewhat variable, peak rates occurring over a relatively short time period.

quadrature The position of the Moon or of a planet when its angular distance from the Sun, as viewed from the Earth, is 90°.

quantum gravity A theory, not yet developed, that aims to unite ➤*Grand Unified Theories* and a theory of the gravitational interaction into a single unified theory in which all interactions in physics are described by a unified set of equations. If achieved, it would unify ➤*General Relativity* and quantum theory. In a rigorous form, the theory would describe conditions in the universe about 10^{-43} seconds after the onset of the ➤*Big Bang*.

quarter day Originally the days of the ➤*solstices* and ➤*equinoxes*, dividing the year into four quarters. In the English legal system, the quarter days (for collecting rents, and so on) are: 24 June, 26 September, 25 December and 25 March.

quasar (quasi-stellar object, QSO) A small extragalactic object that is exceedingly luminous for its angular size and has a high ➤*redshift*. The word originated as a contraction of 'quasi-stellar' radio source, the term given in 1963 to a class of apparently star-like objects, emitting radio waves and having high redshifts. Quasars are now generally believed to be the most luminous type of ➤*active galactic nuclei*. Around a small number, the faint nebulous light of a surrounding galaxy has been detected. Many thousands of quasars have been catalogued.

In general, quasars have a spectrum that shows emission lines, high redshifts (typically from 0.5 to 4, although values higher and lower than these have been recorded), and they are so compact that they appear as sharp as stars on photographs. Although the quasars discovered in the 1960s were all radio sources, most of those now known are not strong radio sources.

Quasars have the largest redshifts found, and their importance in astronomy stems primarily from this feature. If the redshifts follow from the expansion of the universe, ➤*Hubble's law* can be applied. It follows that they are the most distant objects observable, some of them more than 10 billion light years from our Galaxy. The light from quasars has reached us from long ago, so in principle it can tell us about the state of the universe billions of years in the past. The light from distant quasars shows the ➤*Lyman-alpha forest* of numerous absorption lines of hydrogen at a variety of lower redshifts. This absorption occurs in hydrogen clouds near the quasar. If the light from a distant quasar passes through an intervening galaxy, a ➤*gravitational lens* effect may be seen.

The fact that we see such remote objects means that they are intrinsically very luminous, from a few to a hundred times the luminosity of a normal galaxy. The presence of emission lines implies that energy is generated through non-thermal mechanisms. Very-long-baseline interferometry shows that the central energy source in quasars is confined to a volume of space similar to the size of the solar system. This implies that the energy source is probably the infall of matter to a supermassive ➤*black hole*.

quasi-stellar object   ➤*quasar*.

quasi-stellar radio source   ➤*quasar*.

Quetzalcoatl   Asteroid 1915, diameter 0.4 km, discovered in 1953 when it made a close approach to the Earth. It is a member of the ➤*Amor* group.

quiet Sun   The Sun when it is at the minimum level of activity in the ➤*solar cycle*, with little evidence of ➤*solar activity*.

R

RA Abbreviation for ►*right ascension*.

radar astronomy The use of pulsed radio signals in astronomical applications, such as the detection of ►*meteor showers*, the measurement of distances within the solar system and the mapping of surfaces of objects in the solar system. Radar signals transmitted by the 305-metre (1,000-foot) radio telescope at ►*Arecibo Observatory* have successfully been used to map Venus and to characterize the size and shape of ►*asteroids*. The ►*Magellan* spacecraft, placed in orbit around Venus, used ►*synthetic aperture radar* to map the planet's surface, which is concealed by opaque cloud. Radar is of fundamental importance in calibrating the distance scale within the solar system and thereby determining the value of the ►*astronomical unit*.

radial velocity The velocity of an object relative to an observer as measured in the direction of the line of sight. To determine an object's true velocity in space, it is necessary also to know the transverse velocity, which is across the line of sight. For stars, galaxies and other astronomical objects, the radial velocity is often much easier to determine than the transverse velocity because of the operation of the ►*Doppler effect*. ►*proper motion*.

radial-velocity curve A graph in which the ►*radial velocity* of an object is plotted as a function of time. Such curves are used particularly in the analysis of ►*binary star* orbits, where the radial velocities of the components change with a regular periodic cycle.

radian A unit in which angles are measured. It is the angle subtended by an arc of a circle that has the same length as its radius. Thus, 2π radians are equivalent to 360° and one radian is approximately 57.30°.

radiant The point on the celestial sphere from which the trails of meteors belonging to a particular ►*meteor shower* appear to radiate. Meteors entering the Earth's atmosphere from a stream create trails that are almost parallel; the apparent divergence from a radiant is simply a perspective effect.

radiation belt A ring-shaped region around a planet in which electrically charged particles (electrons and protons) are trapped, following spiral trajectories around the direction of the magnetic field of the planet. The radiation belts surrounding the Earth are known as the ►*Van Allen belts*. Similar regions exist around other planets with magnetic fields, such as Jupiter.

radiation era   The period from one second to one million years after the
➤*Big Bang*. During this interval, the behaviour of the universe was dominated
by radiation (i.e. ➤*photons*). The era ended with the ➤*recombination epoch*, when
the temperature of the expanding universe had fallen to a few thousand degrees,
enabling electrons and protons to form the first stable atoms.

radiation pressure   The pressure exerted by a stream of ➤*photons* when they
are transferring momentum to matter. In astronomy, radiation pressure is
important wherever the flux of radiation is extremely high, for example in the
outer layers of a star. In the interstellar medium, radiation pressure on dust
grains can be more important than the local gravitational field. Within the solar
system, the pressure of radiation from the Sun acts to push the smallest particles
outwards. ➤*Poynting–Robertson effect*.

radiative transfer   The process by which the energy associated with ➤*electromagnetic radiation* is transferred as the radiation interacts with matter. In radiative
transfer, photons are continually being absorbed and re-emitted by matter.

In simple terms, the law of conservation of energy requires that the energy
emerging from a medium is equal to the amount entering it, plus any energy
emitted by the medium itself but minus any energy absorbed. This can be
expressed mathematically, but real solutions are difficult because of the complexity of the way in which matter and radiation interact in practice.

radio astronomy   The exploration of the universe through the detection of
radio emission from celestial objects. The principal sources of cosmic radio
emission are: the ➤*Sun*, ➤*Jupiter*, interstellar hydrogen and ionized gas, ➤*pulsars*,
➤*quasars* and the ➤*cosmic background radiation* of the universe itself. The frequencies
used span a vast range, from 10 MHz to 300 GHz. There are several wavebands
protected internationally against interference, such as 1,421 MHz (wavelength
21 centimetres), the natural frequency of atomic hydrogen.

➤*Radio telescopes* are largely either single steerable dishes, up to 100 metres
in diameter, or arrays of dishes linked to form ➤*radio interferometers*. Single
telescopes have poor angular resolution compared with optical telescopes, so
they are mainly used in investigations where positional accuracy is not vital,
such as the timing of pulsar signals or the mapping of large-scale distributions,
such as the microwave background. Where structural detail is required, for
example in the mapping of radio galaxies, it is essential to use an interferometer.

Since its inception in the 1940s, radio astronomy has been directly responsible
for the discovery of pulsars, quasars and the microwave background. ➤*radio
galaxy*.

radio brightness distribution   The distribution across the sky of the radio
emission from an extended radio source, expressed as radio flux density per
unit solid angle as a function of position. It can be displayed as a contour map,
or computer processed to produce an image similar to an optical photograph.

radio galaxy A ►*galaxy* that is an intense source of radio emission. About one galaxy in a million is a radio galaxy. The radio emission is ►*synchrotron radiation* from electrons travelling at speeds close to that of light. In ►*Cygnus A*, often regarded as the prototype radio galaxy, two huge clouds of radio emission, disposed symmetrically on each side of a disturbed elliptical galaxy, span more than three million light years. It seems unlikely that the huge energy output can be generated by normal nuclear reactions in stars. Mechanisms in which black holes act as a kind of 'central engine' have been suggested.

Radio galaxies are closely related to ►*quasars*, many of which have similar radio properties.

radiograph In radio astronomy, a map of the distribution of radio emission processed to make an image similar in appearance to an optical photograph.

radioheliograph A radio telescope designed for mapping the distribution of radio emission from the Sun.

radio interferometer A ►*radio telescope* in which two or more separate antennas observe the same object simultaneously. The signals received by a pair of antennas are fed into a receiver that multiplies the voltages. The amplitude and phase of the correlated output depend on the distribution of radio emission from the source being studied. Little can be learned from a single measurement of this kind but, if the spacing and orientation of the interferometer are changed, the correlated voltage varies and can be analysed by computer to generate maps showing the distribution of radio brightness on the sky. This technique is fully exploited in ►*Earth rotation synthesis*. ►*aperture synthesis, very-long-baseline interferometry*.

radiometer Any instrument for measuring the total amount of electromagnetic radiation received from an object. In infrared astronomy, the term is applied to a device designed to measure only infrared flux. In radio astronomy, a radiometer is a detector able to measure with great accuracy the total radio energy received.

radio source Any natural source of cosmic radio emission. In cosmology, it has a more restricted meaning and refers only to ►*radio galaxies* and ►*quasars*. ►*radio astronomy, source counts*.

radio star An expression used in the early years of ►*radio astronomy* when the resolution of observations was too poor to enable astronomers to match radio sources to visible objects. It was assumed, quite wrongly, that many of these sources were stars; subsequently many were shown to be radio galaxies. True radio stars are very rare.

radio telescope An instrument for the collection, detection and analysis of radio waves from any cosmic source. All such telescopes consist of a radio antenna feeding an amplifier and a detector. The large range of frequencies covered by radio astronomy means that radio telescopes vary greatly because different techniques are used for different parts of the spectrum.

A fundamental problem in radio astronomy is obtaining adequate angular resolution. A telescope with a diameter of 100 wavelengths has a resolving power of only 1°. To reach a resolution of half an arc second, comparable to that of a good optical telescope, a diameter of 50,000 wavelengths constructed to an accuracy of a tenth of a wavelength is required. At a wavelength of 21 centimetres, the diameter of dish needed is 100 kilometres!

Single steerable dishes are used mainly for studies of interstellar matter, through the ►*twenty-one centimetre line*, and variable sources, such as pulsars. Fully steerable dishes are limited to apertures of about 100 metres by the weight of the structure.

The higher angular resolution needed to map structure in objects such as ►*radio galaxies* and ►*quasars* is obtained by linking arrays or networks of telescopes to form a ►*radio interferometer*. ►*aperture synthesis, radio astronomy, very-long-baseline interferometry*.

radius vector   The line joining an orbiting body to its centre of motion at any instant, directed radially outwards. For a circular orbit, the centre of motion coincides with the centre of the circle; for a parabolic or hyperbolic orbit, the centre of motion is the focus, and for an elliptical orbit it is one of the two foci.

Ram   English name for the constellation ►*Aries*.

Ramsden disc   The same as ►*exit pupil*.

Ramsden eyepiece   A telescope ►*eyepiece* consisting of two similar plano-convex lenses, convex sides facing and separated by a distance equal to two-thirds the sum of their focal lengths. This simple design suffers from ►*chromatic aberration* and, for astronomy, the ►*Kellner eyepiece* is preferred.

Ranger   A series of nine American lunar probes, launched between 1961 and 1965. Only the last three, *Rangers 7, 8* and *9*, were successful. *Ranger 7*, launched in July 1964, returned 4,000 images. Thousands more were obtained with *Rangers 8* and *9*, launched in February and March 1965. They were 'hard landers', designed to transmit images during approach to the Moon until they crash-landed. The three successful *Rangers* landed successively in the ►*Fra Mauro* region, the Mare Tranquillitatis and the crater ►*Alphonsus*.

RAS   Abbreviation for ►*Royal Astronomical Society*.

Ras Algethi   (Alpha Herculis; α Her)   The brightest star in the constellation Hercules. It is a binary, consisting of a red supergiant ►*M star* with a sixth magnitude companion of ►*spectral type* F, which appears greenish in contrast.

The secondary is itself a ►*spectroscopic binary*. The primary star is an irregular variable, its magnitude ranging between 3 and 4. The name, of Arabic origin, means 'the kneeler's head'.

Ra-Shalom   Asteroid 2100, diameter 3 km, discovered by E. Helin in 1978. It is the largest known member of the ►*Aten* group of asteroids, whose orbits lie wholly within that of the Earth.

RATAN-600   A radio telescope at the ►*Special Astrophysical Observatory* of the Russian Academy of Sciences, located at Zelenchukskaya in the Caucasus Mountains. The name is an acronym derived from 'radio astronomy telescope of the Academy of Sciences' in Russian. It consists of 900 parabolic plates forming a circle 600 metres (2,000 feet) in diameter. It can be used as a whole, or each quarter can operate as a self-contained unit.

ray   A light-coloured, linear feature extending radially from a crater. A number of lunar craters are surrounded by extensive and conspicuous ray systems that show up particularly at full Moon, when they can be seen with the unaided eye. Ray systems are associated with the youngest craters, such as Tycho and Copernicus. They may be due to rock surfaces exposed in relatively recent impacts; such surfaces would be more reflective than old surfaces that have been subjected to radiation for millions of years. Alternatively, rays may consist of glassy, reflective ejecta.

Rays are also found around craters on several other bodies in the solar system.

R Coronae Borealis   The prototype of a group of peculiar variable stars characterized by sudden and unpredictable dips in brightness of many magnitudes. R Coronae Borealis (R CrB) is normally magnitude 5.8, but every few years becomes up to nine magnitudes fainter when enshrouded in a dust cloud blown off by a strong stellar 'wind'. About forty stars of similar type are known, typically ►*supergiant* stars of ►*spectral type* F or G.

recombination epoch   ►*decoupling era.*

recombination line   A spectral feature produced by electromagnetic radiation at a particular wavelength emitted when an electron in an ionized gas is captured by a positive ion. Energy is released in the form of photons corresponding to discrete wavelengths as the electron drops through the energy levels of the atom.

recurrent nova   A ►*nova* known to have had two or more outbursts. The interval between outbursts typically ranges between 10 and 80 years. It is possible that all novae are recurrent and that only the ones with the shortest periods are recognized. However, the known recurrent novae differ from 'classical' novae in that their amplitudes of six to eight magnitudes are smaller, and their spectra during outbursts are different. At other times their spectra seem to indicate the presence of a ➤*red giant* in the system.

reddening   A change in the distribution of continuum electromagnetic radiation from a source such that the intensity at longer wavelengths is enhanced and that at shorter wavelengths diminished. Reddening occurs when light travels through the interstellar medium. ➤*interstellar reddening.*

red giant   An evolved star that has greatly expanded in size and undergone a change in its surface temperature so that it appears red.

A star becomes a red giant at a phase in the course of ➤*stellar evolution* where the hydrogen fuel for nuclear fusion in the central core has been exhausted. In the internal adjustment that follows, the core collapses until sufficient gravitational energy is released to cause hydrogen burning to restart, but in a shell around the now inert core. The energy generated by shell burning of hydrogen causes the great expansion of the star's outer layers. As the gas expands, it cools. Regardless of the star's original ➤*spectral type*, its surface temperature drops until it reaches 4,000 K. When the Sun becomes a red giant, it will expand until its diameter is roughly the diameter of the Earth's orbit.

Though the light emitted per unit surface area of a star decreases sharply with temperature, a red giant compensates for this effect by the enormous increase in surface area. Thus, red giant stars are relatively very luminous. All the bright red stars visible to the naked eye are giants or supergiants, such as ➤ *Aldebaran* or ➤*Betelgeuse.*

The spectra of red giants show different characteristics according to whether the stellar atmosphere is rich in carbon or oxygen. If oxygen is dominant, carbon monoxide (CO) and metallic oxides, such as titanium oxide (TiO) are evident. If carbon predominates, carbon compounds such as C_2, CH and CN are formed; such giants are called ➤*carbon stars.* The different compositions observed in red giants are presumed to be the result of processes that bring to the surface the products of nuclear reactions that have taken place in the interior.

Red Planet   A popular name for the planet ➤*Mars*, which has a distinctly reddish hue even to the naked eye.

Red Rectangle   A red nebula around the star HD 44179, detected in 1975 during a survey of infrared sources. The emission, which is intrinsic to the nebula and not scattered or reflected light, is in a broad band in the red part of the spectrum, peaking at about 640 nanometres. It is the strongest known source of this red emission, which may be a result of luminescence from hydrogenated amorphous carbon dust.

redshift   (symbol z) The increase in wavelength of electromagnetic radiation caused either by the ➤*Doppler effect*, when the source of radiation is moving away from the observer, or by the presence of a gravitational field. It is quantified

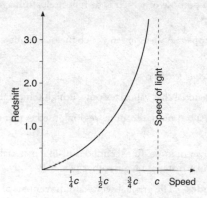

Redshift. The graph shows the relationship between redshift due to the Doppler effect and relative speed of recession.

in terms of the wavelength change $\Delta\lambda$ expressed as a fraction of the rest wavelength λ (measured when the source and observer are not in relative motion): $z = \Delta\lambda/\lambda$. The theory of the Doppler effect gives the relationship between redshift arising from relative motion and the relative velocity of the source and observer (see illustration).

The redshifts of galaxies and quasars are particularly important in astronomy since, through ➤*Hubble's Law*, they are generally regarded as direct indicators of the distances to these objects.

In ➤*General Relativity*, Einstein also predicted that there would be a redshift effect in the presence of a strong gravitational field.

Red Spot ➤*Great Red Spot.*

reflecting telescope A ➤*telescope* in which the main light-collecting element is a mirror.

reflection nebula A cool cloud of interstellar gas and dust that shines because the dust scatters the light from nearby stars; the cloud is not itself luminous. The spectrum of the scattered light is the same as that of the starlight, though blue light is scattered more extensively than red light since the effect depends on wavelength. A notable reflection nebula is that surrounding the stars of the ➤*Pleiades*.

reflectivity ➤*normal reflectivity.*

reflector (1) A ➤*reflecting telescope.*

reflector (2) An element in an optical, radio or other telescope, or any kind of instrument, the purpose of which is to reflect electromagnetic radiation or streams of particles.

refracting telescope   A ➤*telescope* in which the main light-collecting element is an objective lens.

refractor   ➤*refracting telescope*.

refractory   A material that vaporizes only at high temperatures.

regio (pl. regiones)   A term, meaning 'region', used as part of some planetary place names.

regolith   The layer of loose, fine-grained, soil-like material on the surface of the Moon, or similar material on any other planetary body.

regression of nodes   The gradual westward movement of the ➤*nodes* where the Moon's orbit intersects the ➤*ecliptic*. A full circuit takes 18.61 years, and the phenomenon is due to the gravitational influence of the Sun.

regular satellite   A natural satellite of a planet in a ➤*direct* (rather than retrograde) orbit of low or moderate eccentricity in the planet's equatorial plane.

Regulus (Alpha Leonis; α Leo)   The brightest star in the constellation Leo. It is a ➤*B star* of magnitude 1.4; the system is at least triple, the primary having companions of seventh and thirteenth magnitudes. The name is Latin and means 'little king'. Regulus is sometimes known as 'The Royal Star'. ➤Table 3.

relativistic   Having a velocity close to that of light; ➤*Special Relativity* must be used in any calculations involving such velocities and physical theory. The expression *ultrarelativistic* does not mean faster than light; it is used for particles travelling so fast that their kinetic energy is thousands of times higher than their rest mass energy.

Relativity   ➤*General Relativity, Special Relativity*.

remote sensing   The study of the Earth or other astronomical objects by observation or probing from a distance rather than direct contact and exploration. The term is used particularly for the study of the Earth, or other bodies, from orbiting satellites. Methods include both direct, high-resolution imaging and radar.

réseau   A reference grid photographed by a separate exposure on to the same plate as astronomical images.

residual   The difference between an actual observed quantity and its predicted or expected value. Residuals may result from observational error, basic limitations on the accuracy with which the observations can be made, or inaccuracy in the prediction with which the observation is compared.

resolution   The size of the smallest detail that can be distinguished with an imaging instrument such as a telescope or spectrograph.

resolving power   The ability of an optical system to distinguish detail. Theoretically, the ►*resolution* is limited by the size of the aperture, through the effects of ►*diffraction*. Because of diffraction, the image of a point source is actually a disc surrounded by a number of rings. The diameter of the disc, known as the Airy disc, is given in radians by the formula $1.1\lambda/D$, and this is effectively the theoretical resolving power. In practice, however, the resolving power of a large ground-based optical telescope is limited by the quality of ►*seeing* rather than the aperture.

resonance   A situation in which one orbiting body is subject to a regular periodic gravitational disturbance by another. Resonances occur between orbital positions that are linked by periods in whole-number ratios (e.g. 1:1, 2:1, 3:2). They are responsible for phenomena such as the ►*Kirkwood gaps* in the ►*asteroid belt* and divisions in the ►*planetary rings*, such as exist in those around Saturn. ►*commensurability*.

resonance gap   ►*Kirkwood gaps*.

retardation   The difference between the times of moonrise on successive nights.

reticle   Fine lines or wires at the focus of an optical instrument to aid in the measurement of angular distances in the image.

Reticulum (The Net)   A small southern constellation introduced by Nicolas L. de Lacaille in the mid-eighteenth century. Its two brightest stars are of third magnitude. ►Table 4.

retrograde   Motion of an object on the celestial sphere in the east–west direction, or, for orbital motion or axial rotation in the solar system, motion that is clockwise as observed from north of the ►*ecliptic*. ►*direct*.

reversing layer   The outermost layers of a star in which the ►*absorption lines* are presumed to be impressed on the continuous spectrum of the hotter gas beneath. The idea of a reversing layer is, however, an oversimplification since both emission and absorption take place through all the layers of a stellar atmosphere.

RGO   Abbreviation for ►*Royal Greenwich Observatory*.

Rhea   The second-largest satellite of Saturn, discovered by Giovanni Cassini in 1672. ►*Voyager 1* images show that Rhea's light-coloured, icy surface is saturated with craters. At the low temperatures prevailing this far from the Sun, the ice is like rigid rock. There is little evidence that the surface has changed since the era when the impact craters were formed. ▶Table 6.

Rho Ophiuchi cloud (ϱ Oph cloud)   A large nebulous region near the star Rho Ophiuchi (ϱ Oph). It is a mixture of ▶*reflection nebulae*, ▶*emission nebulae*, dark ▶*absorption nebulae* and ▶*molecular clouds*, and is relatively close, at a distance of about 700 light years. Infrared observations reveal the presence of a cluster of at least 40 stars within the dark cloud. This is a region of very active star formation and contains many ▶*T Tauri stars* and ▶*Herbig–Haro objects*.

Rigel (Beta Orionis; β Ori)   The brightest star in the constellation Orion. Its ▶*Bayer letter* is Beta, despite the fact that, at magnitude 0.1, it is slightly brighter than Betelgeuse, which is designated Alpha. Rigel is a ▶*supergiant* ▶*B star*, with a seventh magnitude companion. The name, derived from Arabic, means 'leg of the giant'. ▶Table 3.

right ascension (RA)   One of the coordinates used to define position on the ▶*celestial sphere* in the equatorial coordinate system. It is the equivalent of longitude on the Earth but is measured in hours, minutes and seconds of time eastwards from the zero point, which is taken as the intersection of the celestial equator and the ▶*ecliptic*, known as the ▶*First Point of Aries*. One hour of right ascension is equivalent to 15 degrees of arc; it is the angle through which the celestial sphere appears to turn in one hour of ▶*sidereal time*, as the Earth rotates. ▶*declination*.

Rigil Kentaurus   The star ▶*Alpha Centauri*. This Arabic name, less commonly used, means 'foot of the centaur'.

rille (rill)   An alternative name for a ▶*rima*.

rima (pl. rimae)   A fissure or channel in the lunar surface. Some such features are *graben*, caused by vertical faulting. Others are collapsed *lava tubes*, which tend to be more sinuous. Rimae are also known as *rilles*.

ring antenna   A ▶*radio telescope* in the form of a ring, such as the Russian ▶*RATAN-600*.

ring galaxy   A galaxy shaped like a ring. Such galaxies are thought to result from a collision in which one galaxy passes right through another, triggering rapid star formation in an expanding shock wave of interstellar gas (the ring) within the target galaxy. These objects are very rare.

Ring Nebula (M57; NGC 6720)   A bright ▶*planetary nebula* in the constellation Lyra. It has the appearance of a slightly elliptical luminous ring around a central star. The radius is one-third of a light year, and the nebula is 2,000 light years away. If it has always expanded at its current rate of 19 km/s, it is about 5,500 years old.

ring system ➤*planetary rings.*

Ritchey–Chrétien telescope A telescope in which the optical design is similar to the ➤*Cassegrain telescope* except that both the primary and secondary mirrors are hyperboloids. This results in a wide field of view free of ➤*coma.*

River English name for the constellation ➤*Eridanus.*

Robertson–Walker metric A ➤*metric* or formula for calculating proper intervals in a homogeneous and isotropic universe.

The Robertson–Walker metric is the mathematical formula for time and distance calibration generally adopted for models of the large-scale structure of the universe. It applies to an idealized model, the ➤*Friedmann universe,* that is perfectly homogeneous and perfectly isotropic, i.e. satisfying the ➤*cosmological principle.*

Robertson and Walker showed that, in such an idealized model, ➤*spacetime* could be separated into two concepts common to all observers: curved space and cosmic time. Although this conceptual separation of space and time might seem obvious, it does in fact relate to a highly restricted geometry for spacetime in general. In the vicinity of ➤*black holes,* for example, the distinction between space and time is blurred and the ➤*Kerr metric* or ➤*Schwarzschild metric* is used instead.

Roche limit The minimum distance from the centre of a planet that a fluid satellite can orbit and remain stable against tidal disruption. If the planet and satellite have equal densities, the Roche limit is 2.456 times the radius of the planet.

Anywhere inside the limit, the satellite gets torn apart by tidal forces. Solid satellites can exist inside the Roche limit because the tensile strength of rock holds them together. However, the shattering of satellites by tidal forces could be the means by which ➤*planetary ring* systems are formed. ➤*tides.*

Roche lobes In a ➤*binary star* system, an hourglass-shaped surface on which lie the points where the gravitational force exerted by either star on a small particle is equal. The two segments of the 'hourglass' each enclose a region known as a Roche lobe. Between the stars there is a unique location where the two Roche lobes touch: this is the *inner Lagrangian point,* at which the gravitational force is zero (see illustration (a) on page 318).

When a star in a close binary system expands during the giant phase of its evolution, it may completely fill its Roche lobe (see illustration (b)); such a system is called *semidetached.* Matter then streams through the inner Lagrangian point to the other star (see illustration (c)). In a ➤*contact binary* (see illustration (d)), both stars have completely filled their Roche lobes and mass transfer can

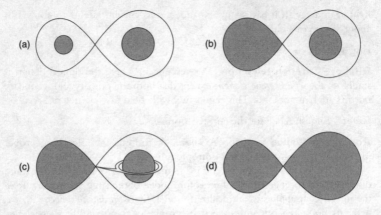

Roche lobe. The Roche lobes of a binary star system form an hourglass shape. At **a**, the stars are completely detached. At **b**, one star fills its Roche lobe and material will stream on to its companion (**c**). At **d**, both Roche lobes are filled and the system is a contact binary.

occur. This mass transfer is important in ➤*dwarf novae*, ➤*X-ray binaries* and ➤*Beta Lyrae stars*, and in the formation of ➤*accretion discs* around stars.

ROE Abbreviation for ➤*Royal Observatory, Edinburgh*.

Ronchi test A test for checking the optical quality of a concave mirror. The mirror is viewed through a grating of fine parallel lines ruled on glass, and the mirror is illuminated with a pinhole light source near the centre of curvature. If the mirror is a perfectly spherical surface, straight bands are seen; for a paraboloid, a curved pattern is obtained.

Roque de los Muchachos Observatory ➤*Observatorio del Roque de los Muchachos*.

Rosalind A small satellite of Uranus, diameter 60 kilometres, discovered in 1986 by the ➤*Voyager 2* spacecraft. ➤Table 6.

ROSAT A German orbiting X-ray astronomy observatory, with participation by NASA and the UK Science and Engineering Research Council. It was launched successfully in 1990. It carried a large X-ray imaging telescope for an all-sky survey and a wide-field camera operating in the XUV region, which bridges the X-ray and extreme ultraviolet regions of the spectrum.

Rosetta A European Space Agency mission to a ➤*comet*, scheduled for launch in 2003. The original plan was to bring back a sample from a comet nucleus for analysis on Earth, but this idea was abandoned in favour of analysis of cometary material by instruments to be carried on the craft, which will rendezvous with the comet for an investigation lasting up to 18 months. The target is Comet 46P/Wirtanen, which would be reached by Rosetta in 2011.

Rosette Nebula (NGC 2237, 2238, 2239 and 2246)   An ►*emission nebula* in the constellation Monoceros, surrounding a young ►*open cluster* of stars (NGC 2244). It is roughly circular in shape, with a central hole cleared of dust and gas by the pressure of radiation from the stars in the cluster. The nebula's distance is estimated to be 4,500 light years.

Rossi X-ray Timing Explorer (RXTE)   A NASA X-ray astronomy satellite launched in 1995. It was named in honour of the X-ray astronomy pioneer, Bruno B. Rossi, in 1996 after launch. The spacecraft was equipped with three instruments. The Proportional Counter Array and the High-Energy X-ray Timing Experiment (HEXTE) working together constituted the largest X-ray telescope flown to date, sensitive to X-rays from 2 to 200 keV. The third instrument, the All Sky Monitor, was designed to record the long-term behaviour of X-ray sources and to monitor the sky.

rotation curve   For a galaxy, a graph showing the variation of orbital velocity about the centre with distance from the centre. The distribution of mass with distance from the spin axis and the total mass of a galaxy may be estimated from its rotation curve.

rotation measure   For a radio source that emits polarized radiation, a parameter indicating the extent to which the polarization vector has rotated while the radiation has been travelling from the source to the observer. The degree of rotation is proportional to the magnetic field across the line of sight and to the electron density along the line of sight. It depends on the square of wavelength, so observations of polarization at several wavelengths enable the rotation measure towards a radio source to be determined. Rotation measures are the main source of information on the strength and direction of the magnetic field in our Galaxy.

Royal Astronomical Society (RAS)   A British organization of professional astronomers and geophysicists, with headquarters in London. It was founded in 1820 by John Herschel and a number of other prominent astronomers and scientists of the day as the Astronomical Society of London. It received its Royal Charter in 1831. During the nineteenth century, many of its Fellows were prominent amateur astronomers, but this situation has now largely changed. It organizes programmes of meetings and publishes research journals.

Royal Greenwich Observatory (RGO)   An astronomical institute in the UK, which was first established at Greenwich near London in 1675 by King Charles II.

In the seventeenth century, the major problem facing navigators was the establishment of longitude at sea. In principle, this could be done by observing the position of the Moon, and Charles II set up a Royal Commission in December 1674 to examine the idea. The Commission sought the advice of the astronomer John Flamsteed, who pointed out many practical difficulties. The King then appointed Flamsteed first ➤Astronomer Royal, with the job of solving the problem. The site in Greenwich Park for the new Royal Observatory was suggested by Christopher Wren.

The Observatory has played an important role in positional astronomy throughout its history. The Washington Conference of 1884 fixed the meridian through Greenwich as the zero point of longitude.

In the twentieth century, the emphasis in the Observatory's work changed to include more astrophysics. The difficulty of observing in London led to a move to Herstmonceux Castle, Sussex, in 1948. By the 1970s, it was clear that first-class observational work could no longer be done at any site in Britain, and a decision was made to move all observational work to La Palma in the Canary Islands. In 1990 the administrative offices of the Observatory were moved to Cambridge. ➤Observatorio del Roque de los Muchachos.

Royal Observatory, Edinburgh (ROE)   A UK national establishment in Edinburgh. Its major role is in the support of the UK's national facilities, especially through instrument development. It was founded as the Astronomical Institution of Edinburgh in 1811 by private citizens. They petitioned King George IV on his visit to Edinburgh in August 1822 with the result that the title was changed to the Royal Observatory, Edinburgh, and the status nominally raised to that of the ➤Royal Greenwich Observatory, though it did not receive funding on the same level in practice. In 1834 the University of Edinburgh agreed to take over the Observatory on condition that the government paid the salary of a professor who would hold the title Astronomer Royal for Scotland. In 1995, that honorary title was separated from the post of Director of the ROE. The Observatory still has close links with the University.

r-process   A process of ➤nucleosynthesis thought to take place when there is a relatively high flux of neutrons. Heavy elements are built up by the capture of more than one neutron by a nucleus in rapid succession, before there has been time for an unstable intermediate isotope to decay. The 'r' stands for 'rapid'. This is the only means be which some of the heaviest known elements and particular isotopes could be made naturally. The r-process is believed to occur in the explosion of a ➤supernova. ➤s-process.

RR Lyrae star   A category of pulsating ➤variable star belonging to ➤Population II (relatively old stars). They are found particularly, but not exclusively, in ➤globular clusters. Though similar to ➤Cepheid variables, they are intrinsically less luminous – by as much as 7 magnitudes – and all have approximately the same absolute magnitude (+0.5). This property makes them useful distance indicators, though the practicability of the method is limited by their modest luminosity. The periods range between a few hours and just over a day, and the typical amplitude of variation is between 0.2 and 2 magnitudes. Their ►*spectral type* is A or F. In some RR Lyrae stars, the amplitude and phase of variability is modulated over a longer period of 20 to 200 days. These are known as Blazhko variables. The brightest known example is RR Lyrae itself, whose basic period of 0.567 days is modulated over 41 days.

RS Canum Venaticorum star   A member of a class of variable stars that are both eclipsing and variable outside eclipse. Both members of the binary system are usually ►*G stars* (sometimes F or K stars) and show extreme solar-like activity. The variation outside eclipse is thought to be caused by extensive 'star-spots' coming in and out of view as the stars rotate. X-ray emission is detected from hot ►*coronae* around these stars, and powerful flares, similar to those that erupt on the Sun, are sometimes seen.

R star   A ►*spectral type* formerly used for stars now incorporated in the class of ►*carbon stars*.

R-type asteroid   A rare type of asteroid with moderately high albedo, of which 349 ►*Dembowska* is an example.

Rudolphine tables   A tabulation of planetary positions based on the ►*Copernican system*, published by Kepler in 1627. It was more accurate than any previous such tabulation and established Kepler's reputation.

Rule   English name for the constellation ►*Norma*.

runaway star   A young hot star travelling through space with an unusually high velocity. It is thought that such stars could originally have been in binary or multiple systems where a companion exploded as a ►*supernova*.

Three well-known examples are Mu (μ) Columbae, AE Aurigae and 53 Arietis. From their speeds and directions of motion it has been calculated that all three were ejected from a common region in the constellation Orion about three million years ago.

runoff channel   A dry, branching river channel with many tributaries, found in the old uplands of Mars. These channels were presumably formed in an era when liquid water fell as rain on Mars.

rupes (pl. rupes)   A term meaning 'scarp', used in the names of certain planetary features of this type.

RV Tauri star   A member of a class of pulsating variable stars. RV Tauri stars are supergiants of ►*spectral types* F, G or K, showing variations of up to four magnitudes, but with minima alternating between deeper and shallower ones. Their periods lie between 30 and 150 days. There is evidence that these stars are losing mass rapidly via a stellar 'wind' and they may be the precursors of ➤*planetary nebulae*. They have the characteristics of old, evolved stars, and some are found in globular clusters.

RXTE Abbreviation for ➤*Rossi X-ray Timing Explorer*.

Ryle Telescope ➤*Mullard Radio Astronomy Observatory*.

S

Sachs–Wolfe effect Anisotropy in the ►*cosmic background radiation* caused by the gravitational field of large-scale structure in the universe.

Sacramento Peak Observatory A solar observatory in New Mexico, forming part of the facilities of the US ►*National Solar Observatory*. The observatory site, known as 'Sunspot', is in the Lincoln National Forest at a height of 2,800 metres (9,200 feet). It was founded in 1951 as the Upper Air Research Observatory of the Air Force Cambridge Research Laboratories to predict disturbances to, for example, communications, caused by solar activity. In 1976, the observatory was transferred to the National Science Foundation and, in 1984, it became part of the National Solar Observatory section of the National Optical Astronomy Observatories.

The largest instrument is a ►*vacuum tower telescope*. It is used to observe small-scale solar features and has a resolving power better than a quarter of an arc second. The John W. Evans Solar Facility (named after the first director) contains various instruments including a 40-centimetre (16-inch) coronagraph and a 30-centimetre (12-inch) ►*coelostat*, which can feed solar radiation to several instruments.

Sagitta (The Arrow) The third-smallest constellation, but nevertheless a rather distinctive little group of stars, included in the 48 constellations listed by Ptolemy (*c.* AD 140). The two brightest stars are third magnitude. It lies in a rich part of the Milky Way, next to Aquila. ►Table 4.

Sagittarius (The Archer) The southernmost constellation of the zodiac, and one of those listed by Ptolemy (*c.* AD 140). The centre of the Galaxy (the Milky Way) lies behind the star clouds in Sagittarius. It is a large constellation, with many bright stars; there are 14 brighter than fourth magnitude. It also contains a large number of star clusters and diffuse nebulae. The ►*Messier Catalogue* lists 15 objects in Sagittarius, more than in any other individual constellation. They include the ►*Lagoon Nebula*, the ►*Trifid Nebula*, the ►*Omega Nebula* and the third-brightest ►*globular cluster* in the sky, M22. ►Table 4.

Sagittarius A The overall name for the complex radio source associated with the galactic centre. It is composed of at least four separate sources, known as Sagittarius A, B, B2 and C. A bright, compact source known as Sagittarius A★ is believed to be the closest to the actual centre of the Galaxy.

Sagittarius Arm   One of the spiral arms of the ➤*Galaxy*. It lies between the Sun and the centre of the Galaxy in the direction of the constellation Sagittarius.

Sagittarius B   ➤*Sagittarius A.*

Saha equations   A set of mathematical expressions, formulated in 1920 by the Indian physicist Megh Nad Saha, from which can be calculated the numbers of atoms in each of the possible ionization states for atoms and electrons in ➤*thermodynamic equilibrium* at a particular temperature. They are important in interpreting stellar spectra and the conditions in stellar atmospheres.

Sail   English name for the constellation ➤*Vela.*

Salpeter function   A simple theoretical expression, named after the theorist Edwin Salpeter (b. 1924), for the numbers of stars of different masses among newly formed stars per unit volume of a galaxy. The Salpeter function, also known as the *initial mass function*, is proportional to $M^{-2.35}$ where M is the mass of a star.

Salpeter process   An alternative name for the ➤*triple-alpha process.*

Salyut   A type of Soviet manned space station. Seven were placed in Earth orbit between 1971 and 1982. Supplies could be brought to the stations, enabling cosmonauts to remain on board for many months.

SAO   Abbreviation for either ➤*Smithsonian Astrophysical Observatory* or ➤*Special Astrophysical Observatory* of the Russian Academy of Sciences.

SAO Star Catalog   Abbreviation for *Smithsonian Astrophysical Observatory Star Catalog*. This general catalogue contains 259,000 stars down to an approximate magnitude limit of 9. It was published in 1966; the ➤*epoch* of the positions given is 1950.0.

SAR   Abbreviation for ➤*synthetic aperture radar.*

saros   The period of time over which the sequence of lunar and solar eclipses repeats. Its length, 6,585.32 days (about 18 years), has been known since antiquity. After this period, the Earth, Sun and Moon will have returned to the same relative positions. Succeeding eclipses in a particular ➤*saros series* occur about 8 hours later and fall nearly 120 degrees of longitude further west. This was known to the ancient Babylonian and Mayan astronomers and to the builders of ➤*Stonehenge*.

saros series   A sequence of lunar or solar eclipses occurring at intervals of one ➤*saros*. Since there are up to seven eclipses every year, there are more than 80 saros series running concurrently. In any one series, an eclipse will recur at a particular longitude after an interval of three saroses (54 years). However, each succeeding eclipse in the series moves systematically in latitude (either north or south) from one pole to the other, until the series is completed.

SAS   Abbreviation for ►*Small Astronomy Satellite.*

satellite   Any body in orbit around a larger parent body. Most planets in the solar system have natural satellites, otherwise known as moons. Artificial satellites are man-made objects launched into orbit around the Earth, or another moon or planet.

satellite galaxy   A dwarf galaxy in orbit around a larger one. A number of satellite galaxies accompany our own Milky Way ►*Galaxy* and the ►*Andromeda Galaxy* in the ►*Local Group.*

satellite laser ranging (SLR)   A technique for measuring with a high degree of accuracy the rotation and gravitational field of the Earth. Trains of laser light pulses 4 centimetres long are beamed to special satellites from which they are reflected back. There are 30 satellite laser ranging stations around the world.

Saturn   The sixth major planet of the solar system in order from the Sun. Saturn is one of the four 'gas giants', second in size only to Jupiter. Its equatorial diameter is 9.4 times the Earth's and its mass 95 times greater. However, its average density is only 0.7 times that of water. Hydrogen and helium make up the bulk of the mass. There is a rocky central core, ten or fifteen times the mass of the Earth, which is surrounded by a thick mantle of liquid hydrogen and helium. In the high-pressure region surrounding the core, the hydrogen takes on the form of a metal. The outermost layers of the planet are gaseous; the visible features of the planet are cloud bands at the top of this atmosphere.

The cloud patterns on Saturn do not normally show much colour contrast. However, storm activity is occasionally observed. In late September 1990, a large white spot developed, expanding over a period of weeks to encircle much of the planet's equatorial region. This apparent eruption of material from the lower atmosphere followed a pattern of such occurrences over a 30-year cycle, corresponding to the orbital period. Similar spots were seen in 1876, 1903, 1933 and 1960, close to saturnian mid-summer in the northern hemisphere. Less extensive eruptions occur from time to time. One was observed by the Hubble Space Telescope in 1994.

Computer processing of images obtained by ►*Voyagers 1* and *2* during their encounters in 1980 and 1981 reveal complex circulation currents, similar to those observed on Jupiter. Saturn rotates rapidly, spinning once every 10 hours 32 minutes on average, though the rate varies with latitude. The resulting flattening at the poles is significant; the polar and equatorial diameters differ by 11 per cent.

The most striking feature is the spectacular ring system. The rings lie in the planet's equatorial plane, which is tilted at an angle of 27° to its orbit round the Sun. They are easily visible in a small telescope. As the relative positions of the Earth and Saturn change, the rings are presented at differing angles, sometimes appearing open, at other times edge-on so that they disappear from view. The rings have the appearance of a series of zones of differing brightness, separated by dark divisions. The most marked divisions are Cassini's and Encke's. The *Voyager* images of the rings showed that they consist of many thousands of narrow concentric ringlets, resulting in a grooved appearance. They are only one kilometre thick and are made up of a huge number of separate rocks and particles, perhaps ranging in size from a hundred metres down to a micrometre.

Before 1980, ten satellites of Saturn were known. More have been discovered since, some telescopically in 1980 when the ring system was edge-on (thus removing the glare) and some by the *Voyager 1* and *2* spacecraft in 1980 and 1981. Eighteen are now known for certain, and there are probably three others and possibly more, subject to confirmatory observations. ➤*Cassini mission*, *planetary rings*, Tables 5, 6 and 7.

Saturn Nebula (NGC 7009) A ➤*planetary nebula* in the constellation Aquarius. Its unusual shape, with a faint partial outer ring, resembles the planet Saturn. The double ring may be the remains of separate shells thrown off by the central star.

Scales English name for the constellation ➤*Libra*.

scattering A process in which all or part of a beam of electromagnetic radiation or particles is deflected from its initial direction of travel, without any absorption or emission.

Light is scattered by fine particles, such as dust, through the mechanism of reflection or diffraction (or both). If the particles are smaller than the wavelength of the light, the effect is purely one of diffraction and the phenomenon is known as *Rayleigh scattering*. The intensity of scattered light, as viewed from any particular direction, varies with wavelength λ as $1/\lambda^4$. This means that blue light is scattered more effectively than red. The daytime sky is blue because the blue component of sunlight is scattered by air molecules, while the rising or setting Sun, viewed through a thick layer of atmosphere near the horizon, is red because blue light is removed by scattering from the rays being observed directly. The same phenomenon produces ➤*interstellar reddening*.

Scattering may also be caused by direct interaction between a beam of radiation and the nuclei or electrons in the material through which it is passing. ➤*Compton effect*.

Scheat (Beta Pegasi; β Peg) The second-brightest star in the constellation Pegasus. It is a ➤*supergiant* ➤*M star* and varies in brightness between magnitudes 2.4 and 2.8. The name is derived from Arabic and probably means 'shoulder'.

Schedar (Schedir; Alpha Cassiopeiae; α Cas) The brightest star in the constellation Cassiopeia. It is a ➤*supergiant* ➤*K star* with a magnitude near 2.2, though it is slightly variable. The Arabic name means 'breast'.

Schickard A large lunar crater, 227 kilometres (141 miles) in diameter, near the Moon's south-west limb. The darkness of the crater floor indicates that it may be flooded by lava.

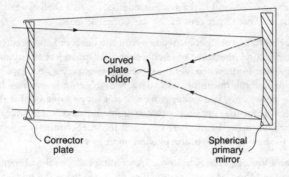

Schmidt camera. A schematic diagram of the optical arrangement.

Schmidt camera A type of wide-field astronomical telescope, designed purely for photographic use. It was invented by Bernhard Schmidt in 1930. The light collector is a spherical mirror (see illustration). Correction for spherical aberration is achieved by means of a thin glass corrector plate with a complex profile, placed at the end of the telescope tube, beyond the focal point. The photographic plate is placed at the prime focus, where the focal plane is curved. A special plate holder is used to bend the photographic plate into the focal plane. By this means, sharp undistorted images can be obtained over very wide fields of view – up to tens of degrees across.

Schmidt–Cassegrain telescope

A design of optical telescope, incorporating features of both the ➤*Schmidt camera* and the ➤*Cassegrain* reflector.

The Schmidt–Cassegrain employs a spherical primary mirror and a corrector plate to compensate for spherical aberration, as in the Schmidt camera (see illustration). However, the prime-focus plate holder is replaced by a small convex secondary mirror, which reflects the light back down the tube and through a hole in the

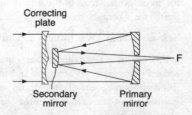

Schmidt–Cassegrain telescope. A schematic diagram of the optical arrangement.

primary mirror. The image can thus be viewed visually, or a camera can be mounted at the back of the main tube.

The resulting telescope is very compact, so the design is particularly suitable for portable telescopes and amateur and educational use.

Schönberg—Chandrasekhar limit The upper limit on the mass of hydrogen that can be converted to helium in the core of a ➤*main-sequence star* before core burning ceases and hydrogen burning moves to a shell around the core. The limit is about 12 per cent of the star's original mass.

Schröter's effect A discrepancy between the observed and predicted phase of Venus around the time of dichotomy (half-phase). Eastern dichotomy (evening apparitions, when Venus is waning) is usually a few days early, and western dichotomy (morning apparitions, when Venus is waxing) a few days late. The cause is not known for certain, but it may simply be that the region of the terminator (the division between night and day) is less bright than the rest of the illuminated hemisphere of Venus. Johann Schröter (1745–1816) was the first to draw attention to the phenomenon in 1793.

Schröter's Valley (Vallis Schröteri) A winding valley in the Oceanus Procellarum on the Moon. It starts in a small crater just outside the northern wall of the crater Herodotus, and extends for about two hundred kilometres.

Schwarzschild metric The ➤*metric* of spacetime used in the presence of a spherical mass. This mathematical description of the gravitational field of a spherical mass, calculated by Karl Schwarzschild in 1915, did not find general application until the 1950s when it was revived by cosmologists. It leads directly to the concept of ➤*black holes*.

Schwarzschild radius The critical radius at which the spacetime surrounding a sphere becomes so curved that it wraps round to enclose the body. An object that has collapsed inside its Schwarzschild radius is a ➤*black hole*, from which nothing can escape into the outside world. The Schwarzschild radius for an object the mass of the Sun is 3 kilometres; for an object the mass of the Earth it is 1 centimetre.

Schwassmann—Wachmann 1, Comet ➤*Comet Schwassmann—Wachmann 1.*

scintillation Twinkling – the rapid variations in the brightness of a star caused by random refraction in turbulent layers of the Earth's atmosphere. A similar phenomenon affecting radio signals from celestial sources occurs in the Earth's ionosphere and also in the ionized gas in both the interplanetary and interstellar media.

scintillation counter A radiation detector used in astronomy to detect gamma rays. The device uses crystals that emit flashes of light when gamma-ray photons strike them. Each flash, or scintillation, is picked up by a photomultiplier, the signal from which is then a measure of the flux of gamma rays on the counter.

scopulus (pl. scopuli)   A term used for a lobate or irregular scarp on a planetary surface.

Scorpio   Alternative (chiefly astrological) name for the constellation ▸*Scorpius*.

Scorpion   English name for the constellation ▸*Scorpius*.

Scorpius (The Scorpion)   A large, bright constellation of the southern part of the zodiac, among those listed by Ptolemy (*c.* AD 140). The brightest star is the first magnitude ▸*Antares*. There are 16 other stars brighter than fourth magnitude. Scorpius is also known as Scorpio, particularly in astrological rather than astronomical contexts. ▸Table 4.

Scorpius X-1   The brightest X-ray source in the sky and the first to be discovered. It is a low-mass X-ray binary star. The X-rays are thought to originate from a ▸*neutron star* and an associated ▸*accretion disc.*

Sculptor (The Sculptor's Workshop)   A faint, inconspicuous constellation of the southern hemisphere introduced by Nicolas L. de Lacaille in the mid-eighteenth century. Its four brightest stars are fourth magnitude. ▸Table 4.

Sculptor's Workshop   English name for the constellation ▸*Sculptor*.

Scutum (The Shield)   A small constellation near the celestial equator, introduced by Johannes Hevelius in the late seventeenth century with the name Scutum Sobieskii in honour of his patron, King John Sobieski III. There are no stars brighter than fourth magnitude though the constellation lies in rich star fields of the Milky Way. Its most notable feature is the star cluster M11, known as the ▸*Wild Duck* Cluster. ▸Table 4.

S Doradus star   Alternative name for ▸*P Cygni* star. S Doradus itself is an irregular variable star in the Large ▸*Magellanic Cloud*, and is currently eleventh magnitude. It has a very high mass, estimated to be about 60 times the mass of the Sun, and is as luminous as a powerful nova.

Sea Goat   English name for the constellation ▸*Capricornus*.

Sea Monster   English name for the constellation ▸*Hydra*.

Seashell Galaxy   A small galaxy interacting gravitationally with a larger galaxy, NGC 5291. The interaction has distorted the shape of the 'Seashell' to resemble a whelk – hence the name.

season   Part of a natural cyclical change in the prevailing environmental conditions on the surface of a planet, over the course of a complete orbit round the Sun. A planet experiences seasons if its rotation axis is not at 90° to the plane of the ►*ecliptic*. Seasonal effects, particularly on the polarice caps, are quite marked on the Earth and Mars.

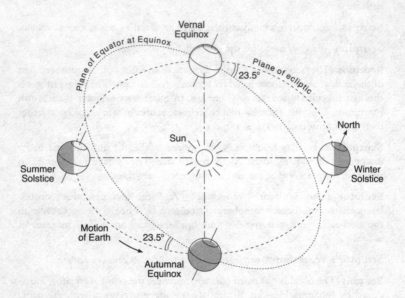

Season. The Earth's seasons arise from the fact that the equator is tilted at an angle of 23.5° to the plane in which it oribts the Sun (the ecliptic). The solstices and equinoxes are labelled here for the seasons in the northern hemisphere.

Conventionally, four seasons are identified – spring, summer, autumn and winter – but there are no strict divisions between them and seasonal conditions may vary considerably from year to year.

second A unit of time measurement defined in the International System of Units (Système International, or SI) as 'the duration of 9,192,631,770 cycles of radiation corresponding to the transition between two hyperfine levels of the ground state of the caesium 133 atom'. ►*arc second*.

second contact In a total or annular ►*eclipse* of the Sun, the point when the edges of the Moon's disc and the Sun's ►*photosphere* are in contact at the start of totality or the annular phase. In a lunar eclipse, second contact occurs when the Moon just enters completely the full shadow (umbra) of the Earth. The term may also be used to describe the similar stage in the progress of a ►*transit* or ►*occultation*.

secular   Continuing, or changing in a non-periodic way, over a long period of time.

secular acceleration   The systematic increase in the velocity of the Moon in its orbit around the Earth, a result of tidal interaction with the Earth and the gravitational attraction of other planets.

secular parallax   The angular displacement over time of a star's position caused by the Sun's motion through space relative to the ►*local standard of rest*. Measurement of the secular parallax provides a method of determining the distance to nearby star groups, on the assumption that the individual motions of the stars in such a group are random with an average of zero. The Sun's relative velocity, also needed in the calculation, is 19.5 km/s, which amounts to 4.11 AU per year.

sedimentary rock   A rock made of fragments of pre-existing rock (or, on Earth, the hard parts of dead organisms) that have been deposited as a sheet, for example on a sea-bed.

seeing   The effect of random turbulent motion in the atmosphere on the quality of the image of an astronomical object. In conditions of good seeing, images are sharp and steady; in poor seeing, they are extended and blurred and appear to be in constant motion. Amateur astronomers sometimes use a scale of seeing quality indicated by Roman numerals. I is the best, II–III typical, IV poor and V extremely bad. The scale is due to Eugenios Antoniadi (1870–1944). Seeing may also be described quantitatively in ►*arc seconds*.

selenocentric   Centred on the Moon.

selenography   The study of the surface features and topography of the Moon.

selenology   The study of lunar rocks and of the surface and interior of the Moon: the lunar equivalent of geology.

semidetached system   A binary star system in which one star has expanded to the extent that its outer layers are pulled away by the gravitational attraction of the companion star. Material streams on to the companion from the larger star, which is described as having filled its ►*Roche lobe*.

semimajor axis (symbol *a*)   Half the length of the maximum dimension of an ►*ellipse*. ► *orbital elements*.

semiregular variable   A member of the group of pulsating variable stars that show some periodicity in their variations but are nevertheless unpredictable. The magnitude range is typically only one or two magnitudes, and the periodicity can range from a few days to several years. A number of subgroups have been distinguished. Types A and B are red giants, type C red supergiants and type

D giants and supergiants of ➤*spectral types* F, G and K. An example is Mu (μ) Cephei, known as the 'Garnet Star'.

separation (sep.)   The angular distance between two stars in a visual ➤*binary star* system, measured in arc seconds. It is one of two standard measurements used to record the relative positions of the members of a binary system, the other being the ➤*position angle*.

Serpens (The Serpent)   One of the 48 constellations listed by Ptolemy (*c.* AD 140) and unique in that it is split into two parts, one either side of Ophiuchus (The Serpent Bearer). The two parts are known as Serpens Caput (the head) and Serpens Cauda (the tail). In all, Serpens contains nine stars brighter than fourth magnitude. ➤Table 4.

Serpent   English name for the constellation ➤*Serpens*.

Serpent Bearer   English name for the constellation ➤*Ophiuchus*.

Serrurier system   A structure for the open tube of a large reflecting telescope, designed to allow equal degrees of flexure when the orientation of the telescope is changed. It is impossible to make a completely rigid tube for the largest telescopes. The solution proposed by Mark Serrurier for the design of the 200-inch ➤*Hale Telescope* tube was to allow flexure in such a way that the optical alignment is not disturbed.

SEST   Abbreviation for Swedish–ESO Submillimetre Telescope. ➤*Submillimetre-wave astronomy*.

SETI   Abbreviation for *Search for Extraterrestrial Intelligence*. It is used as a general term rather than for any particular experiment.

setting circles   Graduated scales attached to the rotation axes of a telescope mounting that indicate the celestial coordinates of the point in the sky at which the telescope is pointed.

Seven Sisters   A popular name for the ➤*Pleiades* star cluster.

Sextans (The Sextant)   A faint constellation of the southern hemisphere introduced in the late seventeenth century by Johannes Hevelius, supposedly to commemorate the instrument he used to make astronomical observations. Its brightest star is magnitude 4.5. ➤Table 4.

Sextant   English name for the constellation ➤*Sextans*.

Seyfert galaxy   A type of galaxy with a brilliant point-like nucleus and inconspicuous spiral arms, first described by Carl Seyfert in 1943. The spectrum shows broad emission lines. About 1 per cent of spiral galaxies are Seyferts. Many are comparatively strong infrared sources; in some the central core is a weak radio source. Brightness variations in the nucleus are common.

Seyfert's Sextet (NGC 6027)   A group of apparently interacting galaxies in the constellation Serpens. It consists of five galaxies, together with a large cloud of gas ejected by the principal galaxy in the group. This galaxy, which is a spiral, and three lenticular galaxies in the group are interacting gravitationally and lie at a distance of 260 million light years. The fifth galaxy is a spiral five times further away, coincidentally lying in the same part of the sky.

shadow bands   A phenomenon sometimes observed briefly just before and just after totality in the course of a total solar eclipse. Irregular bands of shadow, a few centimetres wide and up to a metre apart, are seen to move over the ground. The mechanism is not fully understood but may involve refraction in the atmosphere of the light from the thin crescent of the Sun. They are seen only if the sky is very clear.

Shapley–Ames Catalog   A catalogue of 1,249 galaxies brighter than thirteenth magnitude, prepared from a photographic survey undertaken between 1930 and 1932 at Harvard College Observatory. It was published in the *Annals* of the Observatory, as volume 88 part 2.

Shaula (Lambda Scorpii; λ Sco)   The second-brightest star in the constellation Scorpius, marking the Scorpion's sting. It is a ►*B star* of magnitude 1.6. ►Table 3.

shell star   A ►*B star* with a characteristic spectrum in which sharp absorption lines, flanked by emission 'wings', are superimposed on a normal spectrum of broad absorption lines. The spectrum can be explained by the presence of a ring of circumstellar material, probably created as a result of the star's rapid rotation. ►*Pleione* in the Pleiades is an example.

Shemakha Astrophysical Observatory   A research institute of the Academy of Sciences of Azerbaijan, established in the 1960s. It is located 22 kilometres (14 miles) from the village of Shemakha, at a height of 1,400 metres (4,600 feet) in the Caucasus. The main instrument is a 2-metre (80-inch) reflector.

shepherd satellites   Natural planetary satellites, often in pairs, whose gravitational influence appears to hold a planetary ring in place, preventing it from dispersing. Prometheus and Pandora are the shepherd satellites for the F-ring of ►*Saturn*.

shergottite   A type of stony ►*meteorite*, consisting of basalt-like rock. Together with nakhlites and chassignites, shergottites belong to the class of ►*SNC meteorites*, believed to have originated on the surface of Mars. The name is derived from Shergotty in India where such a meteorite fell.

Shield   English name for the constellation ►*Scutum*.

shield volcano   A large volcano with gently sloping sides, built up from successive lava flows from a single vent. Individual layers may amount to only a few metres but they can build up to create a very high mountain. Typically, the slope of the sides is under 10°. At the top is found a large, shallow, flat-floored crater called a *caldera*.

Ship   English name for the obsolete constellation ➤*Argo Navis*.

shock wave (shock front)   A discontinuity in density and pressure propagating in a solid, liquid or gas at supersonic velocity. Shock waves are normally caused by impacts or explosions.

shooting star   A popular name for a ➤*meteor*.

short-period comet   A ➤*comet* in an elliptical orbit, with a period of several years or decades, comparable with the orbital periods of the planets. Short-period comets have been perturbed into their present orbits by the gravitational influence of the planets, particularly Jupiter, during close encounters. Two-thirds of all short-period comets are in orbits that extend no more than one astronomical unit beyond the orbit of Jupiter. It is suspected that they originate in the ➤*Kuiper Belt*.

short-period variable   A vague term for a regular ➤*variable star* with a relatively short period.

Sickle   The ➤*asterism* formed by the stars Alpha (α), Eta (η), Gamma (γ), Zeta (ζ), Mu (μ) and Epsilon (ε) in the constellation ➤*Leo*, so named because of its shape.

side-lobe   An unwanted response outside the main beam of a ➤*radio telescope*. This fundamental instrumental limitation can cause problems in interpreting radio source maps, but can be minimized in computer processing the data.

sidereal   Pertaining to the stars.

sidereal day   The Earth's rotation period with respect to the stars (considered as a fixed frame of reference for this purpose), defined formally as the interval of time between two consecutive transits of the catalogue equinox (i.e. the zero of right ascension). The length of the sidereal day is 23 hours 56 minutes and 4 seconds.

sidereal month   The orbital period, relative to the stars, of the Moon around the Earth. The length of the sidereal month is 27.321 66 days.

sidereal period   The time taken by a planet or satellite to complete an orbit about its primary, measured relative to the stars.

sidereal rate   The rate at which an equatorially mounted telescope has to be driven around the polar axis in order to compensate exactly for the Earth's rotation and so keep the telescope directed at the same point on the sky. The rate is one revolution in 23 hours 56 minutes and 4 seconds.

sidereal time   Time measured by the rotation of the Earth with respect to the stars (rather than the rate relative to the Sun, which is the basis for civil time). *Local sidereal time* at a particular place is given by the right ascension of the meridian. Thus, the sidereal time is a direct indication of whether a celestial object of known right ascension is observable at that instant. For that reason, observatories are normally provided with *sidereal clocks*.

sidereal year   The orbital period of the Earth around the Sun relative to the stars. Its length is 365.256 36 days, which, because of the effects of ➤*precession*, is 20 minutes longer than the ➤*tropical year*.

Sidereus nuncius   A book by Galileo (1564–1642), published in 1610, in which he announced his first astronomical observations made with the aid of a telescope. The Latin title means 'The Starry Messenger'.

siderite   An alternative, largely obsolete name for an iron ➤*meteorite*.

siderolite   An alternative, largely obsolete name for a ➤*stony-iron meteorite*.

siderostat   A flat mirror mounted and driven so that it can continuously reflect the light of a star on to a fixed piece of equipment, compensating for the apparent motion of the celestial sphere.

Siding Spring Observatory   ➤*Mount Stromlo and Siding Spring Observatories*.

Siegena   Asteroid 386, diameter 204 km, discovered in 1894 by Max Wolf.

Sikhote-Alin shower   A major meteorite fall on 12 February 1947 in eastern Siberia. The largest recovered meteorite weighed 1,745 kg, but it has been estimated that thousands of pieces fell, weighing up to a total of 100 tonnes. Much of it remains unrecovered.

silicate   A rock-forming mineral containing silicon, oxygen and one or more metals, with or without hydrogen.

silicon star   A type of peculiar ➤*A star* in which the absorption lines of silicon are particularly enhanced. ➤*Ap star*.

SIMBAD   An astronomical database compiled at the ➤*Centre de Données Astronomiques*. The acronym is said to be derived from 'Set of Identifications, Measurements and Bibliographic references of Astronomical Data'. The database aims to provide cross-identifications for stars, galaxies, clusters, nebulae and other objects, together with fundamental, observational and bibliographic data.

singularity   A mathematical concept that can be visualized as a warped region of spacetime where quantities may become infinite so that ordinary physical laws cease to apply. The ►*Big Bang* is thought to have originated from such a singularity.

Sinope A small satellite of Jupiter (number IX) discovered in 1914 by S. B. Nicholson. ►Table 6.

sinus Literally 'bay', a term used for certain features with a bay-like appearance along the borders of the ►*mare* areas of the Moon.

Sirius (Alpha Canis Majoris; α CMa) The brightest star in the constellation Canis Major and, at magnitude −1.46, the brightest star in the sky. It is a visual binary with an orbital period of 50 years, the primary (A) being an ►*A star* and the secondary (B) an eighth magnitude ►*white dwarf*. Sirius B was first detected optically in 1862, and its nature was determined from its spectrum in 1925. Sirius lies at a distance of 8.7 light years and is the seventh-nearest star to the solar system.

The name is derived from Greek and means 'scorching', a reference to the star's brilliance. Sirius is also known as 'the Dog Star', from the constellation in which it lies. ►Table 3.

Sirrah An alternative name for the star ►*Alpheratz*.

SIRTF Abbreviation for ►*Space Infrared Telescope Facility*.

Sisyphus Asteroid 1866, diameter 7.6 km, discovered in 1972 by P. Wild. It is notable for its comet-like orbit, inclined at 41° to the plane of the solar system, which occasionally brings it relatively close to the Earth.

Skylab An American space station, launched into Earth orbit in May 1973. Three crews, each of three men, were sent to the station for periods of several weeks between 1973 and 1974. Astrophysical and solar studies were undertaken, as well as experiments on the effects on the astronauts of what were then record periods under weightless conditions in orbit. The station burnt up on re-entering the atmosphere in 1979.

SLR Abbreviation for ►*satellite laser ranging*.

Small Astronomy Satellites (SAS) Name of three NASA spacecraft deployed in the 1970s for X-ray and gamma-ray astronomy.

SAS-1, launched in 1970 from the Italian San Marco platform off the coast of Kenya, was the first satellite dedicated to ►*X-ray astronomy*. After launch, it was given the name Uhuru, which means 'freedom' in Swahili, because the launch date of 12 December coincided with the seventh anniversary of Kenya's independence.

SAS-2, launched in 1972, was the first satellite to carry a gamma-ray detector. SAS-3, which followed in 1975, carried further X-ray experiments.

small circle   A circle on the surface of a sphere that divides the sphere into two unequal parts. Circles of constant latitude or declination, other than the equator, are small circles. ➤*great circle.*

Small Magellanic Cloud (SMC)   ➤*Magellanic Clouds.*

small molecular cloud   ➤*molecular cloud.*

Smithsonian Astrophysical Observatory (SAO)   A research establishment founded by the US Smithsonian Institution in 1890. Initially it had modest facilities in Washington, DC. In 1955, under the directorship of Fred Whipple, its headquarters were moved to the grounds of ➤*Harvard College Observatory* (HCO) in Cambridge, Massachusetts, and its activities were expanded. In 1967, an observatory was established at Mount Hopkins in Arizona, now known as the ➤*Fred Lawrence Whipple Observatory*. In 1973, the Harvard–Smithsonian Center for Astrophysics was formed by combining the resources of the SAO and HCO under one director, George Field.

SMM   Abbreviation for ➤*Solar Maximum Mission.*

SNC meteorite   A member of a small group of unusual basaltic ➤*meteorites*, apparently originating in the mantle of a ➤*parent body*. The initials are taken from the names given to subgroups: shergottites, nakhlites and chassignites. It is possible that these meteorites come from the surface of Mars.

SN 1987A   ➤*Supernova 1987A.*

Sobieski's Shield   Former English name for the constellation ➤*Scutum*. It is no longer used but is encountered in nineteenth-century books.

Socorro   The location in New Mexico, USA, of the National Radio Astronomy Observatory's ➤*Very Large Array* (VLA).

SOFIA   A 2.5-metre (98-inch) telescope mounted in a Boeing 747 aircraft, due to begin operations in 2001. SOFIA is an acronym for 'Stratospheric Observatory for Infrared Astronomy'. It is the successor to the ➤*Kuiper Airborne Observatory* (KAO).

Flying at between 41,000 and 45,000 feet, SOFIA will operate above 85 per cent of the Earth's atmosphere and above 99 per cent of the atmospheric water vapour, which interferes with ➤*infrared astronomy*. Its capability will encompass the visible, infrared, submillimetre and microwave regions of the spectrum.

soft gamma repeater   A class of rare gamma-ray sources that are observed to emit infrequent bursts of radiation. For example, one was seen to have three bursts in 1979 and a further three in 1992. Each burst lasts a maximum of a few seconds. They are thought to be young, rapidly spinning ➤*neutron stars*. Two are known to be located within ➤*supernova remnants*.

SOHO   Abbreviation for ➤*Solar and Heliospheric Observatory*.

Sojourner   The name of the small roving vehicle carried by the Mars mission ➤*Mars Pathfinder*.

sol   A martian 'day', which is 24 hours 37 minutes and 22.6 seconds in length.

solar   Of or pertaining to the Sun.

Solar-A   ➤*Yohkoh*.

solar activity   A variety of energetic phenomena on the Sun that vary cyclically in frequency and intensity over time. The most obvious cycle of change takes about 11 years, though there is also evidence for longer ones. Phenomena such as ➤*coronal mass ejections*, ➤*flares*, ➤*sunspots*, ➤*prominences* and ➤*faculae* are manifestations of solar activity. ➤*solar cycle*.

Solar and Heliospheric Observatory (SOHO)   A scientific satellite launched by the ➤*European Space Agency* on 2 December 1995 with an intended lifetime of two years. It was placed in orbit around the Sun at the point known as the L1 ➤*Lagrangian point*, where the gravitational forces of the Earth and Sun are equal. Its twelve instruments were designed to investigate the solar atmosphere and how it is heated, solar oscillations, how the Sun expels material into space, the structure of the Sun and processes operating within it.

solar apex   ➤*apex*.

solar constant   The total solar power incident per unit area at the top of the Earth's atmosphere, corrected to the mean Sun–Earth distance. Its value is about 1.35 kW/m² but, despite its name, it is not precisely constant. The value changes slightly over the course of a ➤*solar cycle* – a large sunspot group decreases it by about 1 per cent – and it may also be subject to longer-term variations.

solar corona   ➤*corona*.

solar cycle   The periodic variation in the amount of ➤*solar activity*, particularly the number of ➤*sunspots*. The period of the cycle is about 11 years, though it has been closer to 10 years during the twentieth century.

At the commencement of a new cycle there are few, if any, spots on the Sun. The first ones for the new cycle erupt around heliographic latitudes 35°–45° north and south; over the course of the cycle, subsequent spots appear closer to the equator, finishing at around 7° north and south. This pattern can be demonstrated graphically as a ➤*butterfly diagram*.

It is generally thought that the solar cycle is caused by an interaction between the 'dynamo' responsible for the Sun's magnetism and the Sun's rotation. The Sun does not rotate as a solid body: the equatorial regions rotate fastest and this amplifies the magnetic field, which eventually bursts into the photosphere, causing sunspots. At the end of each cycle the overall magnetic field reverses, giving a total period of 22 years, which is known as the *Hale cycle*.

solar day   The same as apparent solar day. ➤*apparent solar time.*

solar eclipse   ➤*eclipse.*

solar flare   ➤*flare.*

Solar Maximum Mission (SMM)   A US satellite launched in February 1980 for studying the Sun during a period of maximum ➤*solar activity.* It failed after nine months, but repairs were successfully made by a Space Shuttle crew in 1984, and the satellite was redeployed. It re-entered the Earth's atmosphere in 1989.

solar nebula   The cloud of interstellar gas and dust that condensed to form the Sun and solar system about five billion years ago.

solar prominence   ➤*prominence.*

solar system   The Sun, together with the planets and moons, comets, asteroids, meteoroid streams and interplanetary medium held captive by the Sun's gravitational attraction. The solar system is presumed to have formed from a rotating disc of gas and dust created around the Sun as it contracted to form a star, about five billion years ago.

The planets and asteroids all travel around the Sun in the same direction as the Earth, in orbits close to the plane of the Earth's orbit and the Sun's equator. The planetary orbits lie within 40 AU of the Sun, though the Sun's sphere of gravitational influence can be considered to be much greater. Comets seen in the inner solar system may originate in the ➤*Oort cloud*, many thousands of AU away.

solar time   ➤*apparent solar time, mean solar time.*

solar tower   A type of telescope used exclusively for observing the Sun. To form an image of the Sun's disc on which detail can readily be distinguished, a long focal length (of order 100 metres) is desirable. Close to the ground, the heating effect of the Sun causes a layer of hot, turbulent air, which makes images formed by mirrors near the ground unsteady. To overcome this problem, the ➤*heliostat* or ➤*coelostat* used to deflect an image of the Sun into the telescope is placed on a tall tower. ➤*vacuum tower telescope.*

solar wind   A stream of particles, primarily protons and electrons, flowing outwards from the Sun at up to 900 km/s. The solar wind is essentially the hot solar ➤*corona* expanding into interplanetary space.

Solis Planum (Solis Lacus)   An ancient volcanic plain on Mars, lying to the south of Valles Marineris. To the visual observer, the area has a variable dark spot (the 'lake') whose appearance has earned the feature the nickname 'the eye of Mars'.

solstices   The two points on the ►*ecliptic* where the Sun reaches its maximum and minimum ►*declination*, or the times at which the Sun is at these points. They lie approximately midway between the vernal and autumnal ►*equinoxes*.

The solstices occur on about 21 June and 21 December. At the summer solstice, the Sun reaches its highest altitude in the sky and the duration of daylight is a maximum. At the winter solstice the Sun's altitude at noon is the lowest as seen from a particular latitude, and the duration of daylight is a minimum. The summer solstice in the northern hemisphere (June) is the winter solstice in the southern hemisphere and vice versa.

solstitial colure   A ►*great circle* on the celestial sphere passing through the two poles and the ►*solstices*.

Sombrero galaxy (M104; NGC 4594)   A spiral galaxy, seen edge-on, in the constellation Virgo. Its marked central bulge and conspicuous lane of dark obscuring dust give it a superficial resemblance to a wide-brimmed hat.

Sonneberg Observatory   A research institute near the German town of Sonneberg in the Thuringia region, noted particularly for the searches for variable stars carried out there. It was opened in 1925.

source   A term applied to a celestial object from which radiation is received when its true nature is unknown, or in order to encompass a variety of objects without being specific as to their nature. Thus, terms such as 'radio source', 'infrared source' and 'X-ray source' are used to mean an unspecified celestial source of the particular kind of radiation.

source count   A graph used as a test of cosmological models, in which the numbers of cosmic radio sources seen are plotted against their apparent luminosities.

Source counts are important for determining the structure of the universe. William Herschel (1738–1822) used star counts in an attempt to elucidate the structure of the Milky Way. He failed because he was unaware of the degree of obscuration by interstellar matter.

The technique was revived in the 1950s and 1960s by radio astronomers, particularly Martin Ryle at Cambridge. Plots were made of the numbers of extragalactic radio sources found within given magnitude ranges over the whole sky. All these sources are ►*radio galaxies* and ►*quasars*. The way the number of sources counted increases as the telescope probes to fainter detection limits can in principle be used to distinguish between different cosmological theories. From the outset, Ryle asserted that source counts favoured the ►*Big Bang* model, in contrast to the ►*steady-state theory* advocated by the Cambridge theorist, Fred Hoyle. It is now accepted that it is difficult to disentangle the results of the evolution of radio sources from cosmological effects.

South African Astronomical Observatory (SAAO)   A national optical astronomy facility located at Sutherland and Cape Town in South Africa. It operates under the auspices of the Foundation for Research Development.

The SAAO was formed in 1972 by merging the old Royal Observatory at Cape Town and the Republic Observatory, Johannesburg. Some of their telescopes were moved to the Sutherland site and, in 1974, a 1.9-metre (75-inch) telescope was purchased from the Radcliffe Observatory, Pretoria, and also moved. Some smaller instruments remain at Cape Town, which is the administrative headquarters.

South Atlantic anomaly   A region over the South Atlantic Ocean where the lower ➤*Van Allen belt* of energetic electrically charged particles is particularly close to the surface, presenting a hazard for artificial satellites.

Southern Cross   English name for the constellation ➤*Crux.*

Southern Crown   English name for the constellation ➤*Corona Australis.*

Southern Fish   English name for the constellation ➤*Piscis Austrinus.*

Southern Pleiades   Popular name for IC 2602, a large and bright ➤*open cluster* of stars in the constellation Carina. The star Theta Carinae (magnitude 2.8) is at the centre of the cluster and several other member stars are also visible to the naked eye. The whole cluster is about one degree across on the sky.

Southern Triangle   English name for the constellation ➤*Triangulum Australe.*

southing   The ➤*transit* of a celestial object across an observer's ➤*meridian*, i.e. the moment at which it is due south.

Soyuz   A Soviet spacecraft, used to carry up to three cosmonauts. ➤*Apollo–Soyuz project.*

space   The regions between the planets and stars, excluding their immediate atmospheres.

Space Infrared Telescope Facility (SIRTF)   A NASA orbiting telescope for infrared astronomy, with a launch planned for late 2001 or 2002. It will have a 0.85-metre (33-inch) mirror and be equipped for imaging and spectroscopy in the wavelength band between 3 and 180 microns.

Spacelab   A small space station, built by the European Space Agency to fit in the payload bay of the ➤*Space Shuttle.*

space research   The branch of scientific endeavour dedicated to all aspects of manned and unmanned space flight.

Space Shuttle A US reusable space vehicle that takes off like a rocket but lands like an aircraft on a runway. The first flight was made by *Columbia* on 12 April 1981. The second Shuttle, *Challenger*, was destroyed in an explosion shortly after its tenth launch in 1986. *Discovery* and *Atlantis*, the third and fourth Shuttles, first flew in 1984 and 1985 respectively.

space telescope A telescope placed in orbit around the Earth so as to be above the atmosphere. ➤*Hubble Space Telescope*.

Space Telescope Science Institute A (US) research institute located in Baltimore, Maryland, operated by the ➤*Association of Universities for Research in Astronomy* under contract to NASA. Its main responsibilities are to manage the science programme of the Hubble Space Telescope, process data from the telescope and coordinate its operations with the Space Telescope Operations Control Center.

spacetime A unified multi-dimensional framework within which it is poss-ible to locate events and describe the relationships between them in terms of spatial coordinates and time. The concept of spacetime follows from the observation that the speed of light is invariant – i.e. it does not vary with the motion of the emitter or the observer. Spacetime allows a description of reality that is common for all observers in the universe, regardless of their relative motion. Intervals of space and time considered separately are not the same for all observers, but the spacetime interval, defined by

$$(\text{spacetime interval})^2 = (\text{time interval})^2 - (\text{space interval})^2$$

is invariant. In ➤*General Relativity*, gravitation is described in terms of curvature of spacetime.

Spacewatch A project to search for ➤*near-Earth objects*, established in the 1980s. It has carried out a full-time survey since September 1990 using the 0.91-metre (36-inch) Spacewatch Telescope of the ➤*Steward Observatory* at ➤*Kitt Peak*.

space weather The variation of physical conditions in the space environment immediately around the Earth and between the Earth and the Sun as the result of variations in the ➤*solar wind*, ➤*coronal mass ejections* and other phenomena related to ➤*solar activity*.

Special Astrophysical Observatory (SAO) The primary observing facility of the Russian Academy of Sciences for optical and radio astronomy. It is located in the Caucasus region between the Black Sea and the Caspian Sea.

The optical observatory at Nizhnij Arkhyz near Zelenchukskaya is the location of the largest single-mirror optical telescope in the world, the Bolshoi Teleskop Azimutalnyi (Large Altazimuth Telescope), which has a 6-metre (236-inch) primary mirror. Completed in 1975, it was the very first large telescope to be designed on an ➤*altazimuth mounting*. There are also a 1-metre (40-inch) ➤*Ritchey– Chrétien telescope* and a 60-cm (24-inch) ➤*Cassegrain* reflector.

The ➤*RATAN-600* radio telescope is located on the outskirts of Zelenchukskaya. There is also a radio branch of the SAO in St Petersburg.

Special Relativity A theory developed by Albert Einstein (1878–1955) and published in 1905, which describes how physical phenomena are observed between unaccelerated (inertial) reference frames in relative motion.

The Special Theory of Relativity was a direct consequence of the fact, proved experimentally, that the speed of light (c) in a vacuum is measured to be the same by all observers, regardless of their state of motion or the motion of the source of light. This had been verified by the famous *Michelson– Morley experiment* in 1887. It was also predicted theoretically (1873) by the equations describing electromagnetic radiation derived by James Clerk Maxwell. The other underlying principle of the theory is the so-called *principle of relativity*. This states that no physical experiment can be devised to detect a uniform state of motion. In other words, neither location in space and time, nor uniform motion, affects the description of physical reality. The concept of motion through some absolute framework of space and time was swept away by the ideas of relativity.

A number of important consequences flow from these principles, relating to the way time intervals, lengths and masses are measured between reference frames in relative motion. Time intervals appear extended (*time dilation*), lengths foreshortened (the *Lorentz contraction*) and masses increase in reference frames moving relative to the observer, though the effects become significant only when the relative speed is a significant fraction of that of light. Using $\beta = $ relative velocity/c, and the subscript o to mean 'in the reference frame of the observer', the transformations are:

$$T = T_0/\sqrt{(1-\beta^2)} \qquad L = L_0\sqrt{(1-\beta^2)} \qquad M = M_0\sqrt{(1-\beta^2)}.$$

The three dimensions of space and time are considered to describe four-dimensional spacetime with the characteristic that the interval,
$\Delta s = \sqrt{(c^2\Delta t^2 - \Delta x^2 - \Delta y^2 - \Delta z^2)}$ is the same for all reference frames.

Another result from the theory is the concept of rest mass and the equivalence of mass and energy as expressed in the relationship $E = mc^2$. This formula gives the amount of energy released when mass is annihilated.

The predictions of Special Relativity have been totally verified, particularly in the physics of atoms and elementary particles. ➤*General Relativity*.

speckle interferometry A technique for combating the blurring of star images caused by turbulence in the Earth's atmosphere. Because of this problem, the resolution theoretically possible with a particular telescope is rarely achieved. Speckle interferometry can be used to improve resolution and to measure the diameters of some stars.

A number of very-short-exposure images are made, typically for 0.02 second. In these, the image of a star appears to be broken up into a pattern of bright speckles by the bending of the starlight as it passes through turbulent cells in the atmosphere. The speckled pattern changes rapidly, so any longer exposure is blurred. The information contained in the individual speckles of a set of images is combined mathematically to extract details about the star.

spectral class ➤*spectral type*.

spectral index A quantity that indicates how the flux density (S) of continuum emission varies with frequency (ν): S is proportional to ν^{α}, where α is the spectral index. It is used particularly in connection with radio sources.

spectral line In a ➤*spectrum*, an absorption or emission feature covering a relatively narrow wavelength range. Line spectra result when there is a transition between two of the discrete energy levels of an atom or ion. A transition to a lower energy state, with the emission of a photon, results in an ➤*emission line*. A transition to a higher energy state, with the absorption of a photon, results in an ➤*absorption line*.

spectral type A classification assigned to a star according to the appearance of its spectrum. The spectral type is based primarily on a temperature sequence; a luminosity class may also be specified. Additional information about a star's spectrum, such as the appearance of emission lines or unusually strong metal lines, may also be indicated.

The present alphabetical notation is a legacy from the first comprehensive attempt at classification, undertaken at Harvard College Observatory, financed from the estate of Henry Draper and published in 1890. The classes, originally designated A–Q, were subsequently rationalized and re-ordered as a temperature sequence, resulting in the final basic types still used: O, B, A, F, G, K and M (see the table). The main classes may be further broken down into up to 10 subdivisions, indicated by the numbers 0 to 9 (e.g. A0, K5).

Spectral type	Temperature range	Principal features of visible spectrum
O	>25,000 K	Relatively few absorption lines. Lines of ionized He, doubly ionized N, triply ionized Si. H lines weak.
B	11,000–25,000 K	Lines of neutral He, singly ionized O and Mg. H lines stronger than in O stars.

Spectral type	Temperature range	Principal features of visible spectrum
A	7,500–11,000 K	Strong H lines. Lines of singly ionized Mg, Si, Fe, Ti, Ca, etc. and some neutral metals.
F	6,000–7,500 K	H lines weaker and neutral metal lines stronger than in A stars. Lines of singly ionized Ca, Fe, Cr.
G	5,000–6,000 K	Lines of ionized Ca most conspicuous features. Many lines of ionized and neutral metals. CH bands.
K	3,500–5,000 K	Neutral metal lines predominate. CH bands present.
M	<3,500 K	Strong lines of neutral metals and molecular bands of TiO.

The classification system continues to be developed and refined as research produces more detailed information. Other classifications include ➤*S stars* and the ➤*carbon stars,* formerly called R and N, now arranged in a sequence from C0 to C9 that roughly parallels the non-carbon stars G4 to M in temperature.

Various prefixes and suffixes are used to give additional information about spectra. Some of the more common ones are:

c	sharp lines
d	dwarf = main-sequence star
D	white dwarf
e	emission (hydrogen emission in O stars)
em	emission in metal lines
ep	peculiar emission
eq	emission with shorter wavelength absorption
f	emission by helium and neon in O stars
g	giant
k	interstellar lines
m	strong metallic lines
n	diffuse lines
nn	very diffuse lines
p	peculiar spectrum
s	sharp lines

sd	subdwarf
wd	white dwarf
wk	weak lines

In 1943, W. W. Morgan, P. C. Keenan and E. Kellman defined spectral criteria for luminosity classes as well as selecting sample stars to act as standards for each of the Harvard subclasses. The luminosity classes are indicated by capital Roman numerals (see illustration):

Ia	Luminous supergiants
Ib	Less luminous supergiants
II	Bright giants
III	Normal giants
IV	Subgiants
V	Dwarfs/Main Sequence

Two further classes were introduced later, but are now rarely used:

VI	Subdwarfs
VII	White dwarfs

These are written after the temperature class and before any suffix letter. For example, a B3 giant with emission lines would be classified as B3IIIe. ►*Hertzsprung–Russell diagram.*

spectrogram   A permanent record of a ►*spectrum*, obtained photographically or digitally by means of an electronic detector.

spectrograph   An instrument for obtaining a permanent record of a ►*spectrum.*

spectroheliogram   A monochromatic image of the Sun produced by means of a ►*spectroheliograph* or by the use of a narrow-band filter.

spectroheliograph   An instrument for obtaining images of all or part of the Sun in monochromatic light. An entrance slit selects the slice of the Sun to be observed. The light is dispersed into a spectrum by means of a diffraction grating, and a second slit is used to select the narrow region of the spectrum to be observed. By using the entrance slit to scan the whole of the solar disc, a full image can be generated.

spectrometer   An instrument for observing a spectrum and measuring features in it by direct observation.

spectroscope   An instrument for observing a ►*spectrum* visually. The term is sometimes also used as an alternative to ►*spectrograph.*

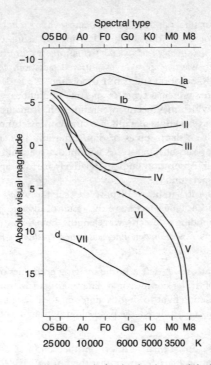

Spectral type. The relationship between the luminosity classes and absolute visual magnitude across the range of temperatures, O to M.

spectroscopic binary A ➤*binary star* whose nature is revealed from its spectrum even though the components cannot be resolved visually. In a *double-lined spectroscopic binary*, two superimposed spectra can be detected. The lines are shifted relative to each other in a periodic way by the ➤*Doppler effect*, as the two stars orbit about their common centre of mass. In a *single-lined spectroscopic binary*, the two stars differ greatly in luminosity so only the spectrum of the brighter component can be discerned. However, the lines are seen to change wavelength in a periodic way when measured in relation to a standard comparison spectrum.

spectroscopy The study and interpretation of ➤*spectra*, especially with a view to determining the chemical composition of and physical conditions in the source of radiation.

spectrum (pl. spectra) The result of dispersing a beam of electromagnetic radiation so that components with different wavelengths are separated in space, in order of increasing or decreasing wavelength. The most familiar example of a spectrum is the natural rainbow, which occurs when sunlight is dispersed into its component colours by the combined prism-like action of raindrops. The full spectrum of ►*electromagnetic radiation* encompasses, in order of decreasing wavelength, radio waves, microwaves, infrared radiation, visible light, ultraviolet radiation, X-radiation and gamma-radiation.

There are three principal types of spectra: continuous, emission line and absorption line, which may occur in combination. If intensity of radiation is plotted against wavelength as a graph, a continuous spectrum shows a smooth distribution, with no sharp spikes or dips. Emission lines show as relatively narrow spikes or peaks of intensity. A continuous spectrum may or may not be present as well. Absorption lines are relatively narrow dips in the intensity of a continuous spectrum.

Continuous spectra result from processes such as ►*black body radiation* or ►*synchrotron radiation*. Line spectra are the result of discrete packets of energy (quanta), corresponding to precise wavelengths, being emitted or absorbed in atoms or molecules. ►*absorption line, absorption spectrum, continuous spectrum, emission line.*

spherical aberration A defect in the imaging properties of a lens or mirror caused by parts of the lens or mirror at different distances from the optical axis reflecting or refracting light to slightly different focal lengths. It is a defect of spherical surfaces, not suffered by paraboloids, though both are subject to ►*coma.*

spherical space Space that is uniform in the same way as ►*Euclidean space*, but with the additional characteristic that there is no parallel to a straight line through a given point. Mathematically, the space has positive curvature, and in it the angles of a triangle add up to more than 180°.

Spica (Alpha Virginis; α Vir) The brightest star in the constellation Virgo, magnitude 1.0. It is an ►*eclipsing binary*, varying by about 0.1 magnitude with a period of 4.014 days. The primary is a blue-white ►*B star* with a mass about eleven times the Sun's. The name means 'an ear of corn'. ►Table 3.

spicules Spike-like structures in the Sun's ►*chromosphere* that may be observed at or near the limb. They change rapidly, having a lifetime of five to ten minutes. Typically, spicules are 1,000 kilometres across and 10,000 kilometres long. They are not distributed uniformly on the Sun but concentrated along the cell boundaries of the ►*supergranulation* pattern.

spider The supporting bars of a secondary mirror in the tube of a reflecting telescope. Diffraction caused by the presence of a spider results in spikes radiating from the photographic images of bright stars.

spin casting A technique developed at the University of Arizona for manufacturing large paraboloid telescope mirrors. Molten glass is rotated as it cools in order to create a paraboloidal surface. Spin cast mirrors do not need grinding – only polishing.

Spindle Galaxy   Popular name for the edge-on galaxy NGC 3115 in the constellation Sextans, the shape of which is reminiscent of a spindle wound with yarn. It is a highly evolved galaxy with no obvious evidence of dust.

spiral galaxy   A ►*galaxy* with spiral arms. Edwin Hubble divided spiral galaxies into two broad groups: those with a central bar (SB galaxies) and those without (S). Each group he further subdivided into three categories, a, b and c. Sa and SBa galaxies have tightly wound arms and a relatively large central bulge. Sc and SBc galaxies have loose arms and a small central bulge, while types Sb and SBb are intermediate between the two extremes.

Our own ►*Galaxy* (the Milky Way) is a spiral, possibly with a small central bar. Its structure is fairly typical: young stars and interstellar material are concentrated in a disc, particularly in the spiral arms, and there is a surrounding spherical halo containing old stars and globular clusters.

The spiral arms are not permanent rigid structures but are thought to have the character of ►*density waves*. As stars and interstellar material orbit around the centre of the galaxy, they create regions of enhanced density in a spiral pattern. The arms are actually made up of different stars and gas clouds at different eras of time. ►*Hubble classification*.

sporadic meteor   A ►*meteor* that does not belong to an identified ►*meteor shower*.

Spörer's law   The tendency for sunspots to appear at latitudes that get lower as a particular ►*solar cycle* takes its course. The phenomenon is illustrated graphically in the so-called ►*butterfly diagram*.

s-process   A process of ►*nucleosynthesis* in which heavy elements are created from light ones by the successive capture of neutrons. The 's' stands for slow. In the s-process there is time for a newly formed nucleus to decay by the emission of an electron (beta decay) before another neutron is captured, in contrast with rapid neutron capture – the ►*r-process* – in which this is not the case. The s-process is believed to occur within stars of less than 9 solar masses during the ►*red giant* phase of evolution.

Some isotopes of heavier elements can be formed only by the s-process, and their abundance in the solar system is strong evidence that the s-process was responsible. These elements were created in a generation of stars that existed prior to the formation of the solar system. When those sufficiently massive to do so exploded as ►*supernovae*, they enriched the interstellar material with the products of nuclear processes from their interiors.

spur   A region of continuum radio emission appearing to protrude from the galactic plane in radio maps of the Milky Way. The most notable is the North

Galactic (or Polar) Spur, projecting northwards at galactic longitude 30°. It may be part of a ►*supernova remnant.*

Sputnik   A series of three unmanned spacecraft launched into Earth orbit by the Soviet Union in 1957 and 1958. *Sputnik 1*, launched on 4 October 1957, was the first man-made object to be launched into orbit around the Earth. For some time, the word 'Sputnik' was popularly used to mean any artificial satellite.

Square of Pegasus   An ►*asterism* in the form of a giant square, made up of the stars Alpha (α), Beta (β) and Gamma (γ) Pegasi and Alpha (α) Andromedae. Alpha Andromedae was formerly known as Delta (δ) Pegasi.

SS Cygni star   A type of ►*dwarf nova* that has outbursts lasting several days.

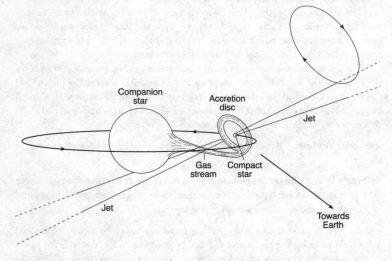

SS 433. Material from a large star pulled towards a companion neutron star forms a wobbling accretion disc. Material blasted from the centre of the disc creates two jets.

SS 433   A peculiar star, number 433 in the catalogue of stars showing hydrogen emission compiled by C. Bruce Stephenson and Nicholas Sanduleak, generally thought to be a binary system in which a ►*neutron star* is accreting material from a more massive normal companion.

SS 433 is located 18,000 light years away inside the old ►*supernova remnant* W 50, believed to be about 40,000 years old. It appears as a fourteenth magnitude star in the constellation Aquila. In 1976 it was found to be a source of X-rays, and radio emission was detected the following year. The optical spectrum reveals a complex situation with periodic variations and evidence for a pair of jets travelling at a quarter the speed of light. Detailed analysis of the spectrum has led to accurate determinations of the masses of the two stars. The compact component is 0.8 solar masses, ruling out the possibility of its being a black hole. The current mass of the companion, which is losing material rapidly, is put at 3.2 solar masses.

The neutron star and its O- or B-type companion orbit each other in a period of 13 days. Material from the companion streams into an ►accretion disc around the neutron star. Excessive heating causes some of the material to be blasted out from the central hole of the ring-shaped accretion disc, forming a pair of narrow jets. The disc precesses, wobbling slightly like a spinning top, over a period of 164 days. This creates a helical pattern in the radio emission from the jets and a regular 164-day cycle in the apparent velocity of the jets as viewed from Earth.

S star A cool giant star of basic ►spectral type K or M, which shows in its spectrum distinct absorption bands of the molecule zirconium oxide (ZrO). S stars also often have bands of lanthanum oxide (LaO), yttrium oxide (YO) and vanadium oxide (VO).

In normal M stars, the most prominent molecular bands are those of titanium oxide (TiO). The dominance of ZrO in S-type stars reflects a higher ratio of carbon to oxygen and a high abundance of zirconium. The zirconium and other heavier elements are the products of nuclear reactions in the star's interior which have been brought up to the surface.

star An intrinsically luminous ball of gas generating energy in its hot core through nuclear fusion processes.

The minimum mass required to form a star is about one-twentieth the mass of the Sun. Below this limit, the gravitational energy released when the mass condenses is insufficient to raise the temperature to the point at which the fusion of hydrogen to form helium can begin. The most massive stars known are about 100 solar masses. Mass is the prime factor determining the temperature and luminosity the star will have during its existence as a ►main-sequence star, when hydrogen in the core is its nuclear fuel.

Stars are predominantly hydrogen, with helium as the other major constituent. In the Sun, which is in many ways a typical star, 94 per cent of atoms are hydrogen, 5.9 per cent helium and less than 0.1 per cent other elements. By weight, 73 per cent is hydrogen, 25 per cent helium, 0.8 per cent carbon and 0.3 per cent oxygen, the remaining 0.9 per cent being all the other elements. ►binary star, Hertzsprung–Russell diagram, spectral type, stellar evolution, variable star.

star atlas A collection of charts showing the positions of the stars and other astronomical objects on the celestial sphere.

starburst galaxy A galaxy in which there is thought to be an exceptionally high rate of star formation. Starburst galaxies are characterized by excessive emission of infrared radiation, which may account for over 90 per cent of their total energy flux. These 'infrared galaxies' were discovered during a survey carried out in 1983 by IRAS, the ➤*Infrared Astronomical Satellite*.

star catalogue A compendium of information on stars, usually giving for each entry a position and magnitude for identification together with physical or observational data of some kind. Catalogues may contain all stars in the area covered down to some magnitude limit or stars selected on the basis of some property such as membership of a binary system.

star cloud An area of sky, particularly in the ➤*Milky Way*, where large numbers of stars are seen close together, giving a cloud-like effect.

star cluster A group of physically associated stars, presumed to share the same origin. There are two main types, ➤*open clusters* and ➤*globular clusters*. Very young stars are often found in loose groupings called ➤*associations*.

star diagonal ➤*diagonal* (1).

Stardust A NASA mission which will fly through the extended coma of the active comet Wild 2, taking images and returning a sample of its cometary dust to Earth. Launch is scheduled for February 1999. Using an Earth flyby for gravity assist, *Stardust* will reach the comet in 2004 and the sample will be returned in 2006.

starquake A sudden cracking in the outer crust of a neutron star, similar to an earthquake. A starquake changes the moment of inertia of the spinning star, leading to an abrupt change in its period, which is observed as a ➤*glitch*.

star tracker A small ➤*equatorial mounting*, motor-driven to compensate for the Earth's rotation, to which a camera may be attached.

star trail A bright streak on a time-exposure photograph of the night sky taken with a camera that has not been driven to follow the apparent motion of the stars. The trails are the elongated images of stars, recorded as the Earth rotates.

stationary point The point on the sky at which the apparent motion of a planet changes from being ➤*direct* to ➤*retrograde*, or vice versa.

statistical parallax The ➤*parallax* – and hence distance – determined for a group of stars by the statistical analysis of their ➤*proper motions*.

steady-state metric A ➤*metric* or formula for calculating proper intervals in the ➤*steady-state theory* of cosmology.

The ➤*Robertson–Walker metric* gives the method of calculating distances in a homogeneous and isotropic universe. The steady-state metric is a special case that accords with the ➤*perfect cosmological principle* and gives the relations for calculating proper distances in an unchanging universe. ➤*de Sitter universe*.

steady-state theory   One of two rival theories of cosmology of the mid-twentieth century, the other being the ➤*Big Bang* theory. The steady-state theory assumes that the universe is the same everywhere for all observers at all times. It accommodates the observed expansion of the universe by postulating that new matter is continuously created to fill the voids left as the already existing galaxies move apart. The discovery of ➤*cosmic background radiation* in 1963 was a major setback for the theory at a time when it had already been shown to be inconsistent with radio ➤*source counts*. A significant achievement was the boost it gave to the theory of ➤*nucleosynthesis* in stars. In the absence of a Big Bang, the heavy elements had to be made in exploding stars. That feature of the research, which is independent of what cosmological model is chosen, survives intact. ➤*perfect cosmological principle*.

stellar   Pertaining to stars.

Stellar Data Centre   ➤*Centre de Données Astronomiques*.

stellar evolution   The process of change that occurs through the lifetime of a star from its birth out of the interstellar medium, through the exhaustion of its usable nuclear fuel, to final extinction.

Stars form in clusters in the clouds of gas and dust of the interstellar medium. The material of the protostar condenses and collapses. The core heats up through the release of gravitational energy until the temperature is high enough for the nuclear fusion of hydrogen into helium to take place. The time taken for this process depends strongly on the mass of the protostar. A star of 10 solar masses takes only 300,000 years, compared with 30 million years for a star the same mass as the Sun.

Hydrogen burning in the core continues until the fuel supplies are exhausted. During this phase, the star is on the main sequence of the ➤*Hertzsprung–Russell diagram*. Again, the timescale is reduced dramatically with increasing mass. The Sun has a main-sequence lifetime of 10 billion years (of which about half have passed), as compared with only 500 million years for a star three times as massive.

When hydrogen burning in the core stops because the fuel is used up, a fundamental change in the star's structure takes place as adjustments are made to compensate for the loss of the energy source. The inert core contracts rapidly. In the process, gravitational energy is released, which heats the surrounding layers of hydrogen to the point where hydrogen burning recommences, but in a shell around the core. The result of the new outpouring of energy is to push the outer layers of the star further and further outwards. As this gas expands, it cools, and the star becomes a red giant. The combined effect of the increase in size and decrease in temperature is to maintain a more or less constant luminosity.

Meanwhile, the helium core continues to contract until a temperature of a hundred million degrees is reached, high enough for the fusion of helium into carbon and oxygen to begin. Helium burning starts.

Eventually, all the helium in the core is consumed. What happens subsequently depends on the mass of the star. In the more massive stars, contraction of the core after each fuel has been exhausted raises the temperature sufficiently to ignite a new, heavier fuel. Ultimately, a situation can be reached in which the central core has been converted to iron, while around the core, in a series of shells, silicon, oxygen, carbon, helium and hydrogen are being burnt simultaneously. Once a star has developed an iron core of about one solar mass no new reactions are possible. At this stage, the core contracts until it implodes catastrophically, setting off a ➤*supernova* explosion. The naked core that remains becomes a ➤*neutron star*.

In lower-mass stars, such as the Sun, the central temperature never gets high enough to progress beyond the burning of hydrogen and helium in concentric shells. Instabilities develop that result in the outer layers of the star being separated from the core to form an expanding shell of gas, called a ➤*planetary nebula*, that gradually disperses into space. In fact, significant mass is probably

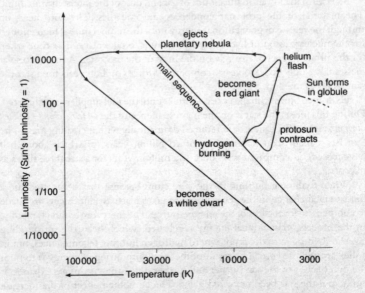

Stellar evolution. A schematic evolutionary track on the Hertzsprung–Russell diagram for a star of one solar mass.

lost from most stars through stellar 'winds', particularly during the later phases of evolution.

The remaining core cools and shrinks, becoming more and more compressed until it is about the size of the Earth. The matter becomes degenerate and a ➤*white dwarf* is formed. There is no internal source of energy and the white dwarfs continue to radiate and cool.

The evolutionary progress of a star is often demonstrated by plotting an *evolutionary track* on a Hertzsprung–Russell diagram. The illustration shows such a track for the Sun. Hertzsprung–Russell diagrams for star clusters illustrate the differential effect of mass on the rate of stellar evolution and can be used to determine the ages of clusters.

The outline of evolution given here is for single stars. Membership of a binary or multiple system may profoundly influence the course of a star's evolution if ➤*mass transfer* takes place.

stellar populations   Two broad classifications into which the stars and associated nebulae in our Galaxy and others are divided according to a number of dynamical and composition criteria. They are known as Populations I and II.

Population I is essentially the younger generation, and its members are located primarily in the arms of spiral galaxies. It includes luminous hot stars, main-sequence stars, open star clusters and associated interstellar clouds. Population I objects are relatively metal rich and they are in roughly circular orbits within the plane of the Galaxy. The Sun and its neighbouring stars belong to Population I.

Population II has the characteristics of an older generation. Its members are typically evolved stars, with low concentrations of heavier elements. They are found in elliptical galaxies and in the centres and haloes of spiral galaxies. Globular clusters are Population II objects.

Sometimes the expression Population III is used for a hypothetical class of objects, belonging to the earliest stage of the life of the Galaxy, which have now completely disappeared. ➤*Population I, Population II.*

stellar wind   Mass loss from a star in the form of a continuous outflow of particles. Many cool ➤*supergiant* stars have low-velocity winds, which can be detected in ➤*binary star* systems where the wind is responsible for ➤*absorption lines* in the spectrum of the companion star. However, the rate of mass loss is highest for the hottest stars which may, by means of a stellar wind blowing at hundreds or even thousands of kilometres per second, lose a significant fraction of their original mass over their lifetimes. ➤*solar wind.*

Stephan's Quintet (NGC 7317, 7318a and b, 7319 and 7320)   A group of five galaxies lying close together in the sky, first noted by M. E. Stephan in 1877. Subsequent measurements of the velocities of recession show that NGC 7320 is much closer than the others, lying along the line of sight by chance.

The other four galaxies appear to be physically associated, sharing a common recession velocity of 6,000 km/s.

Stern   English name for the constellation ➤*Puppis*.

Sternberg State Astronomical Institute   A Russian research institute in Moscow with observing facilities in the Crimea and Kazakhstan.

Steward Observatory   The observatory of the University of Arizona. It operates telescopes at a number of sites in Arizona. The largest is the 2.29-metre (90-inch) Bok Telescope on ➤*Kitt Peak*, opened in 1969 and named in honour of Bart Bok in 1996. There is a 1.54-metre (60-inch) reflector and a 42-centimetre (16-inch) ➤*Schmidt camera* at Mount Bigelow in the Catalina Mountains, 54 kilometres (34 miles) from Tucson at an altitude of 2,515 metres (8,250 feet). 1.5-metre and 1.0-metre telescopes are located at Mount Lemmon, and the ➤*Heinrich Hertz Submillimeter Telescope* together with the 1.8-metre Lennon reflector at the ➤*Mount Graham International Observatory*.

Stickney   The largest crater on the inner martian moon, Phobos. It is 10 kilometres (6 miles) across, over one-third of Phobos's largest diameter of 28 kilometres. Stickney was the maiden name of the wife of Asaph Hall (1829– 1907), the American astronomer who discovered Phobos and Deimos.

Stokes parameters   Four numbers that give a complete mathematical description of any polarized radiation. Radio telescopes can be designed to measure these directly, giving information on the magnetic fields in radio sources. ➤*polarization*.

Stonehenge   A prehistoric stone monument in the UK, thought to be of astronomical significance. Stonehenge, 130 kilometres (80 miles) west of London, is one of the finest of all neolithic sites. It was constructed in three phases, commencing with a bank and ditch around 2800 BC. The surviving group of sandstone megaliths in a circle 30 metres (100 feet) in diameter was erected in about 2000 BC. Some of the stones form foresight and backsight markers that indicate crucial rising and setting points for the Sun and Moon with considerable accuracy. Astronomers have shown how observations made at Stonehenge would have enabled its users to predict solar and lunar eclipses with certainty. If the astronomical interpretation is a correct one, it implies that the designers of Stonehenge had recorded, or remembered through an oral tradition, observations extending over many centuries.

Stonyhurst disc   A printed template used by observers of the Sun to find the ➤*heliographic latitude* and *longitude* of features on the solar surface.

stony-iron meteorite   A major category of ➤*meteorite*, consisting of a mixture of metallic and stony elements. There are two main types: *pallasites* and *mesosider-ites*. Pallasites consist of olivine grains enclosed by metal. Typically there is twice as much olivine as metal by volume. Mesosiderites are agglomerations of silicates and metal, in roughly equal proportions. Stony-iron meteorites are also known as *siderolites* or *lithosiderites*.

stony meteorite   A ➤*meteorite* consisting entirely of stony material. Stony meteorites are divided into two main classes: ➤*chondrites* and ➤*achondrites*. More than 90 per cent of the meteorites seen to fall (as opposed to those found by chance) are stony.

stratigraphy   The study of rock layers (strata). The relative sequence of events that have shaped the surface of a planet or moon may be understood by means of stratigraphy.

stratosphere   The region of the Earth's atmosphere above the ➤*troposphere*. It lies between heights of about 15 and 50 kilometres (9 and 30 miles). From the bottom to the top of the stratosphere, the temperature increases from about 240 K to 270 K.

strewn field   ➤*tektite*.

string theory   A theory in fundamental physics that attempts to construct a model of elementary particles from one-dimensional entities rather than the zero-dimensional 'points' of conventional particle physics. ➤*superstring theory*.

Strömgren radius   The maximum distance at which ultraviolet photons from a star can ionize the surrounding hydrogen completely.

Strömgren sphere   The region of completely ionized gas, typically hydrogen and helium, surrounding a very hot ➤*O star* or ➤*B star*.

S-type asteroid   A category of asteroid of intermediate albedo, thought to be made of silicaceous material, similar to ➤*stony meteorites*. S-type asteroids are relatively common in the inner ➤*asteroid belt*.

Subaru Telescope   An 8.3-metre (27-foot) telescope at the ➤*Mauna Kea Observatories* in Hawaii for the National Astronomical Observatory of Japan. Construction began in 1991 and full operation is expected in 1999. It is designed to operate in both the visual and infrared spectral regions. 'Subaru' is the Japanese word for the ➤*Pleiades*.

subdwarf   A member of a group of stars, mainly of ➤*spectral type* F, G or K, which, when classified according to standard criteria, form a sequence just below (or to the left of) the ➤*main sequence* on the ➤*Hertzsprung–Russell diagram*. These stars have exceptionally low abundances of the heavier elements, compared with the Sun, and belong to the so-called ➤*Population II* of older, 'metal'-deficient stars. The absorption lines in their spectra are relatively weak and, because most such lines occur in the blue and ultraviolet part of the spectrum, the subdwarfs look bluer in colour than their counterparts with solar-like abundances.

sub-Earth point The point on another body in the solar system from which observers would see the Earth at their zenith.

subgiant A star whose luminosity in relation to its spectral type places it between the main sequence and the giant branch on the ➤*Hertzsprung–Russell diagram*. Subgiants are designated as luminosity class I V.

submillimetre–wave astronomy The study of ➤*electromagnetic radiation* from celestial sources in the wavelength band between 0.3 and 3 millimetres. A combination of techniques from radio astronomy and infrared astronomy is required for this transition region of the electromagnetic spectrum. The telescopes have to be located at particularly dry places and high elevations, because water vapour in the Earth's atmosphere absorbs strongly at these wavelengths, and the astronomical signals are mostly weak. However, this region of the spectrum is important for a number of studies in astronomy, including ➤*cosmic background radiation*, regions of star formation and spectral lines of molecules in interstellar clouds.

There are only a few submillimetre telescopes in operation. One is the ➤*James Clerk Maxwell Telescope* situated at the ➤*Mauna Kea Observatories*, Hawaii. The California Institute of Technology's Submillimeter Observatory (CSO) – a 10.4-metre (84-foot) telescope with a segmented mirror – is also at the same site. The Swedish–ESO Submillimetre Telescope (SEST) is at the ➤*European Southern Observatory* (ESO), La Silla, Chile. The reflector has a diameter of 15 metres (49 feet) and consists of 176 panels that can be adjusted separately. The most recent to be built is the ➤*Heinrich Hertz Submillimeter Telescope* at Mount Graham in Arizona.

In 1997, the first orbiting observatory for submillimetre-wave astronomy was launched from the USA – the Submillimeter Wave Astronomy Satellite (SWAS) – with the objective of studying the composition of interstellar clouds.

subreflector A secondary reflector in a radio telescope of a ➤*Cassegrain* design. It receives the signal reflected from the main dish, directing it to the focus behind the main dish, in which there is a central hole. It performs the same function as a secondary mirror in a reflecting optical telescope.

sub-solar point The point on a body in the solar system from which observers would see the Sun at their zenith.

sulci (sing. sulcus) A system of roughly parallel ridges and furrows on the surface of a planet.

Summer Triangle The three bright stars ➤*Vega*, ➤*Altair* and ➤*Deneb*, which are particularly conspicuous in the summer evening sky.

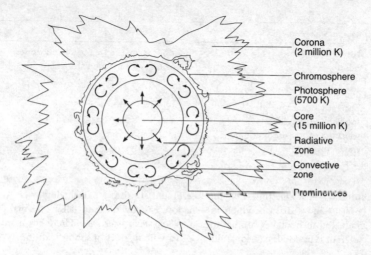

Sun. A schematic cross-section through the Sun, showing the main features of its structure.

Sun The central star of the solar system. On the range stars cover, the Sun is of medium size and brightness, though the vast majority of stars in the solar neighbourhood are smaller and less luminous. It is a dwarf star of ➤*spectral type* G2 with a surface temperature of about 5,700 K. Like all stars, it is a globe of hot gas and its energy source is nuclear fusion taking place in the centre, where the temperature is 15 million K. Four million tonnes of solar material are annihilated each second in the process, in which hydrogen is converted into helium.

Overlying the core is the *radiation zone* where the high-energy photons produced in the fusion reactions collide with electrons and ions to be re-radiated in the form of light and heat. Beyond the radiation zone is a *convection zone* in which currents of gas flow upwards to release energy at the surface before flowing downwards to be reheated. These circulating currents create the Sun's mottled appearance, or ➤*granulation*. The surface layers, or ➤*photosphere*, from which the light we see comes, are some hundreds of kilometres thick. In these layers, manifestations of ➤*solar activity* occur, such as ➤*sunspots* and ➤*flares*. High-speed atomic particles released in flares stream through space, affecting the Earth and its environment. They cause radio interference, ➤*geomagnetic storms* and ➤*aurorae*.

The layer over the photosphere is the ➤*chromosphere*, visible as a glowing pinkish ring during a total solar eclipse. ➤ *Spicules* and ➤*prominences* erupt through the chromosphere. The thinnest, outermost layers, forming the solar ➤*corona*, merge into the interplanetary medium.

sundial

Properties of the Sun	
Mass	1.989×10^{30} kg (332,946 Earth masses)
Radius	6.96×10^{5} km (109 Earth radii)
Effective temperature	5785 K
Luminosity	3.9×10^{26} W
Apparent visual magnitude	-26.78
Absolute visual magnitude	4.79
Inclination of equator to ecliptic	$7° 15'$
Synodic rotation period	27.275 days
Sidereal rotation period	25.380 days

sundial A simple time-keeping instrument consisting of a shadow stick (or *gnomon*) and a dial on which the shadow cast by the Sun falls. The dial is graduated in hours. A sundial measures ➤*apparent solar time*. There are many different types of design for sundials of varying degrees of sophistication.

sundog An alternative name for a ➤*mock Sun* or parhelion.

Sunflower Galaxy (M63; NGC 5055) A spiral galaxy in the constellation Canes Venatici.

sungrazer A ➤*comet* whose distance from the Sun at perihelion passage is so small that it passes through the Sun's outer layers. Around a dozen long-period comets, which have other orbital characteristics in common as well as the small perihelion distance, form a well-established group of sungrazers. They are also known as the *Kreutz group* after the Dutch astronomer Heinrich Kreutz (1854–1907) who, in 1888, was among the first to note the similarity between the orbits of some of the brightest comets ever observed.

sunrise Defined formally as the time at which the apparent upper limb of the Sun is on the astronomical horizon when the altitude of the Sun is increasing.

sunset Defined formally as the time at which the apparent upper limb of the Sun is on the astronomical horizon when the altitude of the Sun is decreasing.

sunspot A region on the Sun where the temperature is lower than that of the surrounding photosphere. Sunspots thus appear relatively dark. The presence of a strong magnetic field, concentrated in a spot area, produces the cooling effect. Sunspots can occur individually but often form groups or pairs of opposing magnetic polarity.

In the dark central part of the sunspot, the umbra, the temperature is about 3,700 K compared with the 5,700 K of the photosphere. The outer and brighter part of a sunspot, the penumbra, consists of aligned bright grains on a darker background which are arranged radially around the spot. ➤*solar activity*, *butterfly diagram*, *Wolf sunspot number*.

Sunspot The location in New Mexico of the ►*Sacramento Peak Observatory*, which belongs to the US ►*National Solar Observatory*.

Sunyaev−Zel'dovich effect Anisotropy in ►*cosmic background radiation* that may be caused by the absorption of radiation passing through rich clusters of galaxies.

supercluster A concentration of clusters of galaxies. About fifty are known, containing on average twelve rich galaxy clusters, though the largest have many more. These structures are hundreds of millions of light years across.

SuperCOSMOS An automated photographic plate processing facility located at the Royal Observatory, Edinburgh. It is the more powerful successor to a previous facility, COSMOS, which ceased operation in 1993. The name 'COSMOS' is contrived from the parameters the instrument can measure: Coordinates, Sizes, Magnitudes, Orientations and Shapes.

supergalactic plane The reference plane of a system of coordinates used for expressing the positions of relatively nearby galaxies. It passes through the Sun, the centre of our Galaxy and the centre of the ►*Virgo Cluster* of galaxies. It is almost perpendicular to the ►*galactic plane*.

supergiant A member of the class of the largest, most luminous stars known. Supergiants can be up to 500 times larger than the Sun and many thousands of times more luminous. There are supergiants of all spectral types. They are massive stars (mass greater than about ten times the Sun's) in an advanced state of ►*stellar evolution*. A supergiant is likely to become a ►*supernova*.

supergranulation A pattern of large-scale ►*convection* cells on the Sun. They are best detected by the horizontal motions they produce in the photosphere away from the centre of the solar disc. They are virtually invisible in integrated (white) light.

supergravity Theories that attempted to incorporate a theory of gravity with all other forces. These theories were found to be flawed because they treated the most basic entities as points of zero size, and have been superseded by the more successful ►*superstring theories*.

superior conjunction The point in the orbit of either Mercury or Venus when the planet lies on the far side of the Sun as viewed from the Earth.

superior culmination The same as upper ►*culmination*.

superior planet Any of the major planets whose orbits lie outside that of the Earth − Mars, Jupiter, Saturn, Uranus, Neptune and Pluto.

Super-Kamiokande A neutrino detector located 1 km underground at the Kamioka mine in Japan. It is the successor to the earlier Kamiokande (Kamioka nucleon decay experiment). The main element of the detector is a tank containing 50,000 tonnes of water. Sensors record ➤*Cerenkov radiation* in the form of visible light, emitted when high-velocity charged particles travel through the water. ➤*neutrino astronomy*.

superluminal motion   Motion at a velocity that apparently exceeds that of light. The angular separation of the components of some double radio sources is increasing at a rate that is apparently equivalent to as much as ten times the speed of light when the distance of the source is taken into account. Speeds in excess of that of light, however, are physically impossible, as shown by ➤*Special Relativity*. In reality, the effect is a purely geometrical one caused by one component travelling almost directly towards us along the line of sight at a velocity nearly as great as that of light. The phenomenon has been observed in the quasar 3C 273.

supermassive star   A very massive star. The term has no precise definition, but the most massive stars have up to about 100 times the Sun's mass.

supernova (pl. supernovae)   A catastrophic stellar explosion in which so much energy is released that the supernova alone can outshine an entire galaxy of billions of stars. In addition to the radiant energy produced, ten times as much energy goes into the kinetic energy of the material blown out by the explosion, and a hundred times as much is carried off by neutrinos.

A supernova explosion occurs when an evolved massive star has exhausted its nuclear fuel. Under these circumstances, the core becomes unstable against collapse.

Two distinct kinds of supernova are recognized, known as *Type I* and *Type II*. They are distinguished by the presence of hydrogen features in the spectrum of Type II supernovae which are absent from Type I. The light curves of Type I supernovae are all very similar: the luminosity increases steadily for about three weeks then declines systematically over six months or longer. The light curves of Type II supernovae are more varied.

Type I supernovae are subdivided into Types Ia and Ib, according to the strength of a particular silicon absorption line in the optical spectrum. The line is strong in Ia and weak in Ib.

Type Ia supernovae are thought to be ➤*white dwarfs* in binary systems, where mass transfer from the companion takes place. A wave of carbon burning through the newly acquired material could account for the energy released. The explosion may represent the total disintegration of the white dwarf. The nuclear reactions create about one solar mass of the unstable isotope ^{56}Ni, which decays to ^{56}Co and finally ^{56}Fe over a period of months. This radioactive decay would take place at a rate consistent with the observed decline in light output. The difference in mechanism between Types Ia and Ib is not yet clear.

Type II supernovae appear to be stars of eight solar masses or more that have run the course of ►*stellar evolution* and totally exhausted the nuclear fuel available in their cores. At this stage their structure is like that of an onion, consisting of concentric spherical shells in which different nuclear reactions are taking place. Once silicon burning starts in the central core, instability develops within a day because the iron created cannot fuse into heavier elements without an input of energy. In the absence of energy generation, the pressure balancing the weight of the overlying layers is removed.

When the crunch comes, the core collapses in less than a second. The rate accelerates as iron nuclei break up and neutrons form. However, implosion cannot continue indefinitely. When the density of nuclear matter is reached, there is a sudden strong resistance to further pressure, the imploding material bounces back and an outward shock wave is generated. The outer layers of the star are blown outwards at thousands of kilometres per second, leaving the core exposed as a neutron star.

The material ejected in the explosion forms an expanding ►*supernova remnant*. The neutron stars can be detected as ►*pulsars* through their radio emission and, in some cases, by pulsed light and X-ray emission as well.

The explosion of supernovae serves to enrich the chemical composition of the interstellar medium from which subsequent generations of stars are created. Very old stars contain much lower quantities of the elements heavier than hydrogen and helium than are found in the Sun and solar system, and many of these heavier elements can be created naturally only in the explosion of a supernova.

Supernovae are fairly rare events: only five have been observed visually in our own Galaxy in the last thousand years. Others have taken place, and radio emission from their remnants has been detected, but the outbursts were concealed behind obscuring dust. However, ►*Supernova 1987A* in the nearby Large Magellanic Cloud provided an opportunity unprecedented in modern times, enabling astronomers to study a supernova at relatively close hand. Numerous supernovae are detected each year in galaxies beyond our own. ►*Crab Pulsar*.

Supernova 1987A (SN 1987A) A ►*supernova* in the Large Magellanic Cloud discovered on 24 February 1987 when it was about sixth magnitude. It was the nearest and brightest supernova observed since 1604. The star that exploded was identified as a twelfth magnitude blue supergiant, known as Sanduleak −69° 202. Maximum magnitude, reached in mid-May, was near 2.8.

supernova remnant The expanding shell of material created by the ejection of the outer layers of a star that explodes as a ►*supernova*. Some supernova remnants are observable visually; others have been detected through their radio and X-ray emission. A shock wave precedes the ejected shell, colliding with and heating the interstellar gas. A reverse shock, moving inwards, is created, which heats the ejected material and the interstellar material, causing it to emit X-rays. Electrons accelerated by the shocks emit radio waves by the ►*synchrotron* *radiation* mechanism. The ejected material breaks up into clumps, so the radiation emitted from the shell often does not make up a uniform ring.

A small proportion of supernova remnants, including the ►*Crab Nebula*, have a rather different appearance. In these, the synchrotron radiation coming from within the shell far outshines any from the shell itself. This type of supernova remnant has been termed a *plerion*. A continuing supply of electrons travelling at relativistic speeds is needed to account for the emission. In the Crab Nebula, the known pulsar can produce the electrons, but for plerions where no pulsar has been detected, it is assumed that we are observing at the wrong angle to pick up the pulses from the central pulsar. Some other well-known examples of supernova remnants are ►*Cassiopeia A*, ►*Kepler's Star*, ►*Tycho's Star* and the ►*Cygnus Loop*.

superstring theory A version of ►*string theory* that incorporates ideas of *supersymmetry*, in which all classes of elementary particles are placed on an equal footing. The astronomical context is that such classes of theory may have applied to the behaviour of matter in the very early universe.

surface gravity The local value of the acceleration due to gravity experienced by a free-falling object at the surface of an astronomical body.

surge A type of solar ►*prominence* consisting of a straight or slightly curved spike of material shot out of a small luminous mound. Surges last about 10 or 20 minutes, and either fade or fall back following the ascent path in reverse.

Surveyor A series of seven unmanned American spacecraft launched between 1966 and 1968 to soft-land on the Moon. Five were successful. They conducted experiments to test the Moon's surface for a subsequent manned landing and returned a large number of close-up images of the lunar surface.

SU Ursae Majoris star A type of ►*dwarf nova* that has outbursts lasting typically for several days and occasional ones that are two magnitudes brighter and five times longer.

Swan English name for the constellation ►*Cygnus*.

Swan Nebula An alternative name for the ►*Omega Nebula*.

Swedish–ESO Submillimetre Telescope ►*Submillimetre-wave astronomy*.

Swordfish English name for the constellation ►*Dorado*.

Sword Handle Popular name for the double star cluster in Perseus, ►*h and χ (chi) Persei*.

Sword of Orion The stars Theta (θ) and Iota (ι) in the constellation ►*Orion*, which are usually shown as forming a sword hanging from the belt of the mythological figure associated with the constellation.

Sylvia   Asteroid 87, diameter 272 km, discovered by N. Pogson in 1866.

symbiotic stars   A term coined in 1928 by P. Merrill for stars with a particular type of unusual combination spectrum. Features of both a cool star and emission lines characteristic of a very-high-temperature gas are present. The normal interpretation is that the cool star is losing mass to a ➤*dwarf* or ➤*white dwarf* companion. Energy from a hot spot or heated accretion disc could ionize a large volume of the infalling gas to account for the nebular-type emission lines.

Such stars are also variable because of the irregularity of the mass transfer and eclipse of emitting material by the large cool star. They are also known as Z Andromedae stars.

synchronous rotation (captured rotation)   The coincidence of the rotation and orbital periods for a satellite, causing the same face always to be presented to the parent planet. Synchronous rotation, such as that of the Moon, is brought about by tidal action over long time periods.

synchrotron radiation   Electromagnetic radiation emitted by an electrically charged particle travelling almost at the speed of light through a magnetic field. The name arises because it was first observed in synchrotron accelerators used by nuclear physicists. It is the major source of radio emission from ➤*supernova remnants* and ➤*radio galaxies*. Much of the light and the X-ray emission from the ➤*Crab Nebula* is produced via the synchrotron process by the very-high-energy electrons from the central ➤*pulsar*.

The spectrum of synchrotron radiation has a characteristic profile very different from that of the thermal radiation emitted by hot gas, making synchrotron sources easy to identify. The polarization of the emission provides a means of estimating the magnetic field in the source.

synodic month   The interval of time between two successive new Moons (or any other specified phase), which is 29.530 59 days.

synodic period   For planets, the mean interval of time between successive conjunctions of a pair of planets, as observed from the Sun; for satellites, the mean interval between successive conjunctions of a satellite with the Sun, as observed from the satellite's parent planet.

synthetic aperture radar (SAR)   A radar technique used, for example, by the ➤*Magellan* mission to Venus, in which the echoes from radar pulsars emitted at a rate of thousands per second are processed by computer to generate a detailed picture of the structure of the reflecting surface.

Syrtis Major Planum (formerly Syrtis Major Planitia)   A cratered volcanic plain on Mars, identified with a dark, triangular feature (Syrtis Major) easily visible in telescopic views of the planet. The name was officially changed from 'Planitia' to 'Planum' in 1982.

syzygy   The rough alignment of the Sun, Earth and Moon, or the Sun, Earth and another planet. Syzygy thus describes both ►*conjunction* and ►*opposition*.

T

Table (Table Mountain)　English name for the constellation ➤*Mensa*.

taenite　A form of iron−nickel alloy found in iron ➤*meteorites*. It contains up to 7.5 per cent of nickel by weight. ➤*Widmanstätten figures, octahedrite*.

TAI　Abbreviation for ➤*International Atomic Time*. (The order of the initials is that of the term in French.)

tail　The extended part of a ➤*comet* that grows from the head when the comet is in the vicinity of the Sun. ➤*ion tail, dust tail*.

Tarantula Nebula (NGC 2070)　A large region of ➤*ionized hydrogen*, 900 light years across, in the Large ➤*Magellanic Cloud*.

T association　➤*association*.

Tau Ceti (τ Cet)　A star of the same type as the Sun, lying at a distance of 11.7 light years. It is the seventeenth-nearest star known, but, at magnitude 3.5, one of only a handful of nearby stars visible to the naked eye.

Taurids　An annual meteor shower, of relatively low activity, the twin radiants of which lie in the constellation Taurus. The peak occurs around 3 November. The meteoroid stream responsible is associated with ➤*Comet Encke*.

Taurus (The Bull)　A conspicuous zodiacal constellation, supposedly representing the head and forequarters of a bull. It was listed by Ptolemy (*c.* AD 140) and is possibly one of the most ancient of constellations. The brightest star is the first magnitude ➤*Aldebaran*, which appears to belong the ➤*Hyades* cluster, though it is in fact in the foreground. In total, there are fourteen stars brighter than fourth magnitude. The ➤*Pleiades* cluster and the ➤*Crab Nebula* also lie within the boundaries of Taurus. ➤Table 4.

Taurus A　The radio source associated with the ➤*Crab Nebula*.

Taurus−Littrow valley　The *Apollo 17* landing site on the Moon, located on the south-eastern border of Mare Serenitatis, in the region of the crater Littrow. The valley is completely surrounded by mountains, some more than 2,000 metres (6,500 feet) high. The centre of the lava-flooded valley was chosen for the manned landing to allow exploration of both the North and South massifs enclosing the area.

Tautenberg   The location in Germany of the ➤*Karl Schwarzschild Observatory*.

Taygeta   One of the brighter stars in the ➤*Pleiades*.

TDB   Abbreviation for barycentric dynamical time. (The order of the initials is that of the term in French.) ➤*dynamical time*.

TDRSS   Abbreviation for ➤*Tracking and Data Relay Satellite System*.

TDT   Abbreviation for terrestrial dynamical time. ➤*dynamical time*.

Teapot   A familiar name sometimes applied to an ➤*asterism* formed by a group of the brighter stars in the constellation Sagittarius because of its supposed resemblance to the shape of a teapot.

Tebbutt's Comet   ➤*Comet Tebbutt*.

technetium star   A star whose spectrum shows the presence of the unstable element technetium. The longest-lived isotope of technetium has a half-life of 2.1×10^5 years, which is short in relation to the typical lifetimes of stars. (The age of the Sun, for example is 5×10^9 years.) The implication is that the technetium is being created within the star itself and subsequently brought to the surface; it could not have been in the material from which the star formed. Technetium has been detected only in a small group of ➤*carbon stars*.

Teide Observatory   An observatory site on the island of Tenerife in the Canary Islands, shared by the Instituto de Astrofisica de Canarias with European partners. The instruments located there include several solar telescopes, a spectroheliograph, a radio telescope for studying ➤*cosmic background radiation* and a 155-cm (61-inch) infrared telescope.

tektite   A characteristic piece of natural glass. Tektites are found distributed on the Earth's surface in four main areas, called *strewn fields*, which are in Australasia, the Ivory Coast, Moravia and Bohemia in the Czech Republic, and Texas and Georgia in the USA. Individual tektites range up to 15 kilograms in mass (though most are much smaller) and their shape and structure suggest that the molten material from which they formed underwent a high-velocity flight through the atmosphere. The most popular theory for their origin is that they were created from terrestrial material when the impacts of large meteorites melted and ejected rock at the impact sites. Their ages and links to known impact structures support this theory.

tele-compressor   A converging lens inserted in the light path of a telescope to reduce the effective focal length.

tele-extender   An optical device in an extension tube used in conjunction with an eyepiece to increase the effective focal length of a telescope.

telemetry   The technique of remote control of a spacecraft, or instruments on it, and the receipt of results, by means of radio signals.

telescope An instrument to collect ➤*electromagnetic radiation* from a distant object, bring the radiation to a focus and produce a magnified image or signal. As technological advances have made it possible for astronomers to study the complete electromagnetic spectrum, telescopes of specialized design, and complementary detectors, have been devised to operate over different wavebands. The term 'telescope', originally coined for an optical instrument, has come to have this broader meaning in astronomy. However, telescopes for use in the radio, visible and X-ray regions, for example, employ widely differing designs and techniques.

Optical telescopes fall into two main categories, refractors and reflectors, according to whether the main light-gathering element is a lens or a mirror. A *refracting telescope* has an objective lens at the front of the telescope tube and either an eyepiece or equipment such as a camera at the back where the image is formed. In a *reflecting telescope*, the objective is a concave mirror at the back of the tube.

The objective of a refracting telescope is usually a compound lens, with two or more elements, of relatively long focal length. The use of a multi-element lens reduces the degree of ➤*chromatic aberration* inherent in lenses. Such a lens is known as an achromatic doublet or triplet. A long focal length also helps minimize both chromatic and ➤*spherical aberration*, but it also means that refractors tend to be long and bulky. In the past, exceptionally long refractors were constructed in an effort to reduce the aberrations. The abbreviation O G (for object glass) is sometimes used to indicate a refracting telescope.

There are inherent difficulties in constructing and mounting glass lenses of large diameter, and large thick lenses absorb too much light for astronomical purposes. The world's largest refractor has an objective lens 101 centimetres (40 inches) in diameter and is at the ➤*Yerkes Observatory*.

All large astronomical telescopes are reflectors, and reflectors are also popular with amateurs, being less expensive than refractors and easier to make. In a reflector, the light converges to a focal point in front of the main mirror, called the ➤*prime focus*. It is usually redirected, by means of a secondary mirror, to a place where detection is more convenient. Several systems are in common use. The ➤*Newtonian telescope*, ➤*Cassegrain telescope*, ➤*coudé focus* and ➤*Nasmyth focus* all have different applications. In a very large telescope, the observer may be able to work directly at the prime focus in a cage suspended within the main tube. The obstruction caused by a secondary mirror or prime focus cage has little effect on the performance of the telescope in practice. Large multi-purpose professional telescopes are usually constructed to offer observers a choice of foci. The Newtonian focus is used only on amateur visual telescopes.

The primary mirrors in reflecting telescopes are made most usually from glass or a ceramic material that does not expand or contract with temperature changes. The surface must be carefully figured to the required shape, either part of a sphere or part of a paraboloid, to an accuracy of a fraction of the wavelength of light. A thin layer of aluminium is then deposited on to the glass to provide the reflecting surface. In early reflecting telescopes, such as those made by William Herschel (1738–1822), the primary mirror was a casting of speculum metal (68 per cent copper, 32 per cent tin). For this reason, the abbreviation 'spec.' is still sometimes used to indicate a reflecting telescope. The earliest glass mirrors were coated with silver, but this has the disadvantage of tarnishing quickly when exposed to the air.

In the design of the most modern large telescopes, techniques known as ➤*active optics* allow thinner, more light-weight mirrors to be kept accurately in shape by means of an array of computer-controlled supports behind. This makes it possible to construct telescopes with mirrors of larger diameter than was previously possible, and mirrors composed of a number of separate segments.

Both the light-gathering power and the ➤*resolving power* of a telescope depend on the size of its objective. Astronomers continually aspire to larger instruments to reach fainter limiting magnitudes and achieve resolution of greater detail, though some of these objectives are also served by developing more sensitive detectors and the application of ➤*interferometers*.

Magnifying power is not of great significance, except with small amateur telescopes for visual use. The magnification for visual observing is changed by employing different ➤*eyepieces*. The maximum magnification is usually governed by ➤*seeing* conditions rather than the limit of performance of the telescope.

The images formed by astronomical telescopes are inverted. Since the introduction of a lens to rectify the image would serve no useful purpose and would absorb valuable light, astronomers prefer to work directly with inverted images.

The mounting of an astronomical telescope is an important part of its structure, since the observer needs to be able to point the instrument easily at selected objects and to follow them as the Earth's rotation causes their apparent movement across the sky. Very small amateur telescopes and modern computer-controlled telescopes employ ➤*altazimuth mountings*. Before the advent of computer control, the most practical method was the ➤*equatorial mounting*. Many existing telescopes are on equatorial mounts, and the system remains popular for amateur instruments. ➤*adaptive optics, New Technology Telescope, radio telescope, Schmidt camera, X-ray astronomy*.

Telescope   English name for the constellation ➤*Telescopium*.

Telescopio Nazionale Galileo   A 3.5-metre (11.5-foot) reflecting telescope at the ➤*Observatorio del Roque de los Muchachos*, in the Canary Islands. It was commissioned by Padua University, Italy, as a national facility for Italian astronomers and was completed in 1997. It is modelled on the European Southern Observatory's ➤*New Technology Telescope*.

Telescopium (The Telescope)   An insignificant southern constellation introduced by Nicolas L. de Lacaille in the mid-eighteenth century. It contains only one star as bright as third magnitude. ➤Table 4.

Telesto   A small satellite of Saturn, discovered in 1980 when the planet's rings were edge-on (and thus invisible) as viewed from Earth. It is co-orbital with Tethys and Calypso. ➤Table 6.

telluric   Pertaining to the Earth. In astronomy, *telluric lines* are ➤*spectral lines* in the spectrum of an astronomical object caused by molecules in the Earth's atmosphere.

terminator   The boundary between the illuminated and unilluminated portions of the surface of a planet or moon.

terra (pl. terrae)   An extensive land mass on a planetary surface. The lighter-coloured highland areas of the Moon are sometimes called 'terrae' in contrast with the darker ➤*mare* areas.

terrestrial   Pertaining to the Earth.

terrestrial dynamical time   ➤*dynamical time.*

terrestrial planet   One of the inner rocky planets (Mercury, Venus, Earth and Mars), which are similar in fundamental structure to the Earth, in comparison with the ➤*jovian planets.*

tessera (pl. tesserae)   A term used in the naming of areas on the surface of Venus that show polygonal patterning.

Tethys   A satellite of Saturn discovered by Giovanni Cassini in 1684. Its low density, only 1.1 times that of water, suggests that at least half of the interior must be ice. Images from the ➤ *Voyager* spacecraft show the surface to be heavily cratered, though there are regions of lower crater density, indicating that geological activity resulting in resurfacing took place in the past. Two notable features are the crater Odysseus, which is 400 kilometres (250 miles) in diameter, and Ithaca Chasma, a valley more than 2,000 kilometres (1,250 miles) long that cuts round three-quarters of the satellite's circumference. It is 100 kilometres wide and several kilometres deep.

Tethys's orbit is shared by two very small satellites, Telesto and Calypso. ➤Table 6.

Thalassa   A satellite of Neptune (1989 N5) discovered during the flyby of ➤*Voyager 2* in August 1989. ➤Table 6.

Tharsis Ridge   A raised volcanic area on Mars, 10 kilometres (6 miles) above the datum level for the planet. Three large volcanoes, rising to 27 kilometres (17 miles) above the datum level, lie in a line along the ridge. They are Arsia Mons, Pavonis Mons and Ascraeus Mons.

Thebe   A small satellite of Jupiter (number XIV), discovered by S. P. Synnott in 1980. ➤Table 6.

Themis   Asteroid 24, diameter 228 km, discovered in 1853 by A. de Gasparis. It is the prototype of a ➤*Hirayama family* of C-type asteroids whose orbits have semimajor axes of 3.13 AU.

Themis family   One of the ➤*Hirayama families* of asteroids, located at a distance of 3.13 AU from the Sun. The members of the family are all of the carbonaceous type, suggesting that they all come from the same parent body.

Theophilus   A large lunar crater to the north-west of Mare Nectaris, overlapping another large crater, Cyrillus. Theophilus is 100 kilometres (60 miles) in diameter and its terraced walls rise 5 km above the floor. A complex central peak rises to 2.2 km.

Theory of Relativity   ➤*General Relativity, Special Relativity*.

thermal radiation   Electromagnetic radiation arising from the thermal state (i.e. temperature) of the emitter, as opposed to non-thermal radiation, which is emitted by energetic electrons that are not necessarily in thermodynamic equilibrium. ➤*black body radiation*.

thermodynamic equilibrium (thermal equilibrium)   The state of a physical system in which there is no net exchange of thermal energy between members, and temperature remains constant.

third contact   In a total or annular ➤*eclipse* of the Sun, the point when the edges of the Moon's disc and Sun's ➤*photosphere* are in contact at the end of totality or the annular phase. In a lunar eclipse, third contact occurs when the Moon starts to leave the full shadow (umbra) of the Earth. The term may also be used to describe the similar stage in the progress of a ➤*transit* or ➤*occultation*.

third quarter   The phase of the Moon when half the visible disc of the waning Moon is illuminated. Third quarter occurs when the celestial ➤*longitude* of the Moon is 270° greater than the Sun's.

Thisbe   Asteroid 88, diameter 232 km, discovered in 1866 by C. H. F. Peters.

tholus (pl. tholi)   A planetary feature like a small dome-shaped mountain or hill.

Thuban (Alpha Draconis; α Dra)   A third magnitude star in the constellation Draco. Despite its designation as Alpha, it is only the seventh-brightest star in Draco. About 5,000 years ago Thuban was the nearest bright star to the north celestial pole. (The location of the north pole among the stars changes slowly over time because of ➤*precession*.) The name, derived from Arabic, means 'dragon'.

Thule   Asteroid 279, diameter 130 km, discovered by J. Palisa in 1888. At a distance from the Sun of 4.26 AU, it is regarded as marking the outer edge of the main ►*asteroid belt*.

tides   The movement of fluids, or stresses induced in solid objects, by a cyclical change in the net gravitational forces acting upon them. On the Earth, ocean tides are produced and modified by the daily, monthly and annual variations in the net gravitational force on the Earth exerted by the combined pull of the Sun and Moon. These variations arise from the Earth's rotation, the Moon's orbital motion around the Earth and the Earth's orbital motion around the Sun.

time   The means by which the intervals between sequential events are measured. ►*civil time, International Atomic Time, Mean Solar Time, sidereal time, spacetime, Universal Time*.

time zone   A geographical region throughout which ►*civil time* is reckoned to be the same. Time zones are based on longitude bands 15° wide, corresponding to a one-hour difference in ►*local time*. There are, however, considerable deviations from regular lines of longitude in the boundaries of time zones in order to take account of the distribution of land over the Earth's surface and the locations of centres of habitation. The difference between most adjacent time zones is one hour, but there are some instances of half-hour differences aimed at minimizing deviations from local time.

Titan   The largest satellite of Saturn and the second-largest in the solar system (after Ganymede). It was discovered in 1655 by Christiaan Huygens.
   Titan is surrounded by a thick atmosphere which consists mainly of molecular nitrogen but also contains methane. The surface pressure is 1.6 times greater than atmospheric pressure at the surface of the Earth. The action of sunlight on the methane and other atmospheric constituents such as carbon monoxide results in chemical changes, producing hydrocarbons and other molecules. These molecules condense in the cold atmosphere, forming a layer of opaque orange-coloured haze 200 kilometres (125 miles) above the surface. It is believed that the conditions are such that liquid methane may exist on the surface, where the temperature is 95 K (−178°C). There could even be methane 'rain' falling from clouds in the lower atmosphere. ►Table 6.

Titania   The largest satellite of Uranus, discovered by William Herschel in 1787. The flyby of ►*Voyager 2* in 1986 showed Titania to be peppered with numerous craters, though there are regions where the crater density is lower. This suggests that activity resulting in changes to the surface has taken place in the past. In addition, the surface is scarred by a large number of valleys and fractures, some of which cut large craters in half. ►Table 6.

Titius–Bode law   (Bode's Law)   A mathematical formula that gives approximations to the distances of the planets from the Sun starting with only the number in order of sequence. The relationship takes the form:

$$D = 0.4 + (0.3N)$$

where D is the distance in ►*astronomical units* (AU) and N takes the values 0, 1, 2, 4, 8 . . . , doubling for each successive planet. The relationship holds to within a few per cent for the seven innermost major planets as long as the value $N = 8$ is taken to represent the largest asteroid, ►*Ceres*. However, it breaks down seriously for Neptune and Pluto.

The formula was devised in 1766 by J. Titius and copied a few years later by J. E. Bode, who published it. At that time none of the ►*asteroids* had been discovered, and the 'gap' at 2.8 AU, where the formula predicted that there should be a planet, convinced astronomers that a small planet would be found there, which indeed proved to be the case.

TLP   Abbreviation for ►*transient lunar phenomenon*.

Tokyo Astronomical Observatory   The former name of a research institute of the University of Tokyo, which, in a reorganization in 1988, was largely incorporated into the new National Astronomical Observatory of Japan, funded by the Ministry of Education, Science and Culture. The new body has its headquarters at the address of the former Tokyo Astronomical Observatory and also incorporates the Okayama Astrophysical Observatory, Dodaira Observatory, Norikura Corona Observing Station and Nobeyama Radio Observatory. Its 8-metre (300-inch) optical/infrared instrument, the ►*Subaru telescope* at the ►*Mauna Kea Observatories* in Hawaii, is due to be opened in 1999.

Tolles eyepiece   A type of telescope ►*eyepiece* constructed from a single solid cylinder of glass.

topocentric coordinates   The coordinates of a celestial body measured from the Earth's surface (in contrast with geocentric coordinates, which are corrected to correspond to hypothetical observation from the Earth's centre).

Tornado Nebula   A radio source in the direction of the galactic centre, the nature of which remains unknown. The name is derived from the appearance of the radiograph of the source.

Toro   Asteroid 1685, diameter 7.6 km, discovered in 1948 by A. Wirtanen. It is a member of the ►*Apollo* group of asteroids and periodically makes exceptionally close approaches to Earth.

torus   A three-dimensional ring shape, such as that of a ring doughnut or quoit; the orbit of ►*Io* around Jupiter is enclosed in a *plasma torus*.

totality   The phase in the course of a solar or lunar ►*eclipse* during which the Sun is totally obscured or the Moon totally in the Earth's shadow.

Toucan   English name for the constellation ➤*Tucana*.

Toutatis   Asteroid 4179, an Earth-crossing asteroid discovered in 1989. Radar studies have shown it be irregular, measuring 4.7 by 2.4 by 1.9 km (2.9 by 1.5 by 1.2 miles) and have revealed the presence of craters and ridges on the surface. There is some suggestion that it may in fact be two bodies in close proximity. It rotates in a very complex manner. Both its shape and rotation are thought to be the result of collisions with other bodies. The plane of Toutatis's orbit is closer to the plane of Earth's orbit than that of any other known Earth-crossing asteroid. It makes a close approach to Earth in 2004, passing at about four times the Moon's distance.

Tracking and Data Relay Satellite System (TDRSS)   A network of four satellites, launched from Space Shuttles in 1983, 1986, 1988 and 1989, used to track NASA spacecraft and relay their data and commands more efficiently than is possible with a ground-based tracking network. It is the first space-based global tracking system.

transient   A phenomenon of brief duration.

transient lunar phenomenon (TLP)   Alleged temporary appearance of coloured patches or obscuration on the surface of the Moon. Reported observations are associated particularly with the regions of the craters Aristarchus, Gassendi and Alphonsus. It remains unclear whether real physical phenomena have been observed.

transit (1)   The passage of a star or other celestial object across the observer's ➤*meridian* in the course of the daily apparent motion of the celestial sphere.

transit (2)   The passage of either of the planets Mercury or Venus across the visible disc of the Sun.

transit (3)   The passage of a natural satellite across the disc of its parent planet.

transit circle   A telescope mounted so that it can rotate about a fixed horizontal axis in a north–south vertical plane. Transit circles are used for accurate measurements of the altitudes of stars and for timing their passage across the ➤*meridian*. The name *meridian circle* is also used.

transition probability   A number that characterizes the likelihood of a particular energy change taking place within an atom. Transition probabilities are a significant factor in determining the strength of spectral lines. Their values need to be known from laboratory measurements in order to calculate the abundances of elements from the spectra of celestial objects.

transneptunian object   A small planetary body in the outer solar system, in an orbit with a semimajor axis greater than that of Neptune's orbit (30 AU). ➤*Kuiper Belt*.

Trapezium   A popular name for the multiple star system Theta^1 (θ^1) Orionis, which lies at the heart of the ➤*Orion Nebula* and illuminates it. The name comes from the pattern formed by four stars, of magnitudes 5.1, 6.7, 6.7 and 8.0, which are visible in a small telescope. A larger telescope reveals the presence of two other stars of eleventh magnitude.

Triangle   English name for the constellation ➤*Triangulum*.

Triangulum (The Triangle)   A small but distinctive northern constellation between Andromeda and Aries, one of the 48 constellations listed by Ptolemy (*c.* AD 140). Its three brightest stars, of magnitudes 3.0, 3.4 and 4.0, form a small, elongated isosceles triangle. Triangulum includes the large spiral galaxy, M33, which is a member of the ➤*Local Group*. ➤Table 4.

Triangulum Australe (The Southern Triangle)   A small but distinctive southern constellation introduced in the 1603 star atlas of Johann Bayer. Its three brightest stars, magnitudes 1.9, 2.9 and 2.9, form what is almost an equilateral triangle. ➤Table 4.

Triangulum Galaxy (M33; NGC 598)   A large, nearby, spiral galaxy in the constellation Triangulum. It lies at a distance of 2.7 million light years, and is a member of the ➤*Local Group*.

Trifid Nebula (M20; NGC 6514)   A large luminous cloud of ➤*ionized hydrogen* in the constellation Sagittarius. Conspicuous dust lanes radiating from the centre appear to divide the nebula into three parts – hence its name.

trigonometric parallax   ➤*parallax*.

triple–alpha process (Salpeter process)   A nuclear fusion process that occurs in the interior of evolved stars. Three helium nuclei – also known as alpha particles – fuse to form carbon, with the release of gamma-ray energy. This can take place only when all the hydrogen in a stellar core has been used up and the temperature has risen to 100 million degrees.

triplet   ➤*doublet*.

triquetrum   ➤*Ptolemy's rules*.

tri–Schiefspiegler   A reflecting telescope with three mutually tilted, curved mirrors arranged in such a way that the optical path is unobstructed and the aberrations introduced by tilting the mirrors are cancelled out. 'Schiefspiegler' is German for 'oblique reflector', and the 'tri' indicates the number of mirrors. The name was coined by Anton Kutter, who was also responsible for many early designs, which he promoted in the 1950s. The modern version is the outcome of work by a Californian, Dick Buchroeder.

   In a typical system, the light falls first on to a concave primary mirror of relatively large focal ratio (f/12, for example) and near-spherical curvature.

The optical axis is inclined at about 3° to the path of the incoming rays. The converging beam is intercepted by a spherical convex secondary with the same radius of curvature as the primary, positioned just outside the entrance aperture of the telescope. This mirror is also tilted with respect to the incoming light, and largely cancels out the ►astigmatism and ►coma caused by the tilt of the primary. This two-mirror combination is virtually optically perfect up to apertures of about 120 mm. However, coma exceeds acceptable limits at larger apertures, and a third optical component must then be added. The result is successful up to apertures of at least 300 mm, and the system is regarded as an ideal planetary telescope with the quality of definition associated with refracting telescopes but no ►chromatic aberration.

Triton The largest natural satellite of Neptune. It was discovered in October 1846 by William Lassell, only 17 days after the discovery of Neptune itself. It circles Neptune every 5.9 days in a retrograde orbit tilted at 23° to the planet's equatorial plane. This unusual orbit has led to speculation that Triton was captured rather than having formed close to Neptune.

The ►Voyager 2 encounter with the Neptune system on 25 August 1989, and its approach to within 4,000 kilometres (2,500 miles) of Triton, revealed a wealth of detail. Triton's diameter was found to be 2,700 kilometres (1,680 miles), slightly less than thought previously. Its gravitational effect on the spacecraft's trajectory suggests that the bright, icy outer crust and mantle must overlie a substantial core of rock (perhaps even metal) containing two-thirds of the satellite's mass. The surface temperature is 38 K, making it the coldest known object in the solar system. Its size, structure and other properties suggest that the planet Pluto may be very similar to Triton.

Triton is surrounded by a tenuous atmosphere (the surface pressure is 15 microbars) of nitrogen with a trace of methane. The south polar cap is coated with a bright frost, possibly of nitrogen ice, which is gradually evaporating. (Owing to the orbital characteristics of Neptune and Triton, this region had been in sunlight continuously for nearly 100 years when it was observed. No impact craters were detected in this region.) The equatorial region exhibited a variety of terrains suggestive of complex volcanic activity still continuing, including ►plume eruptions. Triton's surface is certainly relatively young in astronomical terms. ►Table 6.

Trojan asteroids Two families of ►asteroids that share the orbit of Jupiter, clustered around the ►Lagrangian points 60° either side of the planet. Over 200 are known, the majority being in the preceding group. They do not remain stationary close to the Lagrangian points but oscillate around them over an arc between about 45° and 80° from Jupiter, taking 150–200 years to complete a cycle. The first to be discovered was ►Achilles, and it was decided to use the names of warriors in the Trojan wars for members of the group identified subsequently.

tropical month   The time taken by the Moon to complete one orbit round the Earth, with the ➤*equinox* as the reference point. Its length is 27.321 58 days.

tropical year   The time taken by the Earth to travel once round the Sun, measured from equinox to equinox. It is 365.242 19 days long.

troposphere   The lowest layer of the Earth's atmosphere, up to a height of approximately 20 kilometres (12 miles). It is bounded by the *tropopause*, which marks the transition to the more stable conditions of the stratosphere above.

true anomaly   ➤*anomaly.*

Trumpler classification   A system for classifying the visual appearance of ➤*open clusters* of stars, published by Robert J. Trumpler in 1930. The system describes the degree of central concentration of the cluster (Roman numerals I to IV in order of decreasing concentration and contrast with the background star field), the range of brightness covered by member stars (1 to 3 in order of increasing brightness range), and the apparent number of members (or 'richness') using the codes p ('poor') for fewer than 50 stars, m ('moderately rich') for 50–100 stars, and r ('rich') for more than 100 members.

Tsiolkovskii   A crater on the lunar farside, 180 kilometres (110 miles) in diameter. The crater floor is partially flooded by dark lava, through which a central peak protrudes. In a hemisphere devoid of dark maria, in marked contrast to the Moon's nearside, Tsiolkovskii is one of the most prominent features.

Tsytovich–Razin effect   The suppression of lower frequencies in the ➤*synchrotron radiation* from free electrons in a hot gas. The cut-off frequency depends on the electron density and magnetic field, and measurements of it provide information on those quantities.

T Tauri star   A type of very young star in an early phase of evolution where contraction is still taking place. The prototype, T Tauri, is an irregular variable within a dark dust cloud in the constellation Taurus.

All T Tauri stars vary irregularly. Their absorption line spectra show that their surface temperatures are in the range 3,500–7,000 K. They are found in dense interstellar clouds, usually alongside young, main-sequence O and B stars, but the T Tauri stars are far more luminous than ➤*main-sequence stars* of the same temperature. Strong emission lines also feature in the spectrum; these come from a low-density envelope of gas around the stars.

Large numbers of T Tauri stars have been discovered, notably in the Rho Ophiuchi (ϱ Oph) dust cloud, through the strong infrared radiation they emit. Loose groupings of T Tauri stars are known as ➤*T associations.*

Strong ➤*bipolar outflows* (twin-lobed jets) stream out from T Tauri stars at speeds of several hundred kilometres per second. Where the outflow compresses and heats the interstellar gas, the resulting luminous nebulae are observed as ➤*Herbig–Haro objects.*

T-type asteroid   A type of asteroid characterized by a fairly low albedo.

Tucana (The Toucan)   A southern constellation introduced in the 1603 atlas of Johann Bayer. Its brightest two stars are third magnitude. The Small ►*Magellanic Cloud* lies within its boundaries, and it also includes the large and bright ►*globular cluster*, known as ►*47 Tucanae*, which is on the borderline of visibility to the unaided eye. ►Table 4.

Tully—Fisher relation   The relationship between the width of the ►*twenty-one-centimetre line* radiation from spiral galaxies and their absolute photographic ►*magnitude*. B. Tully and R. Fisher calibrated this correlation in 1977 to derive a new technique for estimating the distances of spiral galaxies.

Tunguska event   The violent impact of what is thought to have been a comet or meteorite in the Tunguska region of Siberia on 30 June 1908. Though the event caused devastation over a large area, no remains of an impacting body or any crater have been discovered, perhaps because the object exploded in the atmosphere before an actual impact; the height of such an explosion has been estimated at 8.5 kilometres. Observers reported seeing a fireball as bright as the Sun. It exploded with a deafening sound and caused a shock wave that shook buildings and caused damage, though there was no loss of human life.

   The first expedition to the remote area of the explosion did not take place until 1927. It was found that a forest of trees had been snapped in half over a region 30—40 kilometres (20—25 miles) in radius. Over a region of radius 15—18 kilometres (9—11 miles) from the apparent 'impact' site, trees had been flattened radially and stripped of their branches. No totally satisfactory explanation of the event has been found.

twenty-one-centimetre line   Characteristic radio emission or absorption at a wavelength of 21 centimetres by neutral hydrogen in interstellar space.

   Neutral hydrogen is a major component of the ►*interstellar medium* and the 21-centimetre line provides an important means of finding its distribution, density and velocity in our own Galaxy and thousands of others. It was the first spectral line to be detected by radio astronomy (in 1951) and is now the main means for exploring the structure of galaxies through radio observations.

   The small energy change in the hydrogen atom responsible for the 21-centimetre emission has an intrinsically low probability of occurring. An individual hydrogen atom excited into the higher energy level typically waits 12 million years to make the energy jump spontaneously. However, the radiation is observed from interstellar hydrogen because of the vast numbers of atoms it contains and because collisions trigger the transitions.

twilight   The period before sunrise and after sunset when the sky is partially illuminated by scattered sunlight. *Civil twilight* is defined as the period when the zenith distance of the centre of the Sun's disc is between 90° 50′ and 96°;

nautical twilight is the interval when it is between 96° and 102°, and *astronomical twilight* that when it is between 102° and 108°.

twinkling The rapid movement and ➤*scintillation* of a star image caused by turbulence in the Earth's atmosphere.

Twins English name for the constellation ➤*Gemini*.

two-colour diagram A graph on which two ➤*colour indices* (such as $(B-V)$ and $(U-B)$) are plotted, one along each axis, for a sample of stars or other objects, such as asteroids. ➤*UBV photometry*.

Tycho A prominent lunar crater in the Moon's southern uplands. It is surrounded by the brightest and most extensive ray system on the Moon, possibly indicating that it is one of the youngest major features. The terraced walls rise to a height of 4.5 km and the central peak to 2.3 km above the floor of the crater, which is 85 km (53 miles) in diameter.

Tycho catalogue ➤*Hipparcos*.

Tycho's Star A ➤*supernova* in the constellation Cassiopeia observed by Tycho Brahe in 1572. At maximum brightness it rivalled Venus and was visible in daylight. The ➤*supernova remnant* is both an X-ray source and an intense source of radio emission; the expanding shell of gas is faintly visible with powerful optical telescopes.

Tyuratam The nearest town to the ➤*Baikonur* space centre in Kazakhstan.

U

UBV photometry   A photometric system introduced in the 1950s by H. L. Johnson and W. W. Morgan. It is based on the measurement of stellar ►*magnitudes* in three wide spectral bands called *U* (ultraviolet), *B* (blue) and *V* (visual), which are centred on wavelengths 350, 430 and 550 nanometres, respectively. *Colour indices* formed by computing (*U*–*B*) and (*B*–*V*) may be used to deduce some of the physical properties of individual stars or groups of stars.

   To extend the usefulness, more bands were added by Johnson in 1965 extending into the infrared. These are called *R*, *I*, *J*, *H*, *K*, *L*, *M* and *N*, ranging from 0.7 to 10.2 micrometres.

UFO   Abbreviation for ►*unidentified flying object*.

U Geminorum star   A ►*dwarf nova*.

Uhuru   ►*Small Astronomy Satellite*.

UKIRT   Abbreviation for ►*United Kingdom Infrared Telescope*.

UKST   Abbreviation for ►*United Kingdom Schmidt Telescope*.

ultraviolet astronomy   The study of electromagnetic radiation from astronomical sources in the wavelength band 10–320 nanometres. Ultraviolet (UV) radiation is strongly absorbed by the Earth's atmosphere, so all observations have to be carried out from satellites. The earliest observations were made during brief rocket flights in the 1940s and 1950s. The first satellite to make systematic ultraviolet observations was the first ►*Orbiting Solar Observatory* (OSO-1) in 1962. The highly successful International Ultraviolet Explorer (IUE) was launched in 1978 and continued to operate until 1996.

   The ultraviolet is often subdivided into the extreme UV (EUV, 10–100 nm), the far UV (FUV, 100–200 nm) and the near UV (NUV, 200–320 nm). The most extreme UV, at the transition to X-radiation in the approximate waveband 6–60 nm, is also known as the XUV. At these wavelengths, the techniques of ►*X-ray astronomy* are required, but the rest of the UV band can be observed and analysed by methods similar to those used in the visible part of the spectrum. The main difficulty is the limited range of transparent materials and reflective coatings suitable for use in the UV. Glass, for example, is strongly absorbent, and quartz or fluorite have to be used. In a UV telescope carried in the ►*Astro-1* observatory on board the Space Shuttle, the reflection problem was tackled by the use of the rare metal iridium, which is effective down to wavelengths of 40 nm.

Ultraviolet astronomy is important because many of the spectral lines most valuable for analysis, of both atoms and molecules, lie in this waveband. Hotter stars, with surface temperatures in excess of 10,000 K, emit most of their energy in the UV. Even for cooler stars such as the Sun, UV studies are needed for energetic phenomena. The ►*interstellar medium* is another important object of study for ultraviolet astronomy though, at wavelengths below 91.2 nm, almost all the UV radiation is absorbed by hydrogen, the most widely distributed element in the universe, making the detection of distant sources difficult at such short wavelengths.

Ulysses   A European Space Agency mission, launched on 6 October 1990, to study the interplanetary medium and the solar wind at different solar latitudes. It provided the first opportunity for measurements to be made over the poles of the Sun. Its trajectory used the ►*gravity assist* technique to take it out of the plane of the solar system. After an encounter with Jupiter in February 1992, the spacecraft swung back towards the Sun to pass over the solar south pole in 1994 and the north pole in 1995. A second encounter with the Sun takes place in September 2000.

umbra (1)   An area of total shadow, such as the zone on the surface of the Earth from which totality is observed during a solar ►*eclipse*.

umbra (2)   The dark central region of a ►*sunspot*, where the magnetic field is vertical and typically has a strength a few thousand times that at the surface of the Earth. The temperature is about 3,500 K, compared with 6,000 K for the surrounding ►*photosphere*. ►*penumbra*.

Umbriel   A satellite of Uranus, discovered by W. Lassell in 1851. Images from the ►*Voyager 2* encounter in 1986 show that Umbriel is much darker than the other four major satellites of Uranus. It appears that the surface has been covered by dark material relatively recently in astronomical terms. It is also covered with craters; one of them, 110 kilometres in diameter, is very bright, in marked contrast to the rest of the surface. ►Table 6.

unda (pl. undae)   A term for dune-like features on Mars.

Undina   Asteroid 92, diameter 194 km, discovered in 1867 by C. H. F. Peters.

Unicorn   English name for the constellation ►*Monoceros*.

unidentified flying object (UFO)   Any phenomenon in the sky for which the observer does not have a ready rational explanation. The term is often used in connection with hypothetical manifestations of unnatural objects from space.

United Kingdom Infrared Telescope (UKIRT)   A 3.8-metre (150-inch) infrared telescope, located at the ►*Mauna Kea Observatories* in Hawaii. It is operated from the Joint Astronomy Center in Hilo, Hawaii, for the UK Particle Physics and Astronomy Research Council. It is the largest telescope dedicated solely to infrared astronomy and operates in the wavelength band between 1 and 30 microns.

United Kingdom Schmidt Telescope (UKST)   A 1.2-metre (48-inch) ►*Schmidt camera* located at the ►*Anglo-Australian Observatory* and currently administered by the Anglo-Australian Telescope Board. It was opened in 1973 and for a time administered by the Royal Observatory, Edinburgh.

United States Naval Observatory   A US government observatory in Washington, DC, the main purpose of which is to provide the astronomical data required to support Navy and other Department of Defense activities. These include astrometry, the preparation of almanacs, time measurement and the maintenance of the Master Clock for the USA. It has astrographic telescopes located at Anderson Mesa, near Flagstaff, Arizona, and Black Birch, New Zealand, as well as in Washington.

The observatory was founded in 1830 and given the title US Naval Observatory in 1844. For fifty years it was located at the site now occupied by the Lincoln Memorial. It was moved to its present site, next to the official residence of the Vice President, in 1893. The largest telescope at the site is the 66-centimetre (26-inch) refractor, dating from 1873, with which Asaph Hall discovered the moons of Mars, Phobos and Deimos, in 1877. Other instruments there include a 30-centimetre (12-inch) Alvan Clark refractor, two 61-centimetre (24-inch) reflectors and a 15-centimetre (6-inch) transit circle. The largest telescope belonging to the observatory is the 1.5-metre (61-inch) astrometric reflector at Flagstaff. Using this instrument, James Christy discovered the moon of Pluto, Charon, in 1978. At the Arizona site, the observatory has also constructed an optical ►*interferometer*, the Navy Prototype Optical Interferometer, which was the largest of its kind when it came into operation in 1995.

The US Naval Observatory also houses one of the world's leading astronomical libraries.

Universal Time (UT)   A measure of time that relates closely to the Sun's daily apparent motion and serves as the basis for civil timekeeping. It is formally defined by a mathematical formula that links it to ►*sidereal time*, and is thus determined from observations of the stars. The timescale taken directly from the stars, designated UT0, depends slightly on the place of observation. UT0 corrected for the shift in longitude of the observing station caused by ►*polar motion* is known as UT1. The use of the abbreviation UT normally implies UT1.

Coordinated Universal Time (UTC) is the time used for broadcast time signals.

It differs from ►*International Atomic Time* (TAI) by an integral number of seconds, and is maintained to within ±0.90 second of UT1 by introducing when necessary one-second steps – leap seconds.

universe The entirety of all that exists. The size of the observable universe is limited to the distance light has had time to travel since the ►*Big Bang*. ►*particle horizon*.

unsharp masking A photographic technique used for processing images to reveal fine detail. A positive contact copy of a glass photographic plate is made on low-contrast film. The copy is blurred because of the glass between the film and the image on the plate. This blurred image contains the large-scale structure of the image and constitutes the 'unsharp mask'. When replaced in contact with the back of the original plate, it effectively cancels out the large-scale structure, leaving fine detail to be seen much more easily.

Urania Asteroid 30, diameter 94 km, discovered in 1854 by J. R. Hind.

Uraniborg The observatory of Tycho Brahe (1546–1601) on the island of Hven, north of Copenhagen. It was completed in 1580 and used by Brahe to make accurate astronomical observations for twenty years. Only ruins now remain.

Uranometria A star atlas compiled by Johann Bayer (1572–1625), published in 1603. In this atlas Bayer introduced the system of labelling stars with Greek letters which is still in use.

uranometry A largely obsolete term for positional astronomy or ►*astrometry*.

Uranus The seventh major planet of the solar system in order from the Sun, discovered by William Herschel in 1781. It is just bright enough to be seen by the naked eye under good observing conditions. From the Earth, it appears as an almost featureless greenish disc, in even the largest telescope. In 1986, the space probe ►*Voyager 2* passed close to Uranus and its satellites, providing close-up images of them. Ten small satellites were discovered by *Voyager 2*. Five larger satellites were already known: Miranda, Ariel, Umbriel, Titania and Oberon, and two more were discovered in 1997.

Uranus is one of the four 'gas giant' planets of the solar system, with a diameter four times the Earth's and a mass fifteen times greater. It is composed almost entirely of hydrogen and helium. It is generally believed that there is a small rocky core at the centre of the planet, which is surrounded by a thick icy mantle of frozen water, methane and ammonia, that merges into the outermost layer, an atmosphere of hydrogen and helium, with small quantities of some molecular compounds.

Even in the *Voyager* close-ups, Uranus presents a bland, nearly featureless appearance, though there is some evidence for faint bands parallel to the equator.

A curious feature of Uranus is that its rotation axis lies almost in the plane of the solar system, rather than being nearly perpendicular to it, as is the case for the other planets. The internal rotation period is 17 hours 14 minutes.

In 1977, a series of narrow rings was discovered around Uranus in its equatorial plane. The rings are each only a few kilometres wide and not visible from Earth. The discovery occurred when Uranus occulted an eighth magnitude star. The rings caused small dips in the observed brightness of the star just before and just after the occultation by the disc of the planet. Later occultations, of Beta (β) Scorpii and Sigma (σ) Sagittarii, confirmed the result. The ring system was subsequently imaged by *Voyager 2* in 1986, when two further rings were discovered, bringing the total to eleven. ►Tables 5, 6 and 7.

ureilite   A rare and unusual type of ►*meteorite* belonging to the ►*achondrite* class. The space between silicate grains is filled by a carbon-rich material, which in some examples has been transformed into diamond.

Ursa Major (The Great Bear)   One of the most familiar constellations of the northern sky and the third largest in area. It contains nineteen stars brighter than fourth magnitude. The seven main stars of the constellation form an asterism known variously as the Plough, the Big Dipper and Charles's Wain. The two stars Merak and Dubhe in the Plough are known as the Pointers since the line between them is effectively an arrow to Polaris, the Pole Star. Ursa Major is one of the ancient constellation listed by Ptolemy (*c.* AD 140). It contains a group of galaxies belonging to the ►*Local Supercluster*, including the relatively bright spiral galaxy, M81. ►Table 4.

Ursa Minor (The Little Bear)   The northern constellation in which the north celestial pole lies. The brightest star of Ursa Minor, the second magnitude Polaris, is within 1° of the pole. The constellation was among those listed by Ptolemy (*c.* AD 140). Though fainter and smaller than the plough in Ursa Major, its main pattern of seven stars is somewhat similar and is known as the Little Dipper. ►Table 4.

US Naval Observatory   ►*United States Naval Observatory*.

UT   Abbreviation for ►*Universal Time*.

UTC   Abbreviation for Coordinated Universal Time. ►*Universal Time*.

Utopia Planitia   An extensive, sparsely cratered plain in the northern hemisphere of Mars. It was the landing site for the ►*Viking 2* spacecraft. Panoramic images returned by the *Viking* lander showed terrain littered with large numbers of boulders made of open-textured rock.

UV   Abbreviation for ultraviolet. ►*electromagnetic radiation, ultraviolet astronomy*.

uvby photometry   A photometric system introduced by B. Strömgren in the late 1950s. It is based on the measurement of ►*magnitudes* through each of four intermediate-band filters centred on the wavelengths 350 (*u*), 410 (*v*), 470 (*b*) and 550 (*y*) nanometres. The *y* band corresponds closely with the *V* band of ►*UBV photometry* and traditional visual magnitudes. The *v* band covers the ►*Balmer line* Hδ and many other lines in the spectra of cooler stars. The *b* band, half-width 18 nanometres, lies close to but excludes the Balmer line Hβ. The *u* band is in the ultraviolet, between the atmospheric cut-off at 320 nanometres and the ►*Balmer decrement*. The use of the system is intended to provide information about the spectra and physical properties of individual stars and groups of stars without the much longer process of studying spectra in detail.

UV Ceti star   A ►*flare star.*

V

V   A symbol used for visual ▸*magnitude*, particularly that defined in the ▸*UBV photometry* system devised by Johnson and Morgan, which is measured in a band centred around 550 nanometres and corresponds closely to traditional visual magnitudes.

vacuum tower telescope   A design of telescope for observations of the Sun. The example at the ▸*Sacramento Peak Observatory* is typical. In that instrument, sunlight enters the tower 41 metres (135 feet) above the ground. A further 67 metres (220 feet) of the telescope lie below ground. The entire optical path is virtually air-free, to avoid distortion of the solar image that would be caused by the presence of hot air. In the observing room an image of the Sun 51 centimetres (20 inches) in diameter is produced with a resolution better than a quarter of an arc second. Sunlight can be directed into spectrographs or other instruments by tilting the main mirror at the bottom of the central tube. The entire optical system is suspended near the top of the tower on mercury float bearings.

Valhalla   A large circular feature on ▸*Callisto*, surrounded by fifteen concentric rings. The radius of the outermost ring is 1,500 kilometres (930 miles). The feature was caused by an impact, but no longer shows any significant relief because of the plastic nature of the crust when the impact occurred. The rings are effectively 'ripples'.

Valles Marineris   A canyon system on Mars, extending more than 5,000 kilometres (3,100 miles) in the east–west direction in the equatorial region. The western extremity is marked by the Noctis Labyrinthus, a complex area of fault valleys (graben) that cut the surface into polygonal shapes. The central section consists of several parallel canyons whose average depth is 6 kilometres. At the centre they join up with Melas Chasma, a depression 160 kilometres (100 miles) across. The eastern end of the system is marked by Capri Chasma. Erosion of the canyon sides has created debris in the flat valley bottom and has revealed a layered structure in the surrounding plateau. Flow channels leading into the canyon suggest that erosion by water took place in the remote past. Valles Marineris is primarily a fault structure, thought to have been created by the uplift of the volcanic Tharsis Ridge to the west.

vallis (pl. valles)   A sinuous valley on the surface of a planet.

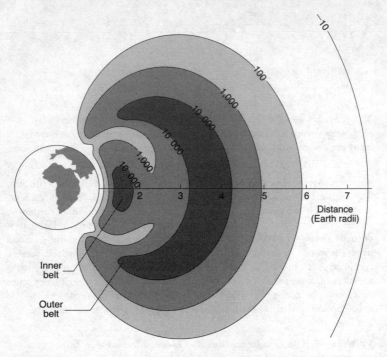

Van Allen belts. Schematic representation of the Van Allen belts around the Earth. The numbers on the contour lines are the average number of particles per cubic metre. The distance scale (in Earth radii) is measured from the centre of the Earth.

Van Allen belts Two ring-shaped regions around the Earth where there are concentrations of high-energy electrons and protons that have been trapped by the Earth's magnetic field. They were discovered by the USA's first successful artificial Earth satellite, *Explorer 1*, which was launched on 31 January 1958. They are named after James Van Allen, the physicist who led the experiment on *Explorer 1*. The inner Van Allen belt lies about 0.8 Earth radii above the equator. The main concentration of the outer belt lies between about 2 and 3 Earth radii above the equator, but a broader region, extending from the inner belt out to as far as 10 Earth radii, contains protons and electrons of lower energy, believed to have come primarily from the ⊳*solar wind*. Because the Earth's magnetic field is offset from the planet's rotation axis, the inner belt dips down towards the surface in the region of the South Atlantic Ocean, off the coast of Brazil. This *South Atlantic Anomaly* presents a potential hazard to the operation of artificial satellites. In 1993, a region within the inner Van Allen belt was found to contain particles that have penetrated from interstellar space.

van Biesbroeck's Star   The companion to the star BD +4°4048, which, at the time of its discovery by George van Biesbroeck (1880–1974), was the least luminous known star. Its ►*absolute magnitude* is +18.

van Maanen's Star   A nearby twelfth magnitude star, discovered by Adriaan van Maanen (1884–1946), identified as a ►*white dwarf*.

variable star   Any star whose light output varies, whether regularly or irregularly. The graph showing brightness as a function of time for a variable is known as the *light curve*.

There are a number of physical reasons why stars vary, and these are used as the basis for broad groupings according to type. Within each group, many more specialized types may be distinguished, often named after a prototype star.

Eruptive and *cataclysmic* variables are characterized by their unpredictability and include a wide variety from ►*T Tauri stars*, in the process of formation, to ►*supernovae*, which have reached an explosive end. This group also includes ►*flare stars*, ►*novae* and ►*dwarf novae*.

Pulsating variables are physically oscillating because of internal instabilities. These include ►*Cepheid variables*, ►*RR Lyrae stars* and ►*Mira stars*. The other major group consists of the *eclipsing binaries*, in which the light variation is caused by one star periodically passing in front of another. Such binaries are often also interacting. The prime example is ►*Algol*.

Some stars, ►*BY Draconis stars* for example, show variability as they rotate because their surface brightness is not uniform.

The method of naming variable stars is somewhat strange. It is mainly due to F. W. A. Argelander (1799–1875), who used the letters R to Z in conjunction with the genitive case of the constellation name for the nine brightest variables in each constellation. After that, pairs of letters, RR to RZ, SS to SZ and so on to ZZ were used. For further variables, the pairs of letters AA to AZ, BB to BZ and so on (J being omitted) were introduced, bringing the number of available designations to 334. Since many more than 334 variables are now known in many constellations, they are designated as V335, V336, and so on.

variation   A term in the mathematical expression for the ecliptic longitude of the Moon which varies with a period of half a synodic month and has a maximum value of 40 arc minutes. It arises because the Sun's gravitational attraction on the Moon varies during the course of a synodic month, being greatest at new Moon and least at full Moon.

vastitas (pl. vastitates)   A widespread lowland area on the surface of a planet.

Vega (1) (Alpha Lyrae; α Lyr)   The brightest star in the constellation Lyra and the fifth-brightest in the sky. It is an ►*A star* of magnitude 0.03. ►Table 3.

Vega (2)   The name of two Soviet spacecraft launched in December 1984 with the dual function of dropping probes into the atmosphere of Venus in 1985 and encountering Halley's Comet in 1986.

Veil Nebula (NGC 6960)   Part of the ➤*Cygnus Loop*, an old ➤*supernova remnant*.

Vela (The Sail)   A large southern constellation, which is one of the four parts into which Nicolas L. de Lacaille divided the ancient constellation Argo Navis. It lies in part of the Milky Way rich with faint nebulosity and contains ten stars brighter than fourth magnitude. The stars Delta (δ) and Kappa (ϰ), together with Iota (ι) and Epsilon (ε) Carinae, make an asterism known as the 'false cross' since it is sometimes confused with the constellation Crux, the true Southern Cross. ➤Table 4.

Vela Pulsar   A ➤*pulsar* in the constellation Vela, associated with a supernova remnant 10,000 years old. It is one of the strongest radio pulsars, and the strongest gamma-ray source in the sky.

It was discovered in 1968 during a general search for pulsars in the southern hemisphere and has a short period, 89 milliseconds, characteristic of young pulsars. The period is steadily increasing at a rate of 10.7 nanoseconds a day as the pulsar loses energy. Since observations of it started, the pulsar has also undergone several major ➤*glitches* in which the period has suddenly decreased by about 200 nanoseconds.

Venera   A series of Soviet spacecraft sent to explore the planet Venus. The first to land successfully was *Venera 7* in 1970. There have been nine further *Venera* probes producing images of the surface of Venus, and data about the atmosphere and the composition of the planet's crust.

Venus   The second major planet of the solar system, in order from the Sun. It is one of the 'terrestrial' planets, similar in nature to Earth and only slightly smaller. Like the Earth, it is surrounded by a substantial atmosphere.

Venus can come closer to the Earth than any other planet and can be the brightest object in the sky (apart from the Sun and Moon). Because its orbit lies inside the Earth's, its position in the sky can never be further than 47° away from the Sun. As a result, Venus can be viewed either in the western sky in the evening, or in the eastern sky in the morning. It is sometimes called the 'morning star' or the 'evening star'.

As a further consequence of its location within the Earth's orbit, Venus appears to go through a cycle of phases similar to the Moon's. At its brightest and nearest, even a small telescope will show that Venus is actually a crescent.

The surface is perpetually covered by dense, highly reflecting clouds which show few features in visible light, though ultraviolet photographs reveal a banded structure, including a characteristic Y-shaped feature. These clouds consist of droplets of dilute sulphuric acid, created by the action of sunlight on the carbon dioxide, sulphur compounds and water vapour present in the atmosphere.

The atmosphere is almost entirely of carbon dioxide, and the surface pressure is more than 90 times that at the surface of the Earth. The exceptionally high surface temperature of 730 K (450°C) is a result of the ➤greenhouse effect.

Venus was the target of a large number of Soviet and American probes in the 1970s and 1980s, notably the Soviet Venera and Vega series and the American ➤Pioneer Venus. The extremely high temperature and pressure present considerable difficulties and many of the probes were destroyed either before returning data or after a relatively short period of operation. Nevertheless, it proved possible to analyse the chemical composition of some surface rocks and to return limited panoramic views of the surface terrain, showing rocky desert landscapes.

The first radar maps produced by spacecraft orbiting the planet showed that most of the surface consists of vast plains, above which several large plateaux rise to heights of several kilometres. The two main highland areas are Ishtar Terra in the northern hemisphere and Aphrodite Terra in the equatorial region. The Maxwell Montes are the highest feature, rising to 11 kilometres above the mean level of the planetary surface.

In 1990, the US ➤Magellan probe arrived in orbit around Venus and commenced a programme of mapping the surface in much greater detail than had been achieved previously, by means of sophisticated radar techniques. Ample evidence has been found of both impact features and volcanic activity in the relatively recent past. By solar system standards, the surface of Venus is young: the oldest craters appear to date from 800 million years ago. However, no direct evidence has yet been found of current volcanism.

The thick atmosphere and high surface temperature mean that impact craters take forms rather different from those on other planets and satellites. Smaller meteorites burn up readily on passing through the atmosphere so there is an absence of smaller craters. The material thrown out in the powerful impacts of larger meteorites did not travel far and tended to spread round the craters in molten form.

Large numbers of volcanic features have been identified: lava flows, small domes 2−3 kilometres across, larger volcanic cones hundreds of kilometres across, 'coronae' and so-called 'arachnoids'. The coronae of Venus are circular or oval volcanic structures surrounded by ridges, grooves and radial lines. They appear to be collapsed volcanic domes and are different from any features seen on other planets or satellites. The 'arachnoids', which get their informal name from their spider-like appearance, are similar in form to coronae, but generally smaller. According to one theory, arachnoids may be precursors to coronae. The bright lines extending outwards for many kilometres indicate formations that may have been created when magma upwelled from the planet's interior, causing the surface to crack. ➤Table 5.

vernal equinox ➤*equinox*.

Very Large Array (VLA) A radio telescope consisting of 27 dishes, each 25 metres (82 feet) in diameter, configured to carry out ➤*Earth rotation synthesis*. Located at Socorro, New Mexico, it is the world's largest ➤*aperture synthesis* telescope. Its elements are arranged in a Y-shape with three arms each 21 kilometres (13 miles) long. The dishes are connected electronically so as to be equivalent to 351 ➤*radio interferometers* observing simultaneously. The maximum resolution obtainable is 0.05 arc seconds at a wavelength of 1.3 centimetres, although in practice most observations are made at 6 centimetres with 1 arc second resolution as this greatly reduces the time needed to construct a radio map.

Very Large Telescope (VLT) The ➤*European Southern Observatory*'s set of four, linked 8-metre (300-inch) telescopes located at ➤*Paranal Observatory* in Chile. The light-gathering power of the four telescopes together is equivalent to that of a mirror 16 metres (52 feet) in diameter. Construction work began in late 1988. The first telescope was finished in 1997 and final completion of all four is scheduled for 2001. The technology employed is the same as for the ➤*New Technology Telescope*.

Very Long Baseline Array (VLBA) A network of radio telescopes in North America for ➤*very-long-baseline interferometry*. It consists of ten antennas, distributed from Hawaii to St Croix in north-east Canada. The effective diameter is 8,000 kilometres (5,000 miles) and the resolution attainable 0.2 milliseconds of arc.

very-long-baseline interferometry (VLBI) A technique in radio astronomy that effectively creates a ➤*radio interferometer* in which the component antennas are separated by very large distances, typically thousands of kilometres. The antennas are not connected electrically or through microwave links. Instead, video signals are registered on magnetic tape, together with very accurate timings, at each observing station. The tapes from each station are brought together later and played through a radio receiver to complete the analysis. The technique provides extremely accurate positions for radio sources, resolutions (but not maps) down to a few milliseconds of arc and the direct detection of ➤*continental drift*.

Antenna separations even greater than the Earth's diameter can be achieved by placing radio telescopes in orbit and using them in conjunction with ground-based telescopes. The launch of the Japanese satellite, HALCA (originally known as MUSES-B), in February 1997 marked the first stage in the development of the international VSOP – the VLBI Space Observatory Programme. HALCA, an umbrella-shaped antenna 8 metres (26 feet) in diameter, was placed in an elliptical orbit, providing a baseline up to three times larger than the Earth.

Vesta Asteroid 4, diameter 576 km, discovered by H. W. M. Olbers in 1802. It is the third-largest asteroid known and the brightest of all, sometimes reaching a visual magnitude of 6, when it is just detectable to the unaided eye under optimum observing conditions. Vesta's brightness is due to its high albedo of 25 per cent. As it rotates every 5.43 hours, regular changes in the colour and spectrum are observed, reflecting the fact that the surface is not uniform.

Vesta appears to be a true mini-planet which has survived largely intact since the solar system formed, rather than being a fragment from a larger body. Hubble Space Telescope images reveal details down to 80 km (50 miles) across, including impact craters. One large crater seems to have torn away part of the crust completely, exposing the mantle below. There is evidence for ancient lava flows dating from 4 billion years ago when the interior was hot and molten. It is thought that Vesta may be the parent body of the ➤*eucrite* type of meteorites.

Victor M. Blanco Telescope A 4-metre (160-inch) telescope at the ➤*Cerro Tololo Inter-American Observatory*.

vignetting The uneven illumination of the image plane in an optical instrument such as a telescope. The usual cause of vignetting is obstruction of the light by parts of the instrument itself.

Viking Two identical American probes sent to the planet Mars in 1975. Both *Vikings 1* and *2* consisted of an orbiter, which remained circling the planet, and a soft lander.

Viking 1 was launched on 9 September 1975 and reached Mars orbit on 19 June 1976. Images were taken to locate a suitable landing site and the landing took place on Chryse Planitia on 20 July 1976. The orbiter's path was adjusted several times in order to obtain close-up images of Mars' satellites, Deimos and Phobos, and to observe different aspects of the martian surface.

Viking 2 was launched on 20 August 1975 and reached Mars orbit on 7 August 1976. The landing took place on Utopia Planitia on 3 September 1976.

The orbiters were equipped with two television cameras, an infrared spectrometer to map the distribution of water vapour and a radiometer for determining temperature distribution. The landers sampled the upper atmosphere during descent, made meteorological measurements and carried out experiments on samples of martian soil. One of the prime objectives was to test for the presence of organic material which might indicate the existence of life, but nothing incontrovertible was found. Thousands of images were returned from both the orbiters and landers. The whole of the martian surface was mapped to a resolution of 150–300 metres.

The *Viking 1* orbiter operated until 7 August 1980, and the *Viking 2* orbiter until 25 July 1978. The landers ceased operating in November 1982 and February 1980, respectively. The mission was regarded as very successful, and it had greatly exceeded its expected lifetime.

Virgin   English name for the constellation ➤*Virgo*.

Virgo (The Virgin)   A zodiacal constellation and the second-largest in the sky. It is one of the 48 constellations listed by Ptolemy (*c.* AD 140). The brightest star is the first magnitude ➤*Spica*, and there are seven others brighter than fourth magnitude. The constellation contains the rich and relatively nearby ➤*Virgo Cluster* of galaxies. Eleven of the brighter galaxies are listed in the ➤*Messier Catalogue*. ➤Table 4.

Virgo A   The strongest radio source in the constellation Virgo, identified with the giant elliptical galaxy M87, which dominates the Virgo Cluster. The radio emission is associated with a jet 4,000 light years in extent and may well be caused by the accretion of matter on to a supermassive black hole in the nucleus of M87.

Virgo Cluster   The nearest rich cluster of galaxies at a distance of about 50–60 million light years and the centre of the ➤*Local Supercluster*. It covers 120 square degrees of sky and contains several thousand galaxies. It is an irregular cluster with no central condensation. The giant elliptical galaxy M87 is the most massive in the cluster. A total of sixteen of the brighter members are included in the ➤*Messier Catalogue*.

virial theorem   The statement that, in a self-gravitating system in equilibrium (e.g. a cluster of stars or galaxies), the gravitational potential energy is twice the kinetic energy of the member objects. From this theorem, the masses of galaxy clusters can be estimated from their size and the average velocities of their luminous members. The values obtained are nearly ten times greater than the total mass of visible galaxies, and this has given rise to the problem of ➤*missing mass* in galaxy clusters. It indicates that there are large quantities of dark matter in clusters of galaxies.

visual binary   A ➤*binary star* in which the two components can be resolved as separate images with a telescope of appropriate size.

visual magnitude   The ➤*magnitude* of a celestial object measured over a wavelength band corresponding to the sensitivity of the human eye. ➤*V*.

VLA   Abbreviation for ➤*Very Large Array*.

VLBA   Abbreviation for ➤*Very Long Baseline Array*.

VLBI   Abbreviation for ➤*very-long-baseline interferometry*.

VLT   Abbreviation for ➤*Very Large Telescope*.

Volans (The Flying Fish)   A small and faint southern constellation introduced in the 1603 atlas of Johann Bayer with the longer name Piscis Volans, which was later shortened. Its six main stars are third and fourth magnitude. ➤Table 4.

volatile   A material that readily evaporates, such as water.

Voyager 1   One of a pair of almost identical planetary probes launched by the USA in 1977. The other was ➤*Voyager 2*.

The *Voyager* missions were possible only because of a chance favourable alignment of the outer planets, Jupiter, Saturn, Uranus and Neptune, occurring just once in more than a hundred years. Between them, the two spacecraft were able to explore these four planets, their environs and satellite systems. The gravitational 'sling-shot' technique was used to accelerate the craft from one encounter to the next. The missions were immensely successful, making numerous discoveries and returning huge quantities of data as well as visual images.

The instruments on the *Voyagers* consisted of two groups. One set was designed to sample the craft's environment and these remained in operation constantly, even between planetary encounters. They measured magnetic field, low-energy charged particles, cosmic rays and the characteristics of the local plasma. The other instruments were a wide-angle (3°) camera, a close-up (0.4°) camera, a ➤*Michelson interferometer* to analyse infrared emission from planetary atmospheres, an ultraviolet spectrometer, a photopolarimeter to measure light intensity and its state of polarization, and a detector for radio emission from planetary magnetospheres. The main communication dish was 3.7 metres in diameter and a plutonium-238 power source was used.

Voyager 1 was launched on 5 September 1977. Its closest encounter with Jupiter was on 5 March 1979 at 350,000 kilometres (217,500 miles) and that with Saturn was on 12 November 1980 at 124,000 kilometres (77,000 miles). It entered the jovian system close to Io and Callisto and the saturnian system near Titan, Rhea and Mimas. After the encounter with Saturn, it left the plane of the solar system to travel on into interstellar space. ➤*Voyager Interstellar Mission*.

Voyager 2   One of a pair of planetary probes launched by the USA in 1977. It was virtually identical to ➤*Voyager 1*, except that its power source was designed to last for much longer in order to survive the extended journey to Uranus and Neptune.

Voyager 2 was launched on 20 August 1977. Its first encounter was with Jupiter on 9 July 1979, passing within 71,400 kilometres (44,000 miles) of the planet. It passed close to Europa and Ganymede, complementing the coverage of the Galilean satellites obtained by *Voyager 1*. Saturn was reached in August 1981. Closest approach was on 25 August, at a distance of 101,000 kilometres (63,000 miles). The trajectory took the probe near the saturnian satellites Tethys and Enceladus.

On 24 January 1986 *Voyager 2* reached Uranus, which it passed at a distance of 107,000 kilometres (67,000 miles) and the mission was completed with the Neptune and Triton encounter of 24 August 1989, when the craft passed within

48,000 kilometres (30,000 miles) of the surface of Neptune. ➤*Voyager Interstellar Mission.*

Voyager Interstellar Mission   The continued operation of the ➤*Voyager 1* and ➤*Voyager 2* spacecraft after their planetary encounters were completed. The power sources on both craft are expected to sustain operation until about 2020. As long as power is available, instruments on board continue to detect particles in the ➤*heliosphere*, and to measure magnetic field. It is anticipated that they will gather the first scientific data from the region of the heliopause, where the solar wind merges with the interstellar medium.

VSOP   Abbreviation for VLBI Space Observatory Programme. ➤*very-long-baseline interferometry.*

V-type asteroid   A class of asteroid of which the only known example is ➤*Vesta.*

Vulcan   A hypothetical planet travelling round the Sun within the orbit of Mercury for which searches were made during the late nineteenth century. It is now known not to exist.

Vulpecula (The Fox)   A faint constellation, next to Cygnus, introduced by J. Hevelius in 1690 with the longer name Vulpecula et Anser – the fox and goose – which was subsequently shortened to its present form. It contains no stars brighter than fourth magnitude, but does include the noted planetary nebula known as the ➤*Dumbbell Nebula.* ➤Table 4.

VV Cephei star   A member of a class of ➤*supergiant* binary stars with emission line spectra. The primary is a supergiant of ➤*spectral type* G or M and the secondary a hot B star. Fewer than twenty are known.

W

walled plain A large, flat-floored lunar ➤*crater*, particularly one that has been flooded by lava.

waning The part of the cycle of the Moon's phases when the illuminated fraction of the visible disc is decreasing. The opposite is 'waxing'.

Water Carrier English name for the constellation ➤*Aquarius*.

Water Jar The group of stars Gamma (γ), Eta (η), Zeta (ζ) and Pi (π) in the constellation Aquarius, normally shown as the Water Carrier's jar in representations of the mythological figure associated with the constellation.

Water Snake English name for the constellation ➤*Hydra*.

wavelength (symbol λ) The shortest distance between two points in a wave train that have the same phase.

waxing The part of the cycle of the Moon's phases when the illuminated fraction of the visible disc is increasing. The opposite is 'waning'.

W C star ➤*Wolf-Rayet star*.

West, Comet ➤*Comet West*.

Westerbork Observatory A Dutch national radio astronomy observatory which is part of the Netherlands Foundation for Research in Astronomy (NFRA, or ASTRON). The administrative headquarters are at ➤*Dwingeloo Observatory*.

The instrument at Westerbork Observatory is called the Westerbork Synthesis Radio Telescope (WSRT). It is a fourteen-element aperture synthesis instrument, and came into operation in 1970. A major extension in 1980 increased the baseline from 1,500 to 3,000 metres (0.9 to 1.8 miles).

Whale English name for the constellation ➤*Cetus*.

Whipple Observatory ➤*Fred Lawrence Whipple Observatory*.

Whirlpool Galaxy (M51; NGC 5194) A face-on ➤*spiral galaxy* in the constellation Canes Venatici at a distance of 13 million light years. It was the first galaxy to be recognized as having spiral structure when it was observed by Lord Rosse in 1845. It is accompanied by a much smaller irregular galaxy, NGC 5195, which is in orbit around it.

white dwarf   A star in an advanced state of ►*stellar evolution*, composed of degenerate matter. A white dwarf is created when a star finally exhausts its possible sources of fuel for thermonuclear fusion. The star collapses under its own gravity, compressing the matter to a degenerate state in which atomic nuclei and electrons that have been completely stripped from atoms are all packed together. The process stops when a quantum mechanical effect, the exclusion principle, comes into play; the electrons cannot be compacted further and so exert a resistance called *degeneracy pressure*. S. Chandrasekhar demonstrated theoretically that the upper mass limit for white dwarfs is 1.4 times the mass of the Sun – larger collapsing masses must become ►*neutron stars* or ►*black holes*.

The first white dwarf to be recognized as such was 40 Eridani B, observed in 1910. It was shown to have a surface temperature of 17,000 K but a total luminosity so low that it could be explained only if the star were smaller than the Earth. Other well-known white dwarfs include van Maanen's star and Sirius B. Sirius B, first seen in 1862, has a mass about the same as the Sun's in a diameter only five times the Earth's, but is 10,000 times fainter than Sirius A, which is a normal ►*A star*. A few hundred white dwarfs are known but they may represent as much as 10 per cent of the stellar population. Their low intrinsic luminosity makes them difficult to detect.

Though called 'white' dwarfs as a group, these degenerate stars actually cover a range of temperatures and colours, from the hottest, which are white and have surface temperatures as high as 100,000 K, to cool red objects at only 4,000 K. Since they have no internal source of energy, white dwarfs are in a long process of gradually cooling off, during which the temperature declines. Their ultimate fate is to become a *black dwarf* – a non-luminous dead star.

The spectra of white dwarfs are bewilderingly complex, reflecting a range of temperature and composition. Typically, the spectrum contains very broad absorption lines, though some show no lines at all. The line-forming region is only a few hundred metres thick. Some white dwarfs show only hydrogen lines, presumably because the helium and heavier elements have sunk to the bottom of the 'atmosphere' under the strong gravitational force. In other stars, helium or metals are seen and no hydrogen remains.

A new classification scheme for white dwarfs was proposed in 1983 by E. M. Sion and collaborators. The designations consist of three capital letters, the first being D for degenerate. The second letter indicates the primary spectrum: A, H only; B, neutral He with no H or metals; C, continuous; O, ionized He with neutral He or H; Z, metal lines only with no H or He; Q, carbon present. The third letter is for secondary spectral characteristics: P, magnetic with polarized light; H, magnetic without polarized light; X, peculiar or unclassifiable; V, variable. The old system was based on the usual ►*spectral class* sequence (O, B, A, F, G, K, M) prefixed by D.

white hole   A hypothetical object, never observed, which mathematically has properties inverse to those of a black hole and would be a place where matter spontaneously appears.

Widmanstätten figures (Widmanstätten pattern)   A characteristic geometrical pattern revealed when certain types of metallic ►*meteorite* are cut, polished and etched with dilute acid. They arise from the crystalline intergrowth of two different forms of iron−nickel alloy, ►*kamacite* and ►*taenite*. ►*octahedrite*.

Wild Duck (M11; NGC 6705)   An open cluster of about 200 stars in the constellation Scutum. Its name comes from its shape, which is fan-like in small instruments and resembles somewhat a flight of wild ducks.

William Herschel Telescope   A 4.2-metre (160-inch) altazimuth reflecting telescope in the Isaac Newton Group at the ►*Observatorio del Roque de los Muchachos*, La Palma, Canary Islands. Observing time is shared between the collaborating countries − the UK, Spain and the Netherlands. It is a general-purpose telescope, equipped with a large range of instruments, and came into operation in 1987.

Willstrop telescope   A design of reflecting optical telescope yielding good images over a field of view of 5° or more. It is a modified version of the ►*Paul−Baker system*. The primary mirror contains a hole, the size of which is 60 per cent of the total diameter and in which lies the focal plane. The shapes of all three mirrors depart substantially from the nearest paraboloid or sphere. The design's advantages are that the telescope is much more compact than a ►*Schmidt camera*, there are no ghost images of the type caused by internal reflections in the corrector lens of a Schmidt camera, and it should be possible to build a telescope to this design larger than any existing Schmidt camera.

Wilson effect   A change in the appearance of a sunspot as it is carried close to the Sun's limb by rotation, in which the penumbra nearest the limb appears wider than that on the other side of the spot. This is because the sunspot is in a depression. The phenomenon was first observed by the Scottish astronomer, Alexander Wilson (1714−86), in 1769.

WIMPs   Acronym for (electrically neutral) weakly interacting massive particles. These have not yet been observed in particle accelerators; if they do exist, they are important in cosmology because they would contribute to the ►*missing mass*.

Winchester   Asteroid 747, diameter 204 km, discovered in 1913 by J. Metcalf.

Winged Horse   English name for the constellation ►*Pegasus*.

winter solstice   ►*solstice*.

WIYN Telescope   A 3.5-metre (138-inch) telescope at ►*Kitt Peak*, opened in 1994. It is operated jointly by the University of Wisconsin, Indiana University, Yale University, and the National Optical Astronomy Observatories.

WLM   Abbreviation for ➤*Wolf–Lundmark–Melotte galaxy*.

WN star   ➤*Wolf–Rayet star*.

Wolf   English name for the constellation ➤*Lupus*.

Wolf–Lundmark–Melotte galaxy (WLM) A small irregular galaxy belonging to the ➤*Local Group*. It lies at a distance of 2.8 million light years.

Wolf–Rayet star   A member of a class of rare, exceptionally hot stars with surface temperatures of 20,000–50,000 K. Their spectra show strong broad-band emission lines: in WC stars, carbon dominates, whereas in WN stars the dominant emission lines are of nitrogen. It is believed that there is a genuine difference in composition between the two subgroups. The emission lines are thought to originate in a rapidly expanding envelope through which the star is losing mass. Some are the central stars of ➤*planetary nebulae*, but their evolutionary status is not fully understood.

The name comes from two nineteenth-century French astronomers, Charles Wolf and Georges Rayet.

Wolf sunspot number   A measure of the sunspot activity on the solar disc, taking into account spot groups as well as individual spots. It was devised by Rudolf Wolf of the Zürich Observatory, and is also known as the *Zürich sunspot number*. The value, R, is calculated from the formula $R = k(10g + f)$ where g is the number of sunspot groups, f the total number of spots and k a weighting factor depending on the instruments used and the observer. The value of k is about 1 for telescopes of 100 mm aperture.

wormhole   A hypothetical tunnel-like structure in the fabric of ➤*spacetime*. Theorists have suggested that on the distance scale of the ➤*Planck length* (10^{-35} metres), spacetime might have a foam-like structure, riddled with wormholes. It might be possible for such a wormhole to 'pinch off' and form a new universe.

wrinkle ridge   A low-relief feature on the lunar surface which may extend for hundreds of kilometres across a mare. Wrinkle ridges appear to be a feature associated with volcanic activity and the flow of lava in the Moon's distant past.

W Ursae Majoris star   A member of a class of eclipsing variable stars in which the two components are stars of ➤*spectral type* F or G that are similar in brightness and almost in contact. The primary and secondary minima of the light curve are nearly equal and the periods are typically several hours.

Würzburg antenna   A type of German radar dish used from 1944 in the UK and the Netherlands for the first systematic observations in radio astronomy.

W Virginis star   ➤*Cepheid variable*.

X

XMM A project of the ▶*European Space Agency* for an orbiting X-ray astronomy observatory to be launched in 1999 with a minimum lifetime of ten years. The large collecting area, and the unprecedented sensitivity and resolution of its detectors, will reduce the time required to make an observation by a factor of over 100 and will allow many more spectra to be obtained than has been possible previously. The satellite will carry three identical telescopes, each consisting of 58 nested precision reflectors.

X-ray astronomy The study of X-radiation from astronomical sources. The X-ray waveband is usually considered to cover the wavelength range from about 10 to 0.01 nanometres, between the extreme ultraviolet (XUV) and gamma-rays. The corresponding energy range is 0.1 – 100 keV.

No X-rays from space can penetrate the atmosphere to the ground, so all X-ray astronomy is carried out with instruments on rockets or satellites. X-rays from the Sun were detected during rocket flights in the 1950s. The first X-ray source beyond the solar system to be discovered was ▶*Scorpius X-1*, found in 1962 by a group at American Science and Engineering, led by Ricardo Giacconi. By 1970 there were more than forty known X-ray sources detected during rocket-borne experiments. However, satellites were needed to conduct more extensive surveys.

US military Vela satellites operating between 1969 and 1979 carried X-ray detectors. The first satellite dedicated to X-ray astronomy was Uhuru (1970), the first of the ▶*Small Astronomy Satellite* series. In 1973, a telescope capable of producing X-ray images was used successfully to image the Sun during the Skylab mission. This X-ray telescope used an array of concentric, cylindrical mirrors to reflect the X-rays at grazing incidence and bring them to a focus, and detectors capable of recording the positions of arrival of the photons over a field of view. Such an imaging X-ray telescope was used for objects other than the Sun for the first time by the ▶*Einstein Observatory*. In 1985, a different type of X-ray telescope, using the 'coded mask' technique, was deployed in orbit on Spacelab 2. This is capable of forming images at higher energies and incorporates a diaphragm with a complex pattern of holes. Other important X-ray astronomy satellites have included ▶*Copernicus* (1971), ▶*EXOSAT*, ▶*Ginga* (1987), ▶*ROSAT* (1990) and ▶*BeppoSAX* (1996).

Thermal radiation in the X-ray band comes from sources at temperatures in excess of one million degrees. Much of the X-ray emission detected from astronomical sources is produced in non-thermal processes, such as the interaction between electrons and ions in plasmas (which can produce both continuous radiation and X-ray spectral lines) and nuclear reactions in interacting binary star systems.

The largest class of bright X-ray sources consists of interacting binary stars in which one component is a degenerate star – a ►*white dwarf*, a ►*neutron star* or a ►*black hole*. There are two categories of such X-ray binary. In high-mass binaries, the companion is a star of 10 or 20 solar masses, and matter from its extended envelope flows directly on to the degenerate star. In low-mass binaries the two components are of similar mass, and mass transfer takes place via an ►*accretion disc*. As it gains gravitational energy, the material flowing between the stars reaches temperatures high enough for the emission of X-rays. Such binaries often show periodic variability attributable to the orbital period of the system, the rotation period of the degenerate star or the precession of the accretion disc. The X-ray luminosity ranges from 100 to 100,000 times the total luminosity of the Sun. Some systems, the ►*X-ray bursters*, show much more dramatic and random variations.

The other main types of source of astronomical X-rays are the hot diffuse gas surrounding galaxies and present between the galaxies in clusters, ►*supernova remnants* and ►*active galactic nuclei*. In 1996, for the first time, X-rays were detected from several ►*comets*. ►*XMM*, Yohkoh.

X-ray binary ►*X-ray astronomy, X-ray burster*.

X-ray burster A stellar X-ray source showing violent and random changes in their emission.

X-ray bursters were discovered with the Dutch satellite ANS in 1976. The bursts may last for several days and may recur, but are not periodic. A rapid burster repeats at intervals no longer than 10 seconds. The numbers known are counted in tens; most are in the galactic plane though some are in globular clusters.

The generally accepted model is that of an interacting binary system, similar to a ►*nova*, except that accretion takes place on to a ►*neutron star* rather than a ►*white dwarf*, and the material transferred is predominantly helium rather than hydrogen. The X-ray burst occurs when the accumulation of accreted material reaches the critical temperature and density to detonate a thermonuclear explosion. ►*X-ray astronomy*.

X-ray Multi-Mirror ►*XMM*.

X-ray star A stellar source of X-rays. ►*X-ray astronomy*.

XTE ►*Rossi X-ray Timing Explorer*.

XUV A term sometimes applied to the short wavelength end of the ultraviolet region of the ►*electromagnetic spectrum* in the range 6–60 nanometres, where it merges with the X-ray band. It overlaps the region also known as ►*EUV*. ►*ultraviolet astronomy*.

Y

Yagi antenna An antenna for receiving radio waves, consisting of a small number of parallel dipoles. The commonest domestic television aerials are of this type. Arrays of Yagis are used in radio astronomy to construct cheap ►*aperture synthesis* telescopes for sky survey work. ►*radio telescope*.

Yarkovsky effect The effect of its rotation on the path of a small particle orbiting the Sun. Rotation causes a temperature variation, so thermal energy is re-radiated anisotropically. ►*Poynting–Robertson effect*.

year The period of time taken for the Earth to orbit the Sun. The exact length of the year depends on the reference point taken. ►*calendar, calendar year*.

Length and definition of years

Year	Reference point	Length in days
Tropical	Equinoxes	365.242 19
Sidereal	Fixed stars	365.256 36
Anomalistic	Apsides	365.259 64
Eclipse	Moon's node	346.620 03
Gaussian	Kepler's law for semimajor axis of 1 AU	365.256 90

Yerkes Observatory An observatory in Williams Bay, Wisconsin, which belongs to the Department of Astronomy and Astrophysics of the University of Chicago. The telescope housed at the observatory is the largest refracting telescope ever built, with an objective lens 1 metre (40 inches) in diameter. It was constructed between 1895 and 1897.

The telescope was largely the brainchild of George Ellery Hale, who later became the driving force behind the 2.5-metre (100-inch) Hooker Telescope at Mount Wilson and the 5-metre (200-inch) Palomar telescope. In association with William Rainey Harper, he persuaded the Chicago millionaire Charles Yerkes to finance the entire project. The lens components were already in existence, having been manufactured for another project that never came to fruition. The mechanical parts of the telescope were constructed by Warner and Swasey of Cleveland, Ohio.

The building is noted for its ornate design by the architect Henry Ives Cobb. The '40-inch' telescope is still used in some of the Department's research programmes.

ylem The ➤*primeval fireball*.

Yohkoh The name given after its launch in August 1991 to a Japanese astronomy satellite originally designated Solar-A. Its prime purpose was the study of X-rays and gamma rays from solar ➤*flares* and other energetic phenomena on the Sun.

yoke mounting A particular form of ➤*equatorial mounting*.

YY Orionis star A member of a subclass of ➤*T Tauri stars* characterized by a particular form of emission line with an absorption wing to the red side. The line profiles are variable over periods of days. About half of all T Tauri stars may belong to this subclass.

Z

z   The symbol normally used for ➤*redshift*.

ZAMS   Abbreviation for ➤*zero-age main sequence*.

Z Andromedae star   ➤*symbiotic stars*.

ZC   Abbreviation for ➤*Zodiacal Catalogue*.

Zeeman effect   The splitting of spectral lines into a number of components when the source is in a magnetic field.

Zelenchukskaya   The location of the ➤*Special Astrophysical Observatory* of the Russian Academy of Sciences.

zenith   The overhead point. The *astronomical zenith* is formally defined as the extension upwards to infinity of a plumb line. The *geocentric zenith* is the continuation of a line from the centre of the Earth through the observer. The *geodetic zenith* is the normal to the geodetic ellipsoid or spheroid at the observer's location.

zenithal hourly rate (ZHR)   The hypothetical rate at which meteors of a particular ➤*meteor shower* would be observed by an experienced observer, watching a clear sky with limiting magnitude 6.5, if the radiant were located in the zenith. The lower the altitude of the radiant, the lower the observed rate. To a first approximation the ratio of the observed rate to the zenithal rate is the sine of the altitude angle of the radiant.

zenith distance   The angular distance from the zenith of a point on the celestial sphere, measured along a great circle.

zenith tube   A telescope mounted to point vertically in order to make positional measurements on stars passing through and near the zenith.

zero-age main sequence (ZAMS)   For a population of stars, such as a cluster, formed at the same time, the main sequence on the ➤*Hertzsprung–Russell diagram* as it would have been at the start of their hydrogen-burning phase. ➤*stellar evolution*.

ZHR   Abbreviation for ➤*zenithal hourly rate*.

zodiac   A belt of twelve constellations through which the Sun's path in the sky – the ➤*ecliptic* – passes. They are Aries, Taurus, Gemini, Cancer, Leo, Virgo,

Libra, Scorpius, Sagittarius, Capricornus, Aquarius and Pisces. Though the ecliptic formerly went only through these twelve constellations, the effects of ►*precession* and the precise definitions of constellation boundaries mean that it now also goes through a thirteenth, Ophiuchus.

Since the orbits of all the planets, apart from Pluto, lie very nearly in a plane, the apparent paths of the planets remain in or close to the zodiacal constellations.

In traditional astrological use, the zodiac is divided into twelve equal 30° portions, each of which is allocated to a 'sign', but these do not correspond exactly to the astronomical constellations, which are of varying sizes. The effect of precession has further contributed to increasing disparity between the true position of the Sun and the astrological signs.

zodiacal band   ►*zodiacal light*.

Zodiacal Catalogue (ZC)   The popular name for a 1940 catalogue of 3,539 of the brighter stars within 8° of the ecliptic. It was compiled by James Robertson and published as Volume X, Part II, of *Astronomical Papers prepared for the use of the American Ephemeris and Nautical Almanac*. It includes all stars brighter than magnitude 7.0 and 313 fainter than magnitude 8.5. The catalogue is used particularly for star positions in the prediction and analysis of ►*occultations*.

zodiacal dust cloud   A tenuous flat cloud of small silicate dust particles in the inner solar system. It is believed to be derived from comets and from collisions in the ►*asteroid belt*.

zodiacal light   A faint cone of light extending along the ecliptic, visible in the sky on clear moonless nights in the west following sunset, and in the east just before sunrise. It is caused by sunlight scattered from micrometre-sized dust particles in the ►*zodiacal dust cloud in* the plane of the solar system. The zodiacal light is dimly present all round the ecliptic, a phenomenon sometimes called the *zodiacal band*, and an enhancement also occurs at the position directly opposite the Sun. This is known as the *gegenschein*, or *counterglow*.

Zond   One of a series of Soviet space probes launched between 1963 and 1970. *Zonds 1* and *2* flew by Venus and Mars, respectively, but returned no data. *Zond 3* photographed the farside of the Moon in 1965. *Zond 4* was a failed mission. *Zonds 5, 6, 7* and *8* made flights around the Moon and were recovered on Earth.

zone of avoidance   A region of sky near the plane of the ►*Milky Way* where absorption by interstellar dust is so great that no galaxies can be seen.

Zürich sunspot number   ►*Wolf sunspot number*.

ZZ Ceti star   A pulsating, variable ►*white dwarf* star. The amplitude of variation of ZZ Ceti stars is between 0.05 and 0.3 magnitude, and their periods range between 30 seconds and half an hour.

FURTHER READING

Allen, Richard Hinckley, *Star Names – Their Lore and Meaning*. Dover Publications, 1899 (reprint).

Audouze, Jean, and Guy Israël (eds), *The Cambridge Atlas of Astronomy*. Cambridge University Press, 1994 (3rd edn).

Bakich, Michael E., *The Cambridge Guide to the Constellations*. Cambridge University Press, 1995.

Beatty, J. Kelly, and Andrew Chaikin (eds), *The New Solar System*. Cambridge University Press/Sky Publishing Corpn, 1990 (3rd edn) & 1998 (4th edn).

Fraknoi, A., D. Morrison, and S. Wolff, *Voyages through the Universe*. Saunders College Publishing, 1997.

Glyn Jones, Kenneth, *Messier's Nebulae and Star Clusters*. Cambridge University Press, 1991 (2nd edn).

Greeley, R., and R. M. Batson, *The NASA Atlas of the Solar System*. Cambridge University Press, 1997.

Green, Robin M., *Spherical Astronomy*. Cambridge University Press, 1985.

Gribbin, John, *Companion to the Cosmos*. Penguin Books, 1996.

Henbest, Nigel, and Michael Marten, *The New Astronomy*. Cambridge University Press, 1996 (2nd edn).

Illingworth, Valerie (ed.), *The Collins Dictionary of Astronomy*. HarperCollins, 1994.

Kaler, James B., *The Ever-Changing Sky*. Cambridge University Press, 1996.

Lang, Kenneth R., *Sun, Earth and Sky*. Springer, 1995.

Lang, Kenneth R., and Charles A.Whitney, *Wanderers in Space: Exploration and Discovery in the Solar System*. Cambridge University Press, 1991.

Levy, David H., *Skywatching*. Time Life Books, 1994.

Maran, Stephen P. (ed.), *The Astronomy and Astrophysics Encyclopedia*. Van Nostrand Reinhold and Cambridge University Press, 1992.

Moore, Patrick, *The Guinness Book of Astronomy*. Guinness Publishing, 1995 (5th edn).

Morrison, David, *Exploring Planetary Worlds*. Scientific American Library, 1993.

Norton, Richard O., *Rocks from Space*. Mountain Press Publishing Co., Missoula, Montana, 1994.

Pasachoff, Jay, *Astronomy: From the Earth to the Universe*. Saunders College Publishing, 1998 (5th edn).

Ridpath, Ian (ed.), *Norton's 2000.0: Star Atlas and Reference Handbook*. Longman, 1989.

Silk, Joseph, *A Short History of the Universe*. Scientific American Library, 1997.

Yeomans, Donald K., *Comets: a chronological history of observation, science, myth and folklore*. John Wiley & Sons, 1991.

TABLES

Table 1. The Greek alphabet. Lowercase letters are used in star designations. ➤ *Bayer letters.*

A	α	alpha	N	ν	nu
B	β	beta	Ξ	ξ	xi
Γ	γ	gamma	O	o	omicron
Δ	δ	delta	Π	π	pi
E	ε	epsilon	P	ϱ	rho
Z	ζ	zeta	Σ	σ	sigma
H	η	eta	T	τ	tau
Θ	θ	theta	Y	υ	upsilon
I	ι	iota	Φ	φ	phi
K	ϰ	kappa	X	χ	chi
Λ	λ	lambda	Ψ	ψ	psi
M	μ	mu	Ω	ω	omega

Table 2. Units, constants and conversion factors.

ångström (Å)	0.1 nanometre★
astronomical unit (AU or a.u.)	$1.495\,978\,7 \times 10^8$ kilometres★
centimetre (cm)	0.3937 inches
electron volt (eV)	1.6022×10^{-19} joules
foot (ft)	30.48 centimetres★
inch (in)	2.54 centimetres★
kilogram (kg)	2.2046 pounds
kilometre (km)	0.6214 miles
light year (l.y.)	9.4605×10^{12} kilometres = 0.306 60 parsec
micron (μm)	1 micrometre = 10^{-6} metres★
mile (mi)	1.6093 kilometres
nanometre (nm)	10^{-9} metres★
parsec (pc)	3.0857×10^{13} kilometres = 3.261 61 light years
radian	$57°.295\,78$
speed of light (c)	299,792.458 kilometres per second
tonne (t)	1,000 kilograms★

★These conversion factors are exact; other values are approximations to the accuracy given.

Table 3. Stars of apparent visual magnitude 2.0 and brighter.

Star	Name	RA h m s	Declination ° ' "	Visual magnitude*	Spectral classification*	Distance (l.y.)†
Alpha Canis Majoris	Sirius	06 45 09	−16 42 58	−1.5	A1 V	8.6
Alpha Carinae	Canopus	06 23 57	−52 41 44	−0.7	F0 I	313
Alpha¹ Centauri		14 39 37	−60 50 02	−0.3	G2 V	4.4
Alpha² Centauri		14 39 35	−60 50 13		K1 V	
Alpha Boötis	Arcturus	14 15 40	+19 10 57	0.0	K0 III	36
Alpha Lyrae	Vega	18 36 56	+38 47 01	0.0	A0V	25
Alpha Aurigae	Capella	05 16 41	+45 59 53	0.1	G8 III	42
Beta Orionis	Rigel	05 14 32	−08 12 06	0.1	B8 I	773
Alpha Canis Minoris	Procyon	07 39 18	+05 13 30	0.4	F5 IV	11
Alpha Eridani	Achernar	01 37 43	−57 14 12	0.5	B5 IV	144
Alpha Orionis	Betelgeuse	05 55 10	+07 24 26	0.5	M2 I	427
Beta Centauri	Hadar	14 03 49	−60 22 22	0.6	B1 II	525
Alpha Aquilae	Altair	19 50 47	+08 52 06	0.8	A7 IV	17
Alpha Tauri	Aldebaran	04 35 55	+16 30 33	0.9	K5 III	65
Alpha¹ Crucis / Alpha² Crucis	Acrux	12 26 36 / 12 26 37	−63 05 56 / −63 05 58	0.9	{B1 IV / B3}	321
Alpha Scorpii	Antares	16 29 24	−26 25 55	1.0	M1 I	604
Alpha Virginis	Spica	13 25 12	−11 09 41	1.0	B1 V	262
Beta Geminorum	Pollux	07 45 19	+28 01 34	1.1	K0 III	34
Alpha Piscis Austrini	Fomalhaut	22 57 39	−29 37 20	1.2	A3 V	25
Alpha Cygni	Deneb	20 41 26	+45 16 49	1.3	A2 I	3,230
Beta Crucis	Mimosa	12 47 43	−59 41 19	1.3	B0 III	352
Alpha Leonis	Regulus	10 08 22	+11 58 02	1.4	B7 V	77
Epsilon Canis Majoris	Adhara	06 58 38	−28 58 20	1.5	B2 II	431
Alpha Geminorum	Castor	07 34 36	+31 53 18	1.6	A1 V	52
Gamma Crucis		12 31 10	−57 06 47	1.6	M3 III	88

Star	Name	RA (h m s)	Dec	Mag	Spectral type	Distance
Gamma Orionis	Bellatrix	05 25 08	+06 20 59	1.6	B2 III	243
Lambda Scorpii	Shaula	17 33 36	−37 06 14	1.6	B2 IV	703
Alpha Gruis		22 08 14	−46 57 40	1.7	B5 V	101
Epsilon Orionis	Alnilam	05 36 13	−01 12 07	1.7	B0 I	1,342
Beta Tauri	Elnath	05 26 18	+28 36 27	1.7	B7 III	130
Beta Carinae		09 13 12	−69 43 02	1.7	A0 III	111
Alpha Persei	Mirfak	03 24 19	+49 51 40	1.8	F5 I	592
Zeta Orionis	Alnitak	05 40 46	−01 56 34	1.8	O9 I	817
Gamma Velorum		08 09 32	−47 20 12	1.8	WC7	841
Alpha Ursae Majoris	Dubhe	11 03 44	+61 45 03	1.8	K0 III	124
Epsilon Ursae Majoris	Alioth	12 54 02	+55 57 35	1.8	A0p	81
Theta Scorpii		17 37 19	−42 59 52	1.9	F0 I	272
Alpha Trianguli Australe		16 48 40	−69 01 39	1.9	K2 III	415
Beta Aurigae		05 59 32	+44 56 51	1.9	A2 IV	82
Eta Ursae Majoris	Alkaid	13 47 32	+49 18 48	1.9	B3 V	101
Gamma Geminorum		06 37 43	+16 23 57	1.9	A0 IV	105
Epsilon Sagittarii	Kaus Australis	18 24 10	−34 23 05	1.9	B9 IV	145
Delta Canis Majoris		07 08 23	−26 23 36	1.9	F8 I	1,792
Epsilon Carinae		08 22 31	−59 30 34	1.9	K0 II	632
Alpha Pavonis	Peacock	20 25 39	−56 44 06	1.9	B3 IV	183
Gamma¹ Leonis / Gamma² Leonis	Algieba	10 19 58	+19 50 30	1.9	{K0 III / G5 III}	126
Beta Canis Majoris	Mirzam	06 22 42	−17 57 22	2.0	B1 II	499
Alpha Hydrae	Alphard	09 27 35	−08 39 31	2.0	K3	177
Alpha Ursae Minoris	Polaris	02 31 50	+89 15 51	2.0	F8 I	431
Beta Ceti		00 43 35	−17 59 12	2.0	K0 III	96
Delta Velorum		08 44 42	−54 42 30	2.0	A0 V	80
Sigma Sagittarii	Nunki	18 55 16	−26 17 48	2.0	B3 IV	224
Alpha Arietis	Hamal	02 07 10	+23 27 45	2.0	K2 III	66

*Values are to the nearest 0.1 magnitude. For variable stars the mean magnitude is given. †Data from the Hipparcos catalogue.

Table 4. The Constellations. The official, internationally recognized constellation names are in Latin and are listed in the first column of the table. Star designations are constructed from Greek or Roman letters, or Arabic numbers, in conjunction with the genitive case of the constellation name, which is listed in the second column (e.g. Alpha Andromedae, AM Herculis, 32 Virginis). Each constellation has a standard three-letter abbreviation (third column). An English translation of the constellation name (except for proper names) is given in the fourth column, though many variations on these are used. Further information on each constellation is given in the individual entries (under the Latin names) in the main body of the dictionary.

Latin name	Genitive	Abbreviation	English name	Area (square degrees)	Rank in size order
Andromeda	Andromedae	And	Andromeda	722	19
Antlia	Antliae	Ant	Air Pump	239	62
Apus	Apodis	Aps	Bird of Paradise	206	67
Aquarius	Aquarii	Aqr	Water Carrier	980	10
Aquila	Aquilae	Aql	Eagle	652	22
Ara	Arae	Ara	Altar	237	63
Aries	Arietis	Ari	Ram	441	39
Auriga	Aurigae	Aur	Charioteer	657	21
Boötes	Boötis	Boo	Bear Driver	907	13
Caelum	Caeli	Cae	Chisel	125	81
Camelopardalis	Camelopardalis	Cam	Giraffe	757	18
Cancer	Cancri	Cnc	Crab	506	31
Canes Venatici	Canum Venaticorum	CVn	Hunting Dogs	465	38
Canis Major	Canis Majoris	CMa	Big Dog	380	43
Canis Minor	Canis Minoris	CMi	Little Dog	183	71
Capricornus	Capricorni	Cap	Sea Goat	414	40
Carina	Carinae	Car	Keel	494	34
Cassiopeia	Cassiopeiae	Cas	Cassiopeia	598	25
Centaurus	Centauri	Cen	Centaur	1,060	9
Cepheus	Cephei	Cep	Cepheus	588	27

Table 4 (cont.).

Latin name	Genitive	Abbreviation	English name	Area (square degrees)	Rank in size order
Libra	Librae	Lib	Scales	538	29
Lupus	Lupi	Lup	Wolf	334	46
Lynx	Lyncis	Lyn	Lynx	545	28
Lyra	Lyrae	Lyr	Lyre	286	52
Mensa	Mensae	Men	Table	153	75
Microscopium	Microscopii	Mic	Microscope	210	66
Monoceros	Monocerotis	Mon	Unicorn	482	35
Musca	Muscae	Mus	Fly	138	77
Norma	Normae	Nor	Rule	165	74
Octans	Octantis	Oct	Octant	291	50
Ophiuchus	Ophiuchi	Oph	Serpent Bearer	948	11
Orion	Orionis	Ori	Hunter	594	26
Pavo	Pavonis	Pav	Peacock	378	44
Pegasus	Pegasi	Peg	Pegasus	1,121	7
Perseus	Persei	Per	Perseus	615	24
Phoenix	Phoenicis	Phe	Phoenix	469	37
Pictor	Pictoris	Pic	Easel	247	59
Pisces	Piscium	Psc	Fishes	889	14
Piscis Austrinus	Piscis Austrini	PsA	Southern Fish	245	60
Puppis	Puppis	Pup	Poop	673	20
Pyxis	Pyxidis	Pyx	Compass	221	65
Reticulum	Reticuli	Ret	Net	114	82
Sagitta	Sagittae	Sge	Arrow	80	86
Sagittarius	Sagittarii	Sgr	Archer	867	15
Scorpius	Scorpii	Sco	Scorpion	497	33
Sculptor	Sculptoris	Scl	Sculptor	475	36
Scutum	Scuti	Sct	Shield	109	84
S......	Serpentis	Ser	Serpent	637	23

Sextans	Sextantis	Sex	Sextant	47	314
Taurus	Tauri	Tau	Bull	17	797
Telescopium	Telescopii	Tel	Telescope	57	252
Triangulum	Trianguli	Tri	Triangle	78	132
Triangulum Australe	Trianguli Australis	TrA	Southern Triangle	83	110
Tucana	Tucanae	Tuc	Toucan	48	295
Ursa Major	Ursae Majoris	UMa	Great Bear	3	1,280
Ursa Minor	Ursae Minoris	UMi	Little Bear	56	256
Vela	Velorum	Vel	Sails	32	500
Virgo	Virginis	Vir	Virgin	2	1,294
Volans	Volantis	Vol	Flying Fish	76	141
Vulpecula	Vulpeculae	Vul	Fox	55	268

Table 5. Characteristics of the major planets and their orbits.

Characteristics of the inner planets

	Mercury	Venus	Earth	Mars
Reciprocal mass[a]	6,023,600	408,524	328,900	3,098,710
Mass[b] (Earth = 1)	0.0553	0.8149	1.0000	0.1074
Mass[b] (g)	3.303×10^{26}	4.870×10^{27}	5.976×10^{27}	6.418×10^{26}
Equatorial radius (Earth = 1)	0.382	0.949	1.000	0.532
Equatorial radius (km)	2,439	6,051	6,378	3,393
Ellipticity[c]	0.0	0.0	0.0034	0.0052
Mean density (g/cm³)	5.43	5.25	5.52	3.95
Equatorial surface gravity (m/s²)	3.78	8.60	9.78	3.72
Equatorial escape velocity (km/s)	4.3	10.4	11.2	5.0
Sidereal rotation period at equator	58.65 days	243.01 days	23.9345 hours	24.6229 hours
Inclination of equator to orbit	(2°)[d]	177°.3[e]	23°.45	25°.19

Characteristics of the outer planets

	Jupiter	Saturn	Uranus	Neptune	Pluto
Reciprocal mass	1,047.355	3,498.5	22,869	19,314	135,300,000
Mass (Earth = 1)	317.938	95.181	14.531	17.135	0.0022
Mass (g)	1.900×10^{30}	5.688×10^{29}	8.684×10^{28}	1.024×10^{29}	1.31×10^{25}
Equatorial radius[f] (Earth = 1)	11.209	9.449	4.007	3.883	0.180
Equatorial radius[f] (km)	71,541	60,268	25,559	24,704	1,150
Ellipticity	0.0649	0.0980	0.0229	0.0170	0.0
Mean density (g/cm³)	1.33	0.69	1.29	1.64	2.13
Equatorial surface gravity (m/s²)	22.88	9.05	7.77	11.00	0.40
Equatorial escape velocity (km/s)	59.6	35.5	21.3	23.3	1.1
Sidereal rotation period at equator	9.841 hours[g]	10.233 hours[h]	17.9 hours[i]	19.2 hours[j]	6.3872 days
Inclination of equator to orbit	3°.12	26°.73	97°.86[e]	29°.6	122°.46[e]

Characteristics of planetary orbits

	Mean distance from Sun (AU)	(10⁶ km)	Sidereal period (years)	(days)	Synodic period (days)	Mean orbital velocity (km/s)	Orbital eccentricity	Inclination to the ecliptic (degrees)
MERCURY	0.3871	57.91	0.24085	87.969	115.88	47.89	0.2056	7.004
VENUS	0.7233	108.20	0.61521	224.701	583.92	35.03	0.0068	3.394
EARTH	1.0000	149.60	1.00004	365.256	—	29.79	0.0167	0.000
MARS	1.5237	227.94	1.88089	686.980	779.94	24.13	0.0934	1.850
JUPITER	5.2028	778.33	11.8623	4,332.71	398.88	13.06	0.0483	1.308
SATURN	9.5388	1,426.98	29.458	10,759.5	378.09	9.64	0.0560	2.488
URANUS	19.1914	2,870.99	84.01	30,685	369.66	6.81	0.0461	0.774
NEPTUNE	30.0611	4,497.07	164.79	60,190	367.49	5.43	0.0097	1.774
PLUTO	39.5294	5,913.52	248.54	90,800	366.73	4.74	0.2482	17.148

ᵃThe mass of the Sun divided by the mass of the planet (including its atmosphere and satellites).
ᵇSatellite masses not included.
ᶜThe ellipticity is $(R_e - R_p)/R_e$, where R_e and R_p are the planet's equatorial and polar radii, respectively.
ᵈValues in parentheses are uncertain by more than 10 per cent.
ᵉBy IAU convention, each planet's north pole is the one lying north of the ecliptic plane; as such, Venus, Uranus, and Pluto are considered to have retrograde rotation.
ᶠSince the outer planets have no solid surfaces, these are the radii at 1-bar pressure level in their atmospheres.
ᵍJupiter's internal (System III) rotation period is 9.925 hours.
ʰSaturn's internal rotation period is 10.657 hours.
ⁱUranus's internal rotation period is 17.240 hours. ʲNeptune's internal rotation period is 16.11 hours.

The data in tables 5, 6 and 7 were taken with permission from *The New Solar System* (3rd edition), edited by J. Kelly Beatty and Andrew Chaikin, Cambridge University Press and Sky Publishing Corp., 1990 © Sky Publishing Corp., and revised in 1997 with data from the National Space Science Data Center.

Table 6. The natural satellites of the planets.

Name	Discoverer(s)	Year of discovery	Magnitude (Vo)	Mean distance from planet (km)	Sidereal period (days)	Orbital inclination (degrees)	Orbital eccentricity	Radius (km)	Mass (g)	Mean density (g/cm³)
Satellite of Earth										
Moon			-12.7	384,400	27.322	18.3–28.6	0.05	1,738	7.35 × 10²⁵	3.34
Satellites of Mars										
Phobos	A. Hall	1877	11.3	9,377	0.319	1.0	0.01	13 × 8	1.08 × 10¹⁹	1.9
Deimos	A. Hall	1877	12.4	23,436	1.263	0.9–2.7	0.00	9 × 5	1.8 × 10¹⁸	1.8
Satellites of Jupiter										
Metis	S. Synnott	1979	17.5	127,960	0.295	(0)	0.00	(20)	?	?
Adrastea	D. Jewitt, E. Danielson	1979	18.7	128,980	0.298	(0)	(0)	12 × 8	?	?
Amalthea	E. Barnard	1892	14.1	181,300	0.498	0.4	0.00	131 × 73 × 67	?	?
Thebe	S. Synnott	1979	16.0	221,800	0.675	0.8	0.00	(50)	?	?
Io	Galileo (S. Marius?)	1610	5.0	421,600	1.769	0.04	0.04	1,821	8.93 × 10²⁵	3.53
Europa	Galileo (S. Marius?)	1610	5.3	670,900	3.551	0.47	0.01	1,565	4.80 × 10²⁵	2.99
Ganymede	Galileo (S. Marius?)	1610	4.6	1,070,000	7.155	0.19	0.00	2,634	1.48 × 10²⁶	1.94
Callisto	Galileo (S. Marius?)	1610	5.6	1,883,000	16.689	0.28	0.01	2,403	1.08 × 10²⁶	1.85
Leda	C. Kowal	1974	20.2	11,094,000	238.72	27	0.16	(5)	?	?
Himalia	C. Perrine	1904	15.0	11,480,000	250.57	28	0.16	(90)	?	?
Lysithea	S. Nicholson	1938	18.2	11,720,000	259.22	29	0.11	(12)	?	?
Elara	C. Perrine	1905	16.6	11,737,000	259.65	28	0.21	(40)	?	?
Ananke	S. Nicholson	1951	18.9	21,200,000	631	147	0.17	10	?	?
Carme	S. Nicholson	1938	17.9	22,600,000	692	163	0.21	15	?	?
Pasiphae	P. Melotte	1908	16.9	23,500,000	735	148	0.38	18	?	?
Sinope	S. Nicholson	1914	18.0	23,700,000	758	153	0.28	14	?	?

Satellites of Saturn

Pan	M. Showalter	1990	?	133,570	0.576	(0)	(0)	(10)	?	?
Atlas	R. Terrile	1980	18.0	137,640	0.602	(0)	(0)	19 × 17 × 14	?	0.27
Prometheus	S. Collins and others	1980	15.8	139,350	0.613	0	0.00	74 × 50 × 34	1.4×10^{20}	0.42
Pandora	S. Collins and others	1980	16.5	141,700	0.629	0	0.00	55 × 44 × 31	?	?
Epimetheus	R. Walker and others	1966	15.7	151,422	0.695	0.34	0.01	69 × 55 × 55	5.5×10^{20}	0.63
Janus	A. Dollfus	1966	14.5	151,472	0.695	0.14	0.01	99 × 96 × 76	2×10^{21}	0.65
Mimas	W. Herschel	1789	12.9	185,520	0.942	1.53	0.02	199	3.8×10^{22}	1.14
Enceladus	W. Herschel	1789	11.7	238,020	1.370	0.02	0.00	249	7.3×10^{22}	1.12
Tethys	G. Cassini	1684	10.2	294,660	1.888	1.09	0.00	530	6.22×10^{23}	1.00
Telesto	B. Smith and others	1980	18.7	294,660	1.888	(0)	(0)	15 × 13 × 8	?	?
Calypso	D. Pascu and others	1980	19.0	294,660	1.888	(0)	(0)	15 × 8 × 8	?	?
Dione	G. Cassini	1684	10.4	377,400	2.737	0.02	0.00	560	1.05×10^{24}	1.44
Helene	P. Laques, J. Lecacheux	1980	18.4	377,400	2.737	0.20	0.01	16	?	?
Rhea	G. Cassini	1672	9.7	527,040	4.518	0.35	0.00	764	2.31×10^{24}	1.24
Titan	C. Huygens	1655	8.3	1,221,850	15.945	0.33	0.03	2,575	1.35×10^{26}	1.88
Hyperion	W. Bond	1848	14.2	1,481,000	21.277	0.43	0.10	185 × 140 × 113	?	?
Iapetus	G. Cassini	1671	10.2–11.9	3,561,300	79.331	7.52	0.03	718	1.59×10^{24}	1.02
Phoebe	W. Pickering	1898	16.5	12,952,000	550.480	175.30	0.16	115 × 110 × 105	?	?

Satellites of Uranus

Cordelia	Voyager 2	1986	24	49,750	0.335	(0.14)	(0)	13	?	?
Ophelia	Voyager 2	1986	24	53,760	0.376	(0.09)	(0.01)	16	?	?
Bianca	Voyager 2	1986	23	59,160	0.435	(0.16)	(0)	22	?	?
Cressida	Voyager 2	1986	22	61,770	0.464	(0.04)	(0)	33	?	?
Desdemona	Voyager 2	1986	22	62,660	0.474	(0.16)	(0)	29	?	?
Juliet	Voyager 2	1986	22	64,360	0.493	(0.06)	(0)	42	?	?
Portia	Voyager 2	1986	21	66,100	0.513	(0.09)	(0?)	(55)	?	?

Table 6 (cont.).

Name	Discoverer(s)	Year of discovery	Magnitude (Vo)[a]	Mean distance from planet (km)	Sidereal period (days)	Orbital inclination (degrees)	Orbital eccentricity	Radius (km)	Mass (g)	Mean density (g/cm³)
Satellites of Uranus cont.										
Rosalind	Voyager 2	1986	22	69,930	0.558	(0.28)	(0)	29	?	?
Belinda	Voyager 2	1986	22	75,260	0.624	(0.03)	(0)	34	?	?
Puck	Voyager 2	1985	20	86,010	0.762	(0.31)	(0)	77	?	?
Miranda	G. Kuiper	1948	16.5	129,780	1.413	4.22	0.00	235	6.59×10^{22}	1.20
Ariel	W. Lassell	1851	14.4	191,240	2.520	0.31	0.00	580	1.35×10^{24}	1.67
Umbriel	W. Lassell	1851	15.3	265,970	4.144	0.36	0.01	585	1.17×10^{24}	1.40
Titania	W. Herschel	1787	14.0	435,840	8.706	0.10	0.00	789	3.53×10^{24}	1.71
Oberon	W. Herschel	1787	14.2	582,600	13.463	0.10	0.00	761	3.01×10^{24}	1.63
1997U2	B. Gladman and others	1997	20	6,460,000	495	153.36	(0.40)	(80)	?	?
1997U1	B. Gladman and others	1997	22	7,770,000	645	146.36	(0.20)	(40)	?	?
Satellites of Neptune										
Naiad	Voyager 2	1989	25	48,227	0.294	4.74	(0)	(29)	?	?
Thalassa	Voyager 2	1989	24	50,075	0.311	0.21	(0)	(40)	?	?
Despina	Voyager 2	1989	23	52,526	0.335	0.07	(0)	74	?	?
Galatea	Voyager 2	1989	23	61,953	0.429	0.05	(0)	79	?	?
Larissa[b]	Voyager 2	1989	21	73,548	0.555	0.20	(0)	104 × 89	?	?
Proteus	Voyager 2	1989	20	117,647	1.122	0.55	(0)	218 × 208 × 201	?	?
Triton	W. Lassell	1846	13.6	354,760	5.877	156.83	0.00	1,353	2.15×10^{25}	2.05
Nereid	G. Kuiper	1949	18.7	5,513,400	360.14	7.23	0.75	170	?	?
Satellite of Pluto										
Charon	J. Christy	1978	16.8	19,636	6.387	96.2	0.008	595	(1.10×10^{24})	(1.3)

[a]The magnitude in visible light at opposition.
[b]Probably detected by H. Reitsema and others during an occultation in 1981.

Table 7. The characteristics of planetary ring systems.

Name	Distance from centre of main planet (km)	Radial width (km)	Thickness (km)	Optical depth	Total mass (g)	Albedo
Rings of Jupiter	(R_J)					
'Halo'	(1.40)–1.72	22,800	(20,000)	6×10^{-6}	?	0.05
	(100,000)–122,800					
'Main'	1.72–1.81	226,400	<30	10^{-6}	(10^{16})	0.05
	122,800–129 200					
'Gossamer'	1.81–(3)	850,000	?	10^{-7}	?	0.05
	129,200–(214,000)					
Rings of Saturn	(R_S)					
D	1.10–1.24	8,500	?	(0.01)	?	?
	66,000–74,500					
C 'Crepe Ring'	1.24–1.52	17,342	?	0.05–0.35	1.1×10^{21}	0.25
	74,658–92,200					
Maxwell Gap	1.45	270				
	87,500					
B	1.53–1.95	25,532	(0.1–1)	0.4–2.5	2.8×10^{22}	0.65
	91,975–117,507					
Cassini Division	1.95–2.02	4,700		0.12	5.7×10^{20}	0.30
	117,500–122,200					
A	2.02–2.27	14,435	(0.1–1)	0.70	6.2×10^{21}	0.60
	122,340–135,775					
Encke Gap[a]	2.216	325				
	133,589					
Keeler Gap	2.263	35				
	136,530					
F	2.329	30–500	?	0.1	?	0.6
	140,374					
G	2.76–2.88	7,000	100–1,000	10^{-6}	6.23×10^{9}	?
	166,000–173,000					
E	(3–8)	(300,000)	(1,000)	10^{-6}–10^{-7}	?	?
	(180,000–480,000)					

[a]Although the discovery of this gap is often ascribed to J. Encke, J. Keeler was probably the first to observe it.

Table 7 (cont.).

Name	Distance from centre of main planet (km)	Radial width (km)	Thickness (km)	Optical depth	Total mass (g)	Albedo
Rings of Uranus	(R_u)					
6	1.637 41,837	~1.5	(0.1)	0.2–0.3	?	(0.02)
5	1.612 42,230	~2	(0.1)	0.5–0.6	?	(0.02)
4	1.625 42,580	~2.5	(0.1)	0.3	?	(0.02)
Alpha (α)	1.707 44,720	4–10	(0.1)	0.3–0.4	?	(0.02)
Beta (β)	1.743 45,670	5–11	(0.1)	0.2	?	(0.02)
Eta (η)	1.801 47,190	1.6	(0.1)	0.1–0.4	?	(0.02)
Gamma (γ)	1.818 47,630	1–4	(0.1)	1.3–2.3	?	(0.02)
Delta (δ)	1.843 48,290	3–7	(0.1)	0.3–0.4	?	(0.02)
Lambda (λ)	1.909 50,020	~2	(0.1)	0.1	?	(0.02)
Epsilon (ε)	1.952 51,140	20–96	<0.15	0.5–2.1	?	(0.02)
Rings of Neptune	(R_N)					
Galle	1.692–1.773 41,900–43,900	~2,000	$\sim 8 \times 10^{-5}$	~0.015	?	(low)
Leverrier	2.148 53,200	~110	~0.002	~0.015	?	(low)
Lassell	2.148–2.310 53,200–57,200	~4,000	1.5×10^{-4}	~0.015	?	(low)
Arago	2.310 57,200	<~100	?	?	?	(low)
Unnamed	2.501 61,950	?	?	?		
Adams	2.541 62,933	~50	$\sim 4.5 \times 10^{-3}$	~0.015	?	(low)

READ MORE IN PENGUIN

In every corner of the world, on every subject under the sun, Penguin represents quality and variety – the very best in publishing today.

For complete information about books available from Penguin – including Puffins, Penguin Classics and Arkana – and how to order them, write to us at the appropriate address below. Please note that for copyright reasons the selection of books varies from country to country.

In the United Kingdom: Please write to *Dept. EP, Penguin Books Ltd, Bath Road, Harmondsworth, West Drayton, Middlesex UB7 0DA*

In the United States: Please write to *Consumer Sales, Penguin Putnam Inc., P.O. Box 999, Dept. 17109, Bergenfield, New Jersey 07621-0120.* VISA and MasterCard holders call 1-800-253-6476 to order Penguin titles

In Canada: Please write to *Penguin Books Canada Ltd, 10 Alcorn Avenue, Suite 300, Toronto, Ontario M4V 3B2*

In Australia: Please write to *Penguin Books Australia Ltd, P.O. Box 257, Ringwood, Victoria 3134*

In New Zealand: Please write to *Penguin Books (NZ) Ltd, Private Bag 102902, North Shore Mail Centre, Auckland 10*

In India: Please write to *Penguin Books India Pvt Ltd, 210 Chiranjiv Tower, 43 Nehru Place, New Delhi 110 019*

In the Netherlands: Please write to *Penguin Books Netherlands bv, Postbus 3507, NL-1001 AH Amsterdam*

In Germany: Please write to *Penguin Books Deutschland GmbH, Metzlerstrasse 26, 60594 Frankfurt am Main*

In Spain: Please write to *Penguin Books S. A., Bravo Murillo 19, 1° B, 28015 Madrid*

In Italy: Please write to *Penguin Italia s.r.l., Via Benedetto Croce 2, 20094 Corsico, Milano*

In France: Please write to *Penguin France, Le Carré Wilson, 62 rue Benjamin Baillaud, 31500 Toulouse*

In Japan: Please write to *Penguin Books Japan Ltd, Kaneko Building, 2-3-25 Koraku, Bunkyo-Ku, Tokyo 112*

In South Africa: Please write to *Penguin Books South Africa (Pty) Ltd, Private Bag X14, Parkview, 2122 Johannesburg*

READ MORE IN PENGUIN

DICTIONARIES

Abbreviations
Ancient History
Archaeology
Architecture
Art and Artists
Astronomy
Biology
Botany
Building
Business
Challenging Words
Chemistry
Civil Engineering
Classical Mythology
Computers
Curious and Interesting Geometry
Curious and Interesting Numbers
Curious and Interesting Words
Design and Designers
Economics
Electronics
English and European History
English Idioms
Foreign Terms and Phrases
French
Geography
Geology
German
Historical Slang
Human Geography
Information Technology

International Finance
Literary Terms and Literary Theory
Mathematics
Modern History 1789–1945
Modern Quotations
Music
Musical Performers
Nineteenth-Century World History
Philosophy
Physical Geography
Physics
Politics
Proverbs
Psychology
Quotations
Religions
Rhyming Dictionary
Russian
Saints
Science
Sociology
Spanish
Surnames
Symbols
Telecommunications
Theatre
Third World Terms
Troublesome Words
Twentieth-Century History
Twentieth-Century Quotations